AF651523

ENVIRONMENT, POPULATION AND DEVELOPMENT

About the Editors

Dr. Prithvish Nag (*b*. 1951) is a M.Sc. (Gold Medalist) and Ph.D. from Banaras Hindu University. He was awarded Commonwealth Geographical Bursary on the 150 anniversary of the Royal Geographical Society. He has been a full member of the Commission on Population Geography of the International Geographical Union. Chairman of the Commission on Population Cartography of the International Cartographic Association; President of the Indian National Cartographic Association and the Indian Association of Special Libraries; and Member of the Editorial Board of the International Journal on GIS.

Dr. Nag has been author of nearly hundred research papers and several publications. His latest books are *Population, Settlement and Development in Zambia, Geography of India* and *Digital Remote Sensing*. He has been UN Consultant in Census Cartography in Oman. At present he is the Director of the National Atlas & Thematic Mapping Organisation, Calcutta. He is also the Vice-chairman of the UNRCC working Group on Institutional Strengthening for Asia and the Pacific.

Dr. Chandra Shekhar Kumar (*b*. 1959) obtained his M.Sc. and Ph.D. degrees from Banaras Hindu University. He has published several research papers and participated in many national and international seminars and brought out a book on *Environmental Problems and Development*. He is a life member of the Indian Science Congress Association, NAGI and Environmental Monitoring Society. He is also recipient of the CSIR fellowship. At present he is attached to the Department of Geography, Banaras Hindu University, Varanasi.

Dr. Smita Sengupta (*b*. 1956) is a M.A. from Calcutta University and Ph.D. from Gujarat University. She has a varied experience of working with university, institute, government and industry. She has served in Anthropological Survey of India in Calcutta, IIT Bombay, Geosoft Topographis (India) Pvt. Lmt. and Tata Infotech, all in Mumbai. At present she is the Manager, GIS in Tata Infotech.

Dr. Sengupta is a recipient of UGC and CSIR fellowship. She has brought out books on *Residential Pattern of Suburbs* and *Geography of India*, apart from over 30 research papers in well-known journals. She is also member of professional bodies like INCA, Geographical Society of India, Association of Population Geographers of India and the Indian Society of Geomatics.

ENVIRONMENT, POPULATION AND DEVELOPMENT

Felicitation volume in honour of

Prof. S. L. Kayastha

Edited by

Prithvish Nag

Chandra Shekhar Kumar

Smita Sengupta

CONCEPT PUBLISHING COMPANY, NEW DELHI-110059

All rights reserved. No part of this work may be reproduced, stored in a retrieval system, or transmitted in any form or by any means, electronic, mechanical, photocopying, microfilming, recording or otherwise, without the prior written permission of the copyright owner and the publisher.

ISBN 81-7022-889-1

First Published 2001

© Editors

Published and Printed by
Ashok Kumar Mittal
Concept Publishing Company
A/15-16, Commercial Block, Mohan Garden
New Delhi-110059 (India)

Phones: 5648039, 5649024

Fax: 091-(11)-5648053

E-mail: publishing@conceptpub.com

Preface

Environment and population have been studied from different angles, by different disciplines, and for a variety of reasons. These topics have been analysed separately or even together. A close relationship between the two is not a formidable conclusion. But, recently, such studies have gained an additional dimension due to greater emphasis on sustainability which in turn is linked with development. Hence, there is a renewed emphasis on 'sustainable development'. Obviously, resources have to be considered for this purpose. In other words, environment population and development has to be studied together keeping in view of the limited resources at local, regional, national or even at global levels. Such integrated studies are common in geography. However, the emphasis on these three aspects may vary from one study to another.

Professor S.L. Kayastha has been studying these aspects with varied emphasis and at various levels. In his recent contributions, such integrated studies are more apparent. With his approach in mind, we the editors, thought to explore further in this direction and have views of others as well. Accordingly we invited potential authors from related disciplines to make an enquiry in their areas of interest. They readily agreed to have a re-look into their research activities and initiatives and to contribute for this felicitation volume in honour of Professor Kayastha. It has been interesting to note that authors took pains to explore how the issues related to environment, population and development have been linked at various spatial levels. Can there be a viable unit for making such an enquiry? Can the concept of 'region' hold good for carrying out such integrated study? Is the intensity of study directly or inversely related to the spatial unit of analysis? What is the model we are developing, explicitly or implicitly? These questions are intricate but have to be considered sooner or later. This volume may not be able to answer all the related issues. But can we make some humble beginning? Professor Kayastha's model (last chapter) may be considered as the beginning.

Organising this volume was more than usual hassles associated with editing. After issuing invitations to the authors, and having received their favourable responses, we the editors found ourself in three different cities– while one is abroad for a long time. However, the publisher was as usual in New Delhi. The circumstances forced us to enter into long distance communication which was not easy for us. This caused delays in bringing out the publication. Nevertheless, we are obliged to the contributors for bearing with us. Our publisher, Shri Ashok Kumar, stood behind us from the very initial stage, and took special interest in bringing out this publication. We are grateful to him.

We are also grateful to Professor Kayastha for providing information about the research work done by him in more than four decades. We are also grateful to his son, Professor Anand Kumar of the Department of Geology, Banaras Hindu University, Varanasi for helping us in this regard.

25 March, 1999

PRITHVISH NAG
CHANDRA SHEKHAR KUMAR
SMITA SENGUPTA

Professor S.L. Kayastha: A Profile

In 1986, when some of our friends and colleagues brought out the felicitation volume in honour of our revered and respected teacher and senior geographer, Prof. S.L. Kayastha, entitled *Geography and Environment: Issues and Challenges,* we thought that it was appropriate to do so. However, his outstanding contributions in a span of ten years compilled us to bring out a new volume to coincide with the completion of fifty years of his academic life spanning: 1944-1994. We wish him a long life so that we may continue to benefit from his contributions, guidance and inspiration.

Professor Kayastha was born on 30th March, 1924 and educated in the famous Government College, Lahore, and later in Aligarh Muslim University. He is the first Ph. D. from the Department of Geography, Banaras Hindu University, Varanasi. Later, he occupied the Chair of the Head of the Department during 1977-79 and 1983-84. For his contributions in geography, the National Geographical Society of India conferred on him the Sandra Vallabh Bhai Patel Medal in 1971.

He has to his credit several pioneering endeavours in geography. He was the first to undertake the study of a river basin as a viable unit for integrated resource development and planning. His comprehensive work, *The Himalayan Beas Basin : A Study in Habitat, Economy and Society*, is a landmark in this direction. His interest in integrated development found further fulfilment in the study of South Mirzapur: An Integrated Area Development Project as major departmental project of the U.G.C., of which he was the Principal Investigator. To Professor Kayastha goes the credit of initiating Environmental Studies in Geography in India. He guided several researches relating to concepts, perceptions and problems of urban pollution, flood plain and human adaptation, rural water supply, regional development planning, population and impact of development on environment. He formulated ecological mapping pilot studies in several universities as part of the U.G.C. Geography Panel of Experts, of which he was the Chairman.

Prof. Kayastha has been a full member of the I.G.U. Commission on Environmental Problems. During various symposia in India and abroad, his contributions and deliberations received high recognition. Prof. Kayastha has visited and lectured in several countries: Romania, Nigeria, Mexico, U.S.A., England, Japan, Brazil, Spain, Nepal, Pakistan, U.S.S.R., France, Australia and China. His contributions during the 24th I.G.U. Congress were truly remarkable; he took part in the U.N. University-IGU Panel discussion, was honoured as a guest speaker at the Centenary Celebrations of Hosie University, Tokyo, presented eight papers, and brought out a special issue of the National Geographical Journal of India on Environmental Perception and Problems. He has been Member Secretary of the Indian National Committee for Geography (IGU/INSA), and National Lecturer of the U.G.C. He was invited to lecture in Cornell University, Ithaca and several other institutions of distinction. His name is familiar in India and abroad due to his long association with the National Geographical Journal of India, since its inception, as its Honorary Editor. His services in the dissemination of geographical knowledge as its editor for about four decades deserve all honour. He undertook this challenging task at a time, when research work found little scope for seeing the light of the day. Many geographers remember him with gratitude seeing their first article in print in the N.G.J.I. The journal in his hands was nurtured with love of labour and it now finds recognition as a publication of original researches.

Varanasi has been a seat of learning since times immemorial. This could be one of the reasons for Professor Kayastha to continue to stay in this oldest living city of the world, even after his retirement from the Department of Geography, Banaras Hindu University. His current contributions attest that his decision was correct, which in turn has benefited his friends, colleagues and other co-professionals. Any one calling on him will find Professor Kayastha cool, polite and gentle as ever. But in return, he will find himself inspired and activated.

Professor Kayastha's significant contributions have provided conceptual models, developed novel ideas and enhanced our knowledge on the environmental forefront. His continuing dynamism in the last decade of his academic career has demonstrated his determination and dedication to the discipline of geography and related sphere of work. He has been senior fellow of the Indian Council for Social Science Research (ICSSR) during 1986-88. On account of his valuable contributions to Environmental Studies, he was unanimously elected as the President of the Geology and Geography Section of the Indian Science Congress Platinum Jubilee Session in Pune in 1988. His Presidential address on Frontiers in Environmental Geography: Reappraisal of Environmental Problems and Policies in India received wide appreciation and provided the spatial dimension of the Environmental Problems in the country. He delivered the Platinum Jubilee Lecture in Cochin in 1990. He was President of the National Association of Geographers of India (NAGI) during 1993-94. His presidential address on Environment, Development and Quality of Life probed deeply into the related issues and processes.

For Prof. Kayastha, the study of population in its spatial dimension is an important aspect of study. As students, we clearly remember how distinctly he used to explain the role of population geographers among other specialists studying population. His address on Sustainability of Metropolitan Centres in the Third World Countries during the International Union of the Scientific Study of Population (IUSSP) Congress in Vigyan Bhawan in New Delhi in 1989 was remarkable. He also participated in the Commission on Population Cartography of the International Cartographic Association (ICA) meeting on Population Mapping Technology which was also held as part of IUSSP Congress. His in-depth knowledge on the Indian population maps was highly appreciated. On the other hand, he actively participated in the activities of the IGU Commission on Population Geography. He attended IGU Commission meetings in Karachi, Kathmandu, Sydney and New Delhi. He presented a paper in the commission meeting in Macquaire University, Sydney on Planning for Population Change, now included in the Professor Moonis Raza Commemoration volume. He also participated in the New Delhi Symposium and presented a paper on population growth in India and its impact on environment and development. It was published in Demographic Transition. His most significant contribution to the activities of the IGU Commission on Population Geography was as the co-editor of its prestigious publication, Population and Disaster, published by Basil Blackwell, Oxford. Its review appeared in the prestigious *Nature.* Prof. Kayastha's Contributions on environment and related themes helped to provide due place in geography for environmental studies. He contributed an important chapter on Environmental Policies in India in C. C. Park edited Environmental Policies: World Overview (Croom Helm, London). He was also entrusted with the responsibility by Indian National Sciences Academy, to prepare a report on Environmental Geography which was published in the National Report: 1988-92. Similarly, in 1994, he was entrusted to contribute a status report on Environment and Development by the University Grants Commission. His other important contribution on the Indian environment scenario is a paper on Forests and Ecology in the Himalaya in the IGU Symposium on Mountain Geosystems Dynamics held in Delhi in 1991, which was published in R.B. Singh edited volume, entitled *Dynamics of Mountain Geosystems* (Ashish, New Delhi).

Professor Kayastha emphasis on Development related issues are equally important. He has studied this aspect in relation to population and environment. His Professor N.K. Bose Memorial Lecture on Population, Disaster and Disaster-Reduction Strategies, delivered during the Indian Science Congress in Indore in 1991 deals with the related issues, was later published in the Geographical Review of India. In 1997 he gave an invited talk in the Indian Science Congress in Delhi on Geography and Sustainable Development : Environmental Scenario and Perspectives. Similarly in 1991, in an IGU seminar on Monitoring Geosystems, he presented a paper on Environmental Transformation and Sustainable Development in the Himalaya. It was later published in R.B. Singh edited Global Environmental Change – Perspectives in Remote Sensing and Geographical Information System (Oxford & IBH, New Delhi/Calcutta). He was delegated as an UNEP Consultant to the International Conference on Environmental Life Elements and Health in Beijing in 1990 and his paper was published in the proceedings published by the Chinese Academy of Sciences. In 1994, he presented paper in IGU/UNDP International Seminar in New Delhi, entitled, Disasters in India : Certain Characteristics, Mitigation and Management, later published in R.B. Singh edited Disasters, Environment and Development (Oxford & IBH, New Delhi/Calcutta).

The population-environment-development provide interrelationships that are reflected in the numerous works of Prof. Kayastha. The analysis of spatial patterns is now emerging in Environmental Science, with which he has dealt in detail. The developing countries have their own perspectives, which have been emphasised by him in various seminars and writings.

Considering his valuable contributions in the front line areas of population, environment and development, we consider it apt to entitle this volume as *Population, Environment and Development.* This is the best way to honour Professor Kayastha by presenting a compendium on the issues which have been his concern for so long and which have established unequivocally a new trend in frontal areas of research in the broad field of Environmental Geography.

PRITHVISH NAG
CHANDRA SHEKHAR KUMAR
SMITA SENGUPTA

CONTENTS

List of Contributors

Kefialew Abate : Department of Geography, Kenyatta University, Nairobi

S.K. Biwas : National Atlas & Thematic Mapping Organisation, Calcutta.

Goutam Bhattacharya : Department of Geography, Gujarat University, Ahmedabad

S. Bandhyopadhyaya : Department of Geography, Vidyasagar University, Midnapur.

N.D. Bhattacharya : National Atlas & Thematic Mapping, Organisation, Calcutta.

Daksha C. Barai : Department of Geography, Bangalore University, Bangalore.

S. Basu : Department of Geography, Calcutta University, Calcutta.

Sukla Basu : Department of Geography, Calcutta University, Calcutta.

S. Bose : Department of Psychology, Calcutta University, Calcutta.

Nandini Chatterjee : Department of Geography, Calcutta University, Calcutta.

G.K. Chakravarty : Department of Anthropology, Vidyasagar University, Midnapur.

Anjana Desai : Department of Geography Gujarat University Ahmedabad.

N.G. Dikshit : Department of Georgraphy, Gujarat University, Ahmedabad.

K.K. Dube : Department of Geography, BHU Varanasi.

A.P. Mishra : Department of Geography, BHU, Varanasi.

S.C. Kharkwal : Department of Geography, H.N.B. Garhwal University, Pauri

Chandra Shekhar Kumar : Department of Geography, BHU Varanasi.

V.K. Kumra : Department of Geography, BHU Varanasi.

Rupali Juyal : Department of Geography, BHU, Varanasi.

Udayan Majumdar : Department of Sociology, University of Calcutta.

P. Nag : National Atlas & Thematic Mapping Organisation, Calcutta.

P. Prabhu : Department of Geography, Bangalore University, Bangalore.

H. Prasad : Department of Geography, Kenyatta University, Nairobi.

S. Priyadarshi: National Atlas & Thematic Mapping Organisation, Calcutta.

V.K. Rai : Department of Geography, BHU Varanasi.

Smita Sengupta : Indian Institute of Technology, Mumbai.

Minati Singh : Department of Geography, BHU, Varanasi.

B.L. Singh : Department of Geography, BHU, Varanasi.

M.B. Singh : Department of Geography, BHU, Varanasi.

R. Singh : Department of Geography, BHU, Varanasi.

S.B. Singh : Department of Geography, BHU, Varanasi.

B.B.K. Sinha : Department of Geography, Visva Bharati, Shanti Niketan.

N. Som : Calcutta Metropolitan Development Authority, Calcutta.

S.K. Thakurta : Calcutta Municipal Corporation, Calcutta.

1

Introduction

PRITHVISH NAG

Any writing on environment, population and development issues in recent times can no skip reference on the United Nations Conference on Environment and Development which was held in Rio-de-Janeiro in 1992. This conference has brought out several issues afresh for discussion at the highest level. The August gathering of the heads of states and governments forced to look at the state of environment in their own countries and the world as a whole. Obviously, it was difficult for such a heterogeneous group to have a common view on the related issues. Every country has its own problems (economic or otherswise), commitments and compulsions. Their economic basic and consumption pattern varies hence their role in exploitation of resources and contribution to pollution can be grouped. Due to these reasons, the environmental issues were related to poverty, energy, shortages and so called green funds on one hand, while on the other global environmental issues, biodiversity, environmental friendly machines, forest and sustainable development. The range of the nature of participants volume of literature, discussions in view, a question obviously strike: What is new in such deliberation, except defining the issues in more scientific terms or with superior statistical data? What are the issues, which have been less acceptable?

Incidentally, the population aspect did not find a proper place in the Rio conference. Twenty years earlier at the first UN Conference on Environment and Development in Stockholm, Indira Gandhi emphasized on this aspect. She clearly said that environment can not be improved in the condition of poverty of the population. This conference broadened the base of the environmental movement. It tied environmental conservation firmly to social and economic development, and also brought togcther people and organizations cf various types (UN agencies, National Governments, NGOs) and various levels (global, national, local). At present, environmental aspects are routinely considered when evaluating all proposed development projects by agencies such as United Nations Development Programme (Gupta, 1997). On the other hand, the consumption capacity of resources and energy of the population of the developed world has been multiple times that of the developing world. An average American uses 500 times the coal equivalent of an African consumes, 1,000 times more oil and eat eight times as much grain. The per capita generation of carbon dioxide in the developed world is ten times more than its counterpart in the developing world. Nevertheless, the developed world do not consider their standard of living negotiable under the pressure of their population probably due to the commitments made in the elections or parliaments or in other platforms. In fact, the U.S. Senate has refused to ratify the 1997 Kyoto protocol on cutting greenhouse gas emission, possibly under the influence of the oil-automobile industry. But a majority of US citizens favour to its rectification. The findings of a survey conducted by the Centre for the Study of Policy Attitudes, the Center for International and Security Studies and School of Public Affairs, University of Maryland, USA show that while the senate harps on meaningful participation by developing countries, a majority in the US favours ratification even if developing countries do not cut emissions (Anonymous, 1998). Daina Cornelius and Jane Covers of Population Reference Bureau observe:

> Developing countries are adding the largest number of human population and account now for 98 per cent of the annual world population increase. Production and consumption in industrialised countries also have a major impact on the global environment. Individually, every person influences the environment many ways from the products he or she chooses to decisions about how many children to have and where to live, collectively, private and public institutions and government policies play a role in determining how societies affect the environment.

In the Rio conference the population issues were ambiguously referred to appropriate demographic policies, which virtually means family planning. Perhaps, this subdued reference has its own implications. It was definitely felt, even much before the conference, that controlled population growth will not solve the world's problems relate to environment and development—a shift from the conventional stand. In nutshell such a conference, has been a more concrete effort to related environment, population with development—exerting more on the former. Population has apparently been considered as a baseline for evolving guidelines for action to save the earth. Further, for the first time the very localised issues like ecological specific species has been linked with the global strategy.. The diversity is in geographic terms as well. The spatial implications in environmental issues were considered important.

Current Issues

There has been a growing awareness about the environment all over. Transnational bodies like World Bank, Governments, NGOs, institutions and individuals are increasingly interested in the state of environment even across the international boundaries. It is not only the depletion of the ozone layer, or climatic changes, or pollution which is a matter of concern, but the governments are interested in environmental issues for bargaining *status quo*, or control of resources, or transfer of environmental friendly technologies, or simply to get funds or edge over other countries in diplomacy. Not only the new organisations have sprung up at the international level, but also the well established institutions have oriented their activities in order to meet the challenges related to environmental issues. Data and literature are freshly exchanged and cross quoted in the research activities. Mostafa K. Tolba, the Executive Director, UNEP says that despite increasing environmental awareness and improved data gathering, the challenges of marine ecological change, burgeoning population growth and decreasing natural resources, still remains awesome (UNEP, 1991).

Though population has been receiving an ambiguous reference in the UNED, it is the basis on which all the discussions are directed to. United Nations Fund for Population Activities (UNFPA) in its subsequent forecasts has been predicting larger and larger population which the earth will have to cater in the near future. In 1980, it predicted that the world population would stabilise at ten billion roughly, 100 years from now. But the new estimate shows that it will be 11.6 billion by 2150 AD. This figure is based on the reduced rates in developing countries. However, if the current rates continue, the population clock will show 12.5 billion fifty years earlier (2100 A.D.). Due to deforestation, pollution and over use of the agricultural land, the environmental condition has deteriorated, particularly in developing countries. The mismanagement of resources and infighting have further aggravated the undesirable population-environmental conditions. It appears that population issues have become very crucial for the environment at the end of the present century. The population in different countries and their regions coupled with the levels of development and ability to control resources have a major impact on the environmental change. Demands for more food, energy and water have direct impact on the environment. It is the environment, which is directly or indirectly harnessed to meet such ever increasing demands of population. As a result, the forests are being vanished, top soils are lost leading towards desertification, water bodies are reclaimed or polluted and pollution are making their impact even in place far away from their sources.

The above descriptions do indicate that at the international level the priorities are changing. It is not always populations that are considered wholly responsible for the so called deteriorating environmental conditions. There are other issues as well, such as consumption patterns, technology, eco-diplomacy and the like. A population may be small but due to its control over resources and technology, it can produce same or more pollution as less fortunate large population groups. With financial control the industries can be located in different parts of the world than the places the products are likely to be used. Hence, the pollutants are generated elsewhere. Under the changing circumstances it will be difficult always to relate the industry-pollution-environment-population issues. Pattern may vary from location to location hence—geographical implications also become relevant.

Basic issue is whether population is always to be considered a problems? Since the beginning of the humanity the population is increasing and so is the environment changing due to such increase or otherwise as well. With the changing circumstances, population-environment-development equations have been modified. What is so strange now? Have we crossed our limits or has environment is unable to sustain more population or we are now better informed then earlier, hence more fear? Do not we require development? Does development not mean that a larger size or greater demand of the population is to be catered? Why are we so hypothetical? Why can't we boldly admit that the process of development is inevitable which is likely to change environmental conditions. Hence there has to be constant modifications in environment-population-development relationship. Such modifications have always taken place, and to denote them always as problem, disaster and crises is unjustified. Obviously the environment, population and development issues resulting into new equtations need relook. There is definitely a scope to have a second opinion on this subject.

The study of environment, population and development should not be synonymous with problems, crises or heading towards doomsday. Man and environment have been studied traditionally by geographers, anthropologists and other social scientists. Population become alarming only after a certain point. However, lesser population is also a problems. It is not only a resource, but also the economy that determines the carrying capacity of population. On the other hand, the environmental conditions become problems only after it crosses certain limit. There can be perception gaps between groups of residents planners, resource managers and political leaders. Further, the identification of problems is associated with how it effects the population. An environmental issue is not a problem if it is not disturbing the population in one way or other (Nag, 1997).

This volume basically concentrated on how the related issues have been interlinked. A comparative picture of India with other regions will provide some idea in this regard. Table 1.1 indicates the population, resource, pollution and infrastructure associated features.

From Table 1.1 it is apparent that the projected population of India (969.7 million) is more than Africa or Europe or North America, Latin America including Caribbean and Oceania put together. Annual increase is also on higher side. But other development indicators show India in lesser favourable condition, which is likely to deteriorate in the years to come. If this population also consumes at the level of developed countries, the impact on the world resources and pollution would be astronomical. Hence India's population and its increase is a matter of global concern and perhaps this is the reason why its population is watched so closely.

Considering the close association of the issues related to environment, population and development and the recent contributions made by Professor S.L. Kayastha, it was thought in the fitness of things to bring out an edited volume taking examples and experiences of his friends, colleagues, students and admirers. Obvioulsy, this group is influenced by the studies and contributions of Professor Kayastha who has extensively work on related themes. According to him (Kayastha, 1998):

> An integrated approach to the study of population, environment and development as a whole remains a significant area of geographical research and facilitates the investigation of integrated problems or interlinked studies. Geography deals with both natural and social phenomena, and hence the field of its concern includes population, environment and development. For a long time, the need for integrated studies of population, environment and development has been felt.

Table 1.1: Environment, Population and Development Indicators

Indicators	World	Africa	Europe	North America	Latin America & Caribbean	Asia	Occania	India
Population (millions) mid-1997	5,840	743	729	298	490	3,552	29	969.7
Natural increase (annual, %)	1.5	2.6	0.1	0.6	1.6	1.6	1.1	1.9
Projected population (millions) 2025 AD	8,036	1,313	706	372	691	4,914	39	1,384.6
Deaths under age 5 per 1,000 children 1995	70	145	15	10	46	75	8	115

(Contd.)

Table 1.1: *(Contd.)*

Indicators	World	Africa	Europe	North America	Latin America & Caribbean	Asia	Occania	India
Percent under age 5 underweight 1990	–	26	–	–	10	33	–	53
Annual renewable freshwater available per capita (cubic m) 1990 2025	9,255 5,896	5,532 2,386	8,699 7,918	19,464 14,211	29,818 18,359	4,767 3,031	36,249 25,960	2,451 1,498
Percent access to safe water 1990-96	–	56	–	–	75	74	–	81
percent with adequate sanitation 1990-96	–	46	–	–	60	36	–	29
Croplant available per capita (ha) 1990 2025	0.26 0.16	0.29 0.13	0.40 0.37	0.84 0.63	0.35 0.22	0.15 0.09	2.45 1.69	0.20 0.12
PPP per capita(inst 1$)1995	6,045	1,961	13,134	26,389	5,824	3,565	18,520	1,400
CO_2 emission per capita (metric tons)1992	3.8	1.3	8.3	18.6	2.2	2.2	14.0	

Source: Modified from Cornelius & Cover (1997).

The potential authors were approached to contribute on such integrated study that could possibly link the issues. The authors kindly agreed to our request. Perhaps this is the better way to felicitate professor, teacher and researcher who has a life long contribution on similar issues.

This Volume

In her Chapter on the Titanic Environment, Minati Singh has made an attempt to give a wholesome view of the study of environment and its current relevance. She discusses about the basic concept, problems, forests resources, pollution, biosphere, ecological evolutionary process, developing activities, population settlements and the like. G.K. Chakravarty in his paper entitled cultural basis for the construction of the environment has given the anthropological view to study environment and development issues of specific populations groups such as tribes. Sukumar Bose has highlighted the emerging frontier in environmental psychology, known as Ecopsychology. The basic concepts are very close to perception research in geography. Recently, these studies have been carried out in this subject under the purview of Perception of the Environment. Professor Kayastha has made a lot of contribution in this branch as well

A.P. Mishra has highlighted the problems of poverty and malnutrition in one of the most backward area of he country, i.e. Sonbhadra district. Here lopsided development has not been beneficial to the tribal population that reside there. But such is not always the case. S.C. Kharakwal in his study of Himalayas, have tried to establish relationship between environmental and developmental issues. After reviewing the literature on the subject, he has indicated the obstacles to sustainable development, and finally dicussed possible strategies for sustainable economic development. Again in Himalayas, in Pauri-Garhwal district, integrated approach consisting of eco-development, micro-level planning and thematic mapping has been suggested by S.B. Singh and Rupali Juyal. Similar approaches have been advocated by the Natural Resources Data Management System of the Union Department of Science & Technology. V.K. Kumra and C.S. Kumar have made a comprehensive study regarding landuse and environmental problems in the country. Landuse changes have been analysed over a period of about four decades (1950–1989) and how environmental problems have effected or could affect such changes have been discussed. These problems are deforestation, floods, droughts, desertification and the like. However, the environmental problems in the coastal areas are entirely different. Anjana Desai and Gautam Bhattacharya have identified coastal problems like marine ingression, erosion, and soil salinity which have led to out migration of population in the Bhal region in the coastal areas of Gujarat. Geographical Information System (GIS), remote sensing and maps have been used for detailed monitoring of the environmental changes. In the eastern side of the country, the environmental problems are similar and here also there has been a loss of agricultural land. Government and semi-government projects have been implemented with different degree of success. This study by S. Bandhyopadhyaya is based on opinion survey of the three village in Sagar island in South 24 Paraganas district. Nandini Chatterjee has carried out another perception study. She has analysed responses regarding social forestry in the six villages in Rajnagar and Suri police station areas of Birbhum district.

Some studies are always better at sub-national or regional levels (Nag, 1997). One of them is related to irrigation water requirements. V.K. Rai has pointed out that the existing irrigation facilities in Ganga-Ghagara *doab* are not sufficient to meet the requirement for agricultural development. Over the years, the traditonal system of irrigation with the help of ponds *bundhs*, masonry wells have been replaced by ground water harvesting through deep tubewells, pumping sets and of course canal networks. He suggests that a balance between surface and ground water irrigation should be achieved for desired development. Since the Noble prize has honoured Professor Amartya Sen, the attention has shifted towards social infrastructure that includes education and health (Bagchi, 1998). B.B.K. Sinha has rightly emphasized on the latter issues. According to him, development of a nation depends on the health level of the people both physical and mental. For this purpose he has surveyed contributions of several authors and finally proposed a model for grading the quality of health. In the same continuation, M.B. Singh, R. Singh and Siddarth Priyadarshi have inquired not only the health care infrastructure in general, but also acceptance of family planning. According to them, lack of education, apathy of the government staff, poverty and inadequate infrastructure are responsible for getting desirable result. Population at micro-level has also attracted geographers. N.D. Bhattacharya has attempted to highlight variations even in a small area of a district, i.e. Murshidabad of West Bengal. He has looked into different population facets historically, spatially, demographically and in terms of composition and migration.

Urban areas have always considered as laboratories for environmental, population and development studies. In fact, environmental issues have rejuvenated urban studies in geography. K.K. Dube has made an attempt to study the dynamics of urban phenomena in India. He has taken into consideration of evolution, distribution, morphological characteristics and fringe development to highlight the emerging urban problems. One of the solutions to urban problems, according to him, is to revitalise the small and medium size towns. S.K. Biswas has been more specific in analysing urban issues. He has taken into consideration of the demographic characteristics such as trends of urbanisation, sex ratio, literacy, working force, growth rate and hierarchy of towns. According to him, qualitative improvement of urban environment and simultaneous development of economic base of the metropolitan areas, which dominate the urban scenario, are absolutely necessary for overall development of the country. Further, in their study on lakes in Bangalore City, Daksha C. Barai and P. Prabhu have discussed the physical and cultural environment. It is interesting to note how the urban development due to population pressure has effected the environment. Slums in urban areas have developed for similar reasons, i.e. increasing population, lack of housing and economic conditions which do not encourage housing for the poorest. V.K. Kumra and B.L. Singh have found these reasons while studying slums in Varanasi. The socio- economic characteristics of these high density housing entities makes this study more interesting. Nevertheless, the case of Kharagpur town its different. Sukla Basu mentions that due to railways and the I.I.T., population migrated from neighbouring states and is now assimilated into one society, almost rejecting the regional, religious and language affinities.

One group of authors have tried to give another view of the urban situation. They have been more concerned with renewal, people's perception and the like. Udayan Majumdar has highlighted different planning and development priorities and activities of planning bodies in their attempt for renewal of Calcutta. According to him, these projects were lacking residents' participation and hence did not meet desirable success. Against largely unplanned Calcutta, Gandhinagar was planned since its inception. In the latter case, the infrastructure and amenities developed slowly. This study by N.G. Dikshit is based on the respondents impressions about the urban environment. They have responded positively irrespective of their social background. In study of Calcutta, Nisit Som, S.K. Thakurta and P. Nag, the prospects of using old garbage in the eastern fringe has been highlighted. After critical evaluating different reports by the experts on the subject, the authors have attempted a new direction towards a viable and less expensive solution of the solid waste management system by utilising the garbage as a good fertilizer after being transformed in a natural process.

Smita Sengupta has made detailed study on the impact of mining activities in the coal fields of Ranigunj. This area has the oldest coal mines of the country and that too operating from very old primitive and manual method to mechanised automatic system. Nevertheless, the mining activities have lead to grave, serious and reached an alarming state. The author has given valuable suggestions for reducing this environmental hazard created by open cast mining. Similarly, in the adjoining Damodar Valley, Swapna Basu has highlighted the impact of wide scale industrialisation that has lead to environmental pollution. The emphasis here is on water pollution. The initiation, causes and results of such pollution have been studied in details. Some suggestions have been made to control this pollution, specifically considering the huge population residing in the new township of Durgapur. Further, Kefialew Abate and H. Prasad, in their study on another developing country of Ethiopia, have given a different opinion of environment, population and development relationship. According to them, crop yield and per capita productivity did not increase over the 36-year (1948-85); and the increasing population pressure must have fallow periods and hence induced desertification *in situ.* This study recommends further investigation regarding the possible relationship between population growth and land degradation.

The Great Expectation

The obvious question out of this study is what new is expected from this volume? How far the recent contributions of Professor Kayastha can be linked with the contributions made here? Is it another edited volume being churning out from various publishing houses in India? What is the focus and what is the necessity? Have we been able to swim across the tide? These issues may pose some basic issues. Perhaps this volume is not linked with the total coverage of the contributions made by this learned scholar on whose name this volume is dedicated. The expectations may be unlimited. But a volume like this can do little justifications to the scholarly work done by Professor Kayastha. At the most we can claim that we will try to toy with the issues and challenges raised by him during his active period stretching for over four years in general, and last one decade in particular. Perhaps several volumes like this will justify his contributions in the study of environmental, population and development issues. We the editors and contributors would closely watch this situation to develop.

References

Agarwal, B. (1992), and the gender and environment debate: Lesson from India, *Feminist Studies* vol. 18 (1).

Agarwal, Bina (1995), Feminist environmentalism or ecofeminism? *The Hindu Survey of the Environment*, pp. 7-14.

Anonymous (1998) US public says yes to Kyoto Protocol, *Down to Earth*, vol. 7, No. 15, December. p. 62.

Bagchi, A.K. (1998), Amartya Kumar Sen and the human science of development, *Economic and Political Weekly*, December 5, pp. 3139-50.

Chandler, William U. (1990), Development and environmental change, *Economic Impact*, vol. 2, pp. 18-25.

Clarke, John I. (1997), *The Future of Population*, Predictions Series, Phoenix, London.

Cornelius, Diana & Jane Cover (1997) *Population and Environment Dynamics*, Population Reference Bureau, Washington (data sheet).

Daly, Herman & John B. Cobb, Jr (n.k.) *For the Common Good : Redirecting the Economy toward Community, The Environment, and a Sustainable Future*, Beacon Press, Boston.

Fri, Robert W. (1990) Energy and environment : A Coming collision, *Economic Impact* Vol. 2, pp.33-35.

Gupta Avijit (1997), Geographers as environment managers : The lost opportunity, in Hazra, Jayati (ed) *Dimensions of Human Geography*, Rawat Publication, Jaipur and New Delhi pp. 54-59.

Handi Gunther (1990), Law and Protection of the atmosphere, *Economic Impact,* Vol. 2, pp. 36-40.

Hema Malani, B. (1996), Influence of biosphere on urban micro-climates, in R.B. Singh (ed) *Research in Geography : Land Use Changes and Sustainable Development,* Volume 1, A.P.H. Publishing Corporation, New Delhi, pp. 155-64.

Kayastha, S.L. (1998), *Geography of Population,* Rawat Publication, Jaipur & New Delhi.

Nag, P. (ed) (1991) *Essays on Environment and Resources : Some Regional Issues,* Deep & Deep, New Delhi.

Nag, P. (1986) Environmental perception, in H.H. Singh *et al* (eds), *Geography and Environment : Issues and Challenges,* Concept Publishing Company, New Delhi, pp. 31-50.

Nag, P. (1997) Emerging environmental issues: Do we care for a second opinion? in P. Nag V.K.Kumra and J. Singh (eds), *Geography and Environment : Local Issues,* Vol. 3, Concept Publishing Company, New Delhi, pp. 285-92.

Nag, P. & M. Kudrat (1998) *Digital Remote Sensing,* Concept Publishing Company, New Delhi.

Reid, Walter V.C. (1990), Sustainable development : Lessons from success, *Economic Impact,* vol, 2,pp. 26-32.

Repetto, Robert (1990), The case for national resource accounting *Economic Impact* vol. 2, pp 41-46.

Sen, Amartya (1999), Hunger in the modern world, *Yojana,* vol. 43, No. 1, January, pp. 9-14.

Singh, H.H., P. Nag, V. K. Kumra *et al* (eds) (1986), *Geography and Environment: Issues and Challenges,* Concept Publishing Company, New Delhi.

Singh, L.R. (ed) (1996), *New Frontiers in Indian Geography,* Professor R.N. Dube Foundation and Geography Department, Allahabad University, Allahabad.

Singh, Savindra (1994) *Environmental Geography,* Prayag Pustak Bhawan, Allahabad.

Singh, S.K. & P. Nag (eds) (1999), *Tourism and Trekking in Nainital Region.* Concept Publishing Company, New Delhi.

UN (1989), *Prospects of World Urbanisation,* 1988, Population Studies Report No. 112 UN Department of International Economic and Social Affairs, New York.

UN Centre for Human Settlements (Hivitat) (1987), *Report on Human Settlements,* Oxford University Press, Oxford.

UNEP (1998), *The United Nations System-wide Medium-Term Environmental Programme,* 1990-1995, United Nations Environmental Programme, Nairobi.

World Commission on Environment and Development (1988), *Our Common Future,* Oxford University Press.

2

The Titanic Environment

Minati Singh

One of the most important and the foremost problems of environmental management is that of understanding the concept of environment. It is perceived differently by different people, *viz.* environmentalists, geographers, planners, decision-makers, administrators and residents. However, its real image is quite comprehensive and complex because the elements of the environmental system are highly interconnected. Man comes across with only his limited surroundings in his every day life. The familiarity with environment and its perception depends on individuals interaction with it and his exposure to it; and therefore, he does not become aware of the ecological factors beyond his limited zone of action. Therefore, the concept of environment with its ramifications is not clear to many individuals working within a limited area of geographical space and narrow specialisation. Different individuals manifest different levels of concern to environmental problems and adopt different approaches.

Concept of Environment

Although, there is much awareness of the present day environmental issues, the idea of environment is not a new specially in geography and to geographers. Geographers pioneered the concept of man-environment relationship in 19th century and inherited the tradition of teaching environment in geography. However, the study of man-environment relationship in geography was more or less discarded when public awareness was being aroused.

Ratzel in Germany, Ellen Churchill Sample and Ellsworth Huntigton in the United States and Giffith Taylor in Canada wrote for re-examination of human history that tended to make man the creature of his spatial environment; rather than its modifier. Little of this determinism could withstand criticism, and it was soon shot-down by heavy handed methodologists, mostly Americans. This led geographers into a wilderness of ideologies from which the relation of man to nature was excluded. The discipline could still flourish, but it lost the opportunity that should have been seized to lead the universities towards a sound treatment of the subject of environment.

The study of man-environment relationship has crossed a long way from the classical to the modern period. The conceptual and methodological approaches have also gone through several changes and have proceeded in many directions. Notions of environmental determinism and possibilism tried to explain man's role in changing the face of the earth. Both the concepts have separated man from the physical system. To a determinist, nature was supreme, while to a possibilist, man was the main force in the interaction system. Neither determinism nor possibilism could bring out the real image. None of the approaches could go far to explain human behaviour over the geographical space.

Later, ecological approach accepted man as an element within the framework of man-environment relationship. Geography was defined as the science of human ecology. Man-environment relationship was replaced by a more powerful concept of an eco-system in which oganisms and environment are one interacting entity. The eco-concept could explain well the physical system of environment e.g. hydrological cycles. The over use or misuse of air, water and space in urban environment are examples of pollution of the eco-system by mankind which is difficult to rectify (Desai, 1985). In this context, the environment as defined by Webster is the aggregate of all the external conditions and influences affecting the life and development of an organism. As per report of UNEP the environment is defined as that whole outer physical and biological system in which man and other organisms as one unit; ae beit a complicated one, with many interacting components (UNEP 1976). These external conditions can be social and physical; the setting can be rural or urban. Since man is an integral part of the environmental system, any change in the system will certainly affect the life of man as well as other living organisms.

To some the term environment conjures up thoughts of woodland scenes with fresh and clean air and pristine waters. To others, it means their man-modified neighbourhoods or immediate surroundings. Still others relate environment to ecology and think of plant-animal interrelationship, food chains, threatened species and so forth. Actually the word environment is a combination of all these concepts plus many more. It included not only the areas of air, water, plants and animals but also other natural and man modified features (transportation system, landuse characteristic, community structure and economic stability etc.), which constitute the totality of our surroundings. In other words, the environment is made up of both biophysical and socio-economic element. Once the importance of ecological and environmental studies was realised by intellectuals and scientists, they started thinking about the ways of management of ecological factors and methods of preventing their pollution and decay. The wise management of the environment depends upon an understanding of mother earth and its components, of its rocks, minerals and waters, of its animal life and potential for livestock, husbandry and of its climate. It depends on positive and realistic planning for balancing human needs against the potentiality of environment. Some people think only of pollution when they consider environmental problems. Good management avoids pollution, erosion and the wastage of resources by irreversible damage.

The natural environment contains our whole patrimony of resources both renewable and non-renewable. The man made environment is composed primarily of shelter, and the supporting infrastructures such as energy, conventional and non-conventional communications (transport, road network etc.) services (water supply, waste disposal etc.) and community facilities (catering to the social cultural recreational connected activities of men). The cultural activities may disturb the physical environment by accelerating the process of environmental degradation. Different types of pollution, directly or indirectly, related to use and utilisation of natural resources may be effective. These changes in the ecological system would have far reaching consequences on the survival of the human race itself.

Economists, industrialists and financiers still adhere to the growth principle and argue that economic growth can raise the standards of living and a fair distribution of employment and income be achieved. Economists in particular, find it hard to accept the reality of a finite earth, or a stable society. Marx argued that environmental degradation and the resource depletion are the results of capitalism and imperialism. They would vanish in a socialist million. And lastly the environmental activists would plead the case for conservation for more stridently than others. Therefore, we must seek political solutions to environmental issues and resource depletion. Problems like burgeoning population, depleting resources, weakening institutions, loss of biosphere equilibrium and several others clamour for easily solutions.

Environmental Problems

Environmental problems, which are currently receving attention, are concerned mainly with the adverse impact of human activities and specially of the careless technology. And these adverse impacts damage the environmental resources and ecological foundations of the life support system. Environmental problems can be generally classified into three different groups:

(a) Environmental problems by destruction of ecological substance,
(b) Environmental problems by discharge of ecologically subversive material into the ecological system, and
(c) Obstructions to the ecological evolutionary process.

Problems by Destruction of Ecological Substance

Ecology deals with that part of the nature which is essential for the survival, existence and the improvement of the human species as a whole. Air, water, land, flora and fauna are the elements that constitute the life support system for the species. These are interconnected, interrelated and interdependent and also co-evolved and are co-adopted. Therefore, degradation of one of its elements may mean degradation of the total ecological system itself. The importance of a system approach to the problem is thus clearly evident. The nature of deterioration is of two type i.e. short-term or long-term. Depending upon the resilience of the system, the short-term ones can be repaired. However, in the long-term process, a continual deterioration of ecological environment posses a threat to the very existence of life on the earth. With the process of technology improvement and development, human activities assume such enormous dimensions that the life support system can no longer sustain these. Accordingly, the waste generated through human activities is much more than the system can absorb or assimilate. This has resulted in the problems of pollution.

Pollution is the most important aspect of the environmental problem which originates from natural and man made sources. "Every year millions of tons of gaseous and particulate pollutants are injected into the atmosphere both through natural processes and as the direct result of human activities. Scientists have pointed out however that the earth's atmosphere can not absorb unlimited amounts of pollutant materials without undergoing changes which may be of an adverse nature with respect to human welfare. (UNCHE, 1972)". This being an unsatisfactory situation, it needs to be changed by taking corrective measures. In India, the problems related to environmental degradation and pollution are very serious and alarming.

Forests

Forests have variety of ecological roles to play and deforestation is one of the most important matters which has attracted the attention of ecologists all over the world. Destruction of forests results in changes in the behaviour of monsoon, soil, erosion by rain water and floods, destruction of various species of birds and animals, changes in the composition of atmosphere due to absence of gases and other materials contributed by the trees and plants.

The problem of deforestation is quite staggering in India. About 16.8 per cent of the total land was covered by forest during 1974-75 and this percentage came down to 13.9 per cent during 1980-82, a drop of over 17 per cent in the total area under forests in eight years. Satellite data have revealed that India is loosing 1.3 million ha of forests every year (i.e. eight times more than the official figures released by the forest department). It is nothing short of tragedy that even at the current rate of afforestation (0.4 million ha annually) the country is losing over one million ha of forests every year. The per capita forest land according to official figures works out to 0.11 ha, which is obviously low.

The demand of wood for fuels, timber and industrial uses including paper, plywood, railway sleepers, buildings, and furniture is growing faster than its supply. The problems of deforestation remains quite staggering. The net area under forests is shrinking. The reduction in forest cover has caused changes in hydrologic cycle including surface run off and ground water recharge. Larger areas are now subject to drought and flooding. For example, Kerala which usually receives 3,200 mm of annual rainfall, lost approximately Rs. 1,000 crores in 1982-83 as crops failed and industries closed down.

About 100 million ha are classified as waste lands, i.e. areas seriously affected by salinity and wind and water erosion while 8,000 ha are annually lost to ravines. If the current rate of forest loss continues the country would be heading for a disaster after two decades. The run-off increases, because of deforestation, and about 16 million ha metres of water flow unharnessed into the sea every year. The loss of top soil (4,700 million tonnes), quantity-wise is the maximum in India, being 18.5 per cent of the total soil loss at the global level (25,400 million tonnes). This is indeed very serious, considering the fact that India has only 2.4 per cent of the land area of the world.

The grass lands are being degraded because of over grazing. The country has the world's largest number of livestock but their productivity level is very low. In fact, parallel to human population, there is explosion of livestock population. Over population of livestock, in ecologically fragile habitat, leads to very serious environmental consequences.

The scheme of 'mother nature' is so perfect that it maintains a dynamic equilibrium in the ecosystem. Nature has her own arrangement to convert waste generated by one species into food for other species. The system of recycling ensures that this equilibrium is maintained continuously. If any of the participants in this system is destroyed, it may upset the recycling systems, and may create problems for the human being.

A wide range of flora and fauna are fast disappearing, as their natural habitat are distroyed. This further improverishes the remaining Indian forests. The ecological balance of the habital system is being disturbed by a rapid increase in human population, growing urbanisation and industrialisation. These processes add foreign elements in the natural ecosystem and thus destroying its efficiency and stability.

Aquatic Resources Problems

The aquatic resources are facing depletion due to over exploitation, habitat destruction and pollution of water all due to human intervention. Pollution owing to siltation and chemical and organic wastes is harmful. Effective control of land based pollution, therefore, be a top priority, and the sea should not be treated as dumping ground even for nuclear wastes. One inexpensive way to monitor pollution is through bio-indicators involving plant and animal species. However, there is little information available about the genetic diversity that these habitats support which of course, would be of great value for aquatic culture programmes. The introduction of exotic plant or animal species could be very hazardous for the native species as it may disturbs the biotic environment.

There are certain aquatic plants which are efficient in absorbing nutrients from water. These can be used for decentralising the effluent resulting from enerobic decomposition of raw sewage before it is let into a fresh water body. Some microscopic plants fix atmospheric N, which is of great agricultural significance, while other absorb tonnes of pollutants. There are still others that are sources of food and feed materials to human and animals. In fact, without aquatic plants, and consequently, the aquatic animals, most water bodies would be just wet deserts. But for some reasons, if the natural balancing amongst the components of an aquatic ecosystem is disturbed it leads to over growth of one or more of its plant components. This may be a negative factor as well. Thus, over abundance of aquatic plants is no doubt a matter of serious concern.

Mining and Minerals

Quarrying and extraction of minerals and energy developments also induce environmental changes by destabilising the ecosystem. Mining in an indiscriminate manner may cause a lot of ecological damages. The environmental implications range from serious water and air pollution, degradation of land and deforestation to noise and ground vibration problems with their after effects.

Exploration, mining and drilling, may degrade the environment:

- by destablising a large tract of land,
- by adverse impact on plants and wild life,
- contamination of fresh water bodies, and discharging dust and fumes into the air.

These activities damage the quality of the atmosphere and exposes the miners to serious health hazards owing to the noxious nature of emissions. Most of the damages from mining are local in nature, but these have adverse effect on the environment. Acid drain mostly from coal mines, sulfur dioxide from metal smelters and seepages or blowouts of offshore oil wells are not exceptions. Once mining activity starts, other associated activities follow, and these together lead to the destruction of the environment at a large scale.

The Damodar river provides a good example. Mining and major mineral based industries are located in the Durgapur-Asansol belt in Bengal and in the contiguous Chotanagpur plateau in Bihar. Damodar is the principle source of water, yet it is highly polluted.

The environmenal impact of mining may include the following:

- Loss of vegetation (forest, agriculture and pasture)
- Loss of top soil
- Surface water pollution
- Lowering of ground water table
- Hazards of ore transport, in the form of

(i) damage of vegetation
(ii) Soil, drainage
(iii) Water quality
(iv) Sediment production and discharge
(v) Fire hazards (in coal mine where inflammable gases are released)
(vi) Air pollution, and
(vii) Noise pollution.

These environmental impacts could be broadly classified into five categories i.e. (a) water pollution, (b) air pollution, (c) despoliation of land, (d) deforestation, and (e) noise and ground vibration problems.

(a) *Water pollution:* Mining operations contribute in two ways towards water pollution; firstly through the discharge of acid mine water into streams or water bodies and secondly, by additing toxic substances like selebium, nickel, fluorine and other elements, which are radio active to the water bodies.

(b) *Air Pollution* : Mining operations contribute towards the air pollution in two ways : (i) addition of gaseous pollutants to the atmosphere, and (ii) the emission of dust particulates. The sources of gaseous pollutants which include sulphur dioxide, oxides of nitrogen and carbon monoxide are mainly the waste dumps. The dust particulates are produced during ore, and coal handling, blasting and transportation. Mine fires, and solid waste disposals are to a great extent responsible for polluting the atmosphere. This hazard has been quite visible in Tharia coalfields which have been ravaged by these fires.

(c) *Despoliation of land* : Some problems associated with mining which results in despoliation of land are :

- Subsidence of land, due to underground mining and large scale excavations.
- Land degradation due to surface mining methods which produce large scale excavations and the resultant overburden and mineral wastes spreading over large areas, making them unusable. Open cast mining also leads to rapid erosion of land because of slope in stability problems and denundation of vegetation. Detrimental effects of soil erosion are ultimately felt in silting and degradation of streams and water bodies.
- The waste dumps which are created as a result of open cast and underground mining could, if not constructed and treated properly, lead to slips or erosion.

(d) *Deforestation* : In many cases the valuable ore bodies are located in either dense forests or adjacent to such areas. During the construction of infrastructure and pit site, the vegetation is destroyed, several mining areas are getting bereft of dense forests posing tremendous problems. Massive deforestation drastically changes the climatic conditions like rain fall, temperature and humidity. This may also result in death of wild animals and birds living in these forest areas.

(e) *Noise and Ground Vibration Problems* : With the increase in the scale of mining operations heavy blasting has to be resorted to in the large open caste mines. This results, in ground vibrations, which may cause damage to the structures in the vicinity besides causing irritation to the people living nearby.

Destruction of mountains due to quarry or other projects, disturbs the flow of air currents which significantly changes the wind pattern of monsoons around that locality. Indiscriminate destruction of mountains might change the pattern of season and monsoons so much that it might pose a great problem for the people.

Biosphere

The biosphere is a thin envelope over this globe which is known to sustain the only known life in the universe. The equilibrium of the biosphere is very delicate and many actions induced by the man affect it in many ways. It is due to the biosphere that we are protected from the harmful radiation heat and other energies that continuously hit our planet. Even an insignificant act of using aerosol sprays for scents, soaps or paints eats a way the ozone layer of the biosphere which protects us from ultra violet rays.

Within the biosphere one can identify the organic components (plants, animals including human being and micro organisms) and the inorganic environmental components. Both these components are intimately interconnected and interdependent and are interrelated to a series of large scale cycling mechanisms which involve transfer of energy, water, chemicals and sediments throughout the biosphere. In a natural state, the biosphere can attain a state of equilibrium which is self sustaining and ecologically efficient. With environmental modifications, this state of equilibrium can be disrupted either partially or fully and may lead to widespread environmental and ecological damages.

There are three sources of energy in a biosphere, namely gravity, internal forces within the earth and solar radiation. Among these the solar radiation is the most important as it gets converted into chemical energy by plants through a deceptively simple but a very important and powerful process of photosynthesis.

Only about 51 per cent of solar radiation reaches earth, the remaining 49 per cent is either absorbed in the atmosphere or reflected upwards. By altering the earth's surface through urban development, it is possible to modify the energy balance over a localised area.

Solar energy is considered to be the main source of life on this earth. It has been estimated that 'of the total solar radiation entering the outer limit of the atmosphere about 30 per cent is reflected back into space by clouds and the rest of the atmosphere, about,13 per cent is absorbed by the ozone layer of the stratosphere and 7 per cent is absorbed by the remaining atmosphere which is warmed up in the process (Ryabchikov, 1975). It is estimated that less than half of the solar radiation is absorbed by the earth's surface, which becomes the energy base. The difference in solar heat between the tropics and the polar regions, between the ocean and the land creates the circulation of the atmosphere (Fedorov, 1980). The circulation forms various ways of interconnection leading to a complex system where each and every mobile and over changing element maintains a balance in respect to one another. This state of biospheric balance is referred to as 'natural equilibrium'. But in time-span man has disturbed and is still disturbing the equilibrium of the natural environment, of which three main forms are clearly visible: (a) changing the structure of the earths surface (ploughing the steppes, cutting down forests, land improvement, creating man-made lakes and seas, and influencing the surface water system in other ways), (b) changing the composition of the biosphere which means in effect changing the balance and circulation of constituent substances through extraction of minerals, creation of spoil banks discharge of various substances into the atmosphere and into water bodies, and through changes in moisture circulation, and (c) changing the energy balance (including the heat balance) of individual areas and of our planet as a whole (Fedorov, 1980).

Industrial Pollution

There is an ever increasing encroachment on natural greenery resulting in the shortage of forest lands, pastures and cultivable land. Increasing industrialisation has led to the pollution of atmosphere by emission of poisonous gases, pollution of water bodies by disposal of untreated industrial wastes into them and pollution of natural surroundings by the noise of machines and vehicular traffic. There is reduction of flora, fauna and wild life all around.

Industries pollute the atmosphere in different ways:

(a) Water is polluted by :
  - Chemicals,
  - Solids, and
  - Liquid.

(b) Air is polluted by :
  - Poisonous vapour,
  - Smoke,
  - Dust and other particles.

(c) The general atmosphere is polluted by noise beyond permissible limits.

(d) Land is polluted by the dumping of wastes, making of bricks, use of fertilisers and pesticides.

We shall now consider these pollution factors one by one.

(a) *Water Pollution*

The large and medium scale industrial units contribute 70 per cent by volume of the total industrial wastes in India. The remaining 30 per cent is generated by the units in small and cottage sectors. Any success in controlling this 70 per cent of the total industrial wastes would however take care of only 7 to 10 per cent of the total volume of waste waters generated in the country. There are, in India, about 2,700 large and medium scale industrial units excluding those in Maharashtra, Tamil Nadu, Orissa and the northeastern states. According to a rapid survey there are 1,700 problem industries, however, the number of such industries may range from 2,500 to 3,000 in the entire country.

Only 353 large and medium such industrial units have been operating industrial waste water treatment plants. In addition, 174 industrial effluent treatment plants are under construction and another 190 units are in the planning stage.

About 70 per cent of water resources in India are polluted mostly as a result of the effluents from mines, industrial and municipal wastes, which are drained into them. The problem of industrial pollution exists in various pockets of the country. This is rather due to the inadequacy of measures adopted for effluent treatment, than to the intensity of industrial activities. Therefore, industrial effluents have a major share in polluting the water bodies. The rivers are particularly vulnerable to an indiscriminate discharge of industrial effluents, leading to mass mortality of fishes and fauna in different parts of the country. Industrial units located near the river banks are disposing industrial waste and debris into the river. Oil leakage and discharge from petrochemical plants have been known to cover the river water for miles at a stretch and thus suffocating all organisms living in water. Some times these floating sheets of oil catch fire and play havoc. Effluents from pulps and paper mills, synthetic rubber factories and chemical factories, fly ash from coal washeries and thermal plants cause extensive pollution. Some of the effluents consist of dangerous chemicals like carcinogens, arsenic nickel, asbestos and vinyl chloride. In most of the industrial cities in India the volume of industrial liquid wastes was about 8 to 16 per cent of the total volume of the city waste generated during 1978-80. By 2000 A.D., the volume of waste water discharged by the Indian industry is estimated to be half of the volume of domestic sewage generated in cities. Whatever may be the ratio of industrial waste to domestic sewage, the stupendous volume of waste water generated in major cities needs to be disposed off hygenically. Delhi consumes about 1.20 million kilolitres of water every day and pours back 0.96 million kilolitres of toxic water into River Yamuna, Ganga and Pandu rivers are the sources of water supply as well as dump for urban sewage. About 40 per cent of the water supply in contaminated. About 40 mgd of waste water is discharged into River Ganga through the sewers alone. In Calcutta too, the situation is very critical. The discharge of waste water from hundreds of industrial units (paper mills, jute mills, textile mills, chemical factories, engineering factories) measures 77 mgd through 350 out falls as against the total of 252 mgd (Haldar, 1984).

Progress in urban sewarage is distressingly slow while that in sewage treatment is extremely unreliable. Hardly 114 towns are partly sewered covering about 26 per cent of urban population. During the water supply and sanitation decade (1981-1990), it has been presumed that all class-1 towns would be provided with sewarage and sewage treatment systems, and all towns and villages would have proper water supply.

The sewage water in Chennai and Mumbai is directly discharged into open sea through many open drain. The total quantity of sewage discharge in Chennai is estimated as 30 mgd. About 445 million litres of untreated sewage is discharged in Mumbai every day in the Arabian Sea.

Marine Pollution

With the growth of cities ports industrial establishments and tourist infrastructures along the sea coasts, vast expanses of the coast areas have been altered. The degradation and multiutilisation of beaches are affecting the aesthetic and environmental values of the coasts. These could be avoided through prudent coastal developments and management based on assessment of ecological values and potential damages.

The oceans have been one of the main thoroughfares of commerce and a principal area for disposal of the industrial and domestic wastes. These will increase the probability of future environmental damage by pollution. Various agents that may cause damage to marine life are: (a) petroleum and its products, (b) halogenated hydro carbons, (c) organic chemicals such as marine biotoxins and detergent, (d) nutrient chemicals like those found in domestic sewage and organic wastes, (e) inorganic chemicals especially, heavy metals like lead and mercury, (f) radioactive substances, (g) thermal wastes and (h) suspended solids.

The numerous sources of marine pollution are : (a) disposal of domestic sewage, agricultural or industrial wastes, (b) operation and accidental or even delibarate discharge of shipborne pollutants like fuel or oil, (c) interference of the marine environment by exploitation, (d) disposal of radio active wastes from nuclear power plants, and (e) military uses of the ocean.

As a large population of India lives in coastal areas, the health of the seas is critically influenced. The rate of sewage production in metropolitan cities is about 120 litres/person/day and about 60 litres/person/day in non-metropolitan areas of the country. The total addition of sewage waste to coastal waters would be of the order of 35 km.

The industrial waste going into the coastal waters have been estimated to be about one-tenth of the total sewage outfall. This would mean that another 3.5 km of waste per year comes from the industries located along the coast. Industries discharge about 0.41 km of wastes in the delta of the River Ganga which has an annual run-off of 370 km^2 Using almost a similar rate of waste disposal for all the Indian rivers, the total industrial wastes of inland origin going into the sea would be 0.1 km.

The production of pesticides in India during the period 1970-76 was about 2,50,000 tonnes. Similarly 1,10,000 tonnes of synthetic detergents are being used annually in the country. Assuming that about one-fourth of these will finally reach the sea, the total addition of pesticides and detergents to the sea would be 62,500 and 27,500 tonnes respectively (Khoskoo, 1986). In 1978, about 1,681 million tonnes of crude oil and its products were transported by sea and approximately half of this was shipped along the tanker routes across the Indian Ocean. Such enormous quantities of pollutants when added to the sea around India will undoubtedly effect the water quality and plant life of marine environment. Therefore, monitoring of marine pollution is necessary.

The coastal water near Mumbai and Chennai the southern half of west coast (Cochin, Ernakulam, Trivandrum) and the sea shores of almost all the coastal states suffer from severity of coastal pollution due to the discharge of waste water (sewage) from the class-I and class-II cities. The use of the ocean as a commercial medium for transport of man and material has also contributed to the pollution of sea-coasts. In the areas surrounding big harbours like Khandala, Mumbai, Chennai, Vishakhapattanam and Calcutta, pollution is abundant due to discharge of rubbish from the ships in harbour and also due to Jettisoning of goods the ships do not want to carry.

(b) *Air Pollution*

The quality of human health greatly depends on the quality of air available to man for breathing, wherever the quantity of atmospheric oxygen decreases the people fall victim to various health ailments.

Apart from this, if poisonous gases, dangerous dust and particles or chemicals are released into the atmosphere these are likely to add to the misery of the people. Pollution creating industries are rapidly comming up in more so in developing countries giving proper thought to environmental hazards. A large proportion of population in cities like Calcutta, Kanpur and Ahmedabad with large scale industrial development are victim of environmental pollution related diseases.

Atmospheric pollution in industrial sector occurs mainly through smoke and vehicular traffic. Smoke pollution results mainly from combustion of fossil fueles especially coal, and is closely related to population density climate and topography. Sulphur dioxide pollution in urban and industrial areas has a major impact on the respiratory condition of the population and on the crops and other vegetation of the surrounding rural areas.

The fluoride pollutants, where fluoride containing ores or clay are smelted or fixed e.g. aluminum smelters, and the brick kilns, are more localised than other forms of pollution. The air near major traffic junctions in Delhi is highly poisionous. Inspite of the benefits accruing from increasing transportation i.e. accessibility distribution of goods and services, trade and commerce, there is the potential danger of causing undesirable and harmful effects on quality of environment, specially air. With more than 70,000 heavy vehicles entering Delhi daily, the discharge amounts to about 170 tonnes of carbon monoxide, 70 tonnes of hydrocarbons, eight tonnes of oxides of nitrogen two tonnes of sulphur dioxide and huge quantities of particulate matter including lead. (HT, 1985). Atmospheric pollution in Indian cities is worsening. Mumbai emits 1,500 tonnes while Calcutta 1300 tonnes of pollutants every day. Acid rain (emission of nitrogen oxides and sulphur dioxides which are transformed into nitric and sulphuric acids and washed down by rain) has destroyed forests and aquatic ecosystems across the world.

(c) *Noise Pollution*

Noise pollution originates from motor vehicles, air crafts and machineries (most common health hazard in the industrial complexes). Mumbai has been characterised as the noisiest city in India with an average load varying between 75 to 90 db. In this respect along with this city, Calcutta, Delhi and Chennai are worst affected by incoming and outgoing aircrafts. The principal sources of noise in Indian metropolitan areas are vehicular traffic, industries such as printing presses and passing trains, loudspeakers etc.

In a nutshell the continuously increasing network of transportation by road, rail, air and water crisscross the ecological environment. All these transportation vehicles greatly contribute to noise pollution as well as churn the atmosphere throwing out a lot of dust. The vibrations of these transports are of various intensities (from sonic boon to horn of trucks) and disturb the ecological factors seriously and stunt their growth. Another source of noise pollution is the factories with their roaring furnaces rattling machines, noises and thud of loading and unloading, bleating of the sirens, release of stream from safty valves etc. Mining and quarrying create great disturbance to the atmosphere by their explosives, milling seiving and machines, loading and unloading trucks etc. Building activity adds to the noise of high decibels by pile driving, brick loading lifts, concrete mixers, and general raise in noise level due to other activities.

The society itself adds to noise pollution in a large measure through amplifying systems. Televisions, festivals, rituals, processions, animals and pets. The scientists have found out that increase in noise levels has very deterimental effects on physical growth and mental capabilities of human beings.

(d) *Fertilisers and Pesticides*

Fertilisers and pesticides are also factors of ecological subversive materials. In agriculture the excessive use of fertilisers and pesticides has led to many serious problems as the pesticide residues which enter the food chain, effect the health of human being, fish, fowl and cattle. The use of pesticides pollutes directly the air, water and land as well as plants and animals.

In India, the consumption of pesticides has increased over eight times during the last two decades from the base figure of about 8,000 tonnes in 1960-61, it rose to about 65,000 tonnes during 1980-81. It was estimated that India would require about 1,19,000 tonnes of pesticides by 1989-90 and about 1,44,000 tonnes by the year 2000.

In the absence of precautionary measures and due to

indiscriminate use, the spraying of harmful pesticides and herbicides has created serious problem for the ecology inspite of small quantity used and relatively limited areas covered.

Actually, the agricultural run-off which contains residues of fertilisers and pesticides and agricultural wastes will continue to mix into surface waters. Some of the residues will accumulate and get into the body systems of fishes, birds, mammals, vegetables and fruits etc. causing ecological disturbance and quite likely leading to extinction of some of the animal and plant species. On the other hand, some varieties may undergo environmental muta-genesis. Moreover, pesticide-pollutant resistant varieties may also develop. This may distort ecological balance.

Plastic

Plastic is a material that does not get recycled in the natural process. It means that whatever plastic material that enters the soil will continue to remain there, heaven knows for how many thousand years. Blatant, consumerism has introduced 'throw away' consumable items where plastic play a major part. Thus, plastic gets deposited over all and sundry places and interferes with growth of vegetation or gets into the stomach of grazing animals. Apart from this, plastic has replaced metal and wherever metal junk used to accumulate the plastic junk accumulates. Being very much cheaper than metal, plastic is not collected for reutilisation and this remains distributed in the soil. Unless plastic waste material is collected separately and not mixed with natural waste, pollution by plastic will become a major problem in the age to come.

Nuclear

Environmental pollution from nuclear radiation has assumed significance in areas near atomic installations. The nuclear arsenal, is one of the major components of the modern war machinery. Establishment of nuclear plants has been a matter of controversy because radiation effect from the leakages of the plants on living organism is much greater than its benefits. Some fear that nuclear plants wastes will foul up the atmosphere, poison the earth and expose future generations to lung cancer and other dreaded diseases.

But these radioactive wastes are dumped into the oceans. Though they are sealed in containers, and the latter dropped to greater depths so as to rest on the ocean floors, the possibility of the containers giving way under enormous pressure and contaminating the surrounding waters can not altogether be ruled out. Disposal of radioactive wastes will pose greater problems as a lot of countries will turn to nuclear power in future. Any nuclear conflict would have global environmental effects which could be so dramatic and drastic that all life forms on our plannet might be totally annihilated. For example Bhopal in India, and Chernobyle in Russia should act as reminder of the consequences of industrial hazard and nuclear pollution.

War

The modern warfare technique has designed so many unusual weapons which attack not only the defence establishments but annihilate all living beings in the areas. Atomic and hydrogen bombs, nuclear fission bombs, napalm bombs are only some examples. There are bombs which destroy only life but does not explode nor does damage any structure or machinery. The area covered in a serious war is limited not only to the borders but also the whole country. One can imagine the damage to ecology on account of a war.

Space Debris

The vast space around the earth at zero magnetic level has become a target for various kinds of sattelites for scientific experiments, international espionage and other purposes. Hundreds of sattelites have been launched over the years and most of them along with their stage I, stage II, rockets are shuttling around the earth at breakneck spead. If such space debris continues to accumulate without taking steps to bring them down safely, two kinds of pollution arise. Firstly, accumulation of space debris beyond a limit will interfere with the heat being received from the sun and might upset earth's magnetic activity. Secondly, the debris when they descend on the earth after some time will cause a lot of destruction.

Aviation Pollution

With the continuously increasing air traffic many kinds of pollution arise. The speed, noise and the vibrations, will seriously affect the migratory birds. The exhaust fumes of the aircraft engines leave poisonous gases in the air which might upset the equilibrium of the biosphere. The supersonic jets create the 'Sonic Boom', which is a kind of explosion. The noise level is so high and powerful that it is harmful for human beings not to talk of other animals.

Genetic Engineering

At the end of the spectrum is the use of science of genetics in the conservation of biological as also genetic diversity. The spectrum of the role of genetics in environmental sciences deals with the whole area of genetic engineering which now gives us hope to manipulate living systems, plants, micro-organisms and animals to suit degraded environments. Accordingly, a variety of micro-organisms have now become relevant not because these are capable of surviving polluted environments but can also utilise pollutants and thus help to control pollution, including oil spills.

Waste generation is an integral part of most human activities. Waste can be an environmental hazard. Therefore, waste utilisation for the benefit of human kind is indeed a priority area. The use of genetically engineered micro-organism for waste utilisation seems to be the most positive aspect of growth.

Genetic diversity is heritable diversity within and between the species of a genes. Genetic diversity is a very critical input to our agriculture, horticulture, forestry and animal husbandry (including fisheries). While dealing with biological diversity, the basic rational has to ensure sustainability in the utilisation of species and eco-systems by maintaining essential ecological processes in the life support system. This is necessary because biological diversity is expected to make an important contribution to social and economic development.

Biological experiments and genetic engineering might soon reach a state where the creation of new species of organisms would be at the call of man. Thus, new living things might be created with combination of physical and mental factors which is at present beyond the scope of imagination. They might vary from new varieties of bacteria and virus to monsters with many heads, hands and tails! The introduction of a single new species would vastly affect the ecological balance. What would happen when a new species is created?

Obstructions to Ecological Evolutionary Process

The ecological system is in a continuous process of adjustment and readjustment between various species of living organisms. The present ecological balance has arrived after millions of years, so as to suit the emergence of mankind. It is a system in which all species play their part. No single species can live along without the other species. If man encroaches upon nature and eliminates some or other of the species, a vacuum is created, and in the process of readjustment by nature, who knows, one day man might himself be eliminated. In recent times the human civilisation has expanded so much that its activities have started interfering with the ecological system in many ways. Let us consider some of the matters arising out of this issue.

Expansion of human settlement without taking adequate steps to preserve ecological sources causes grave danger to the continuity of human race itself. Such activities can be broadly classified into the following categories:

(a) Activities developing from commercial ambition of people.
(b) Activities emerging from political ambition of governments.
(c) Expanding development activities due to increasing population.

(a) *Activities developing from commercial ambition of people*

The people, who are involved in business and commerce, are guided mostly by profit motive and that too short by term goals. To put it plainly, how to make money in the shortest possible time with the least effort. Such people set their target on continuous expansion of their companies and projects. Due to these expansion of large areas of forest are not only razed to the ground, but the future possibility of developing a forest in the area is completly ruled out. Lakes and marsh lands are filled up for creating space for factories and settlement colonies, leaving no scope for another lake or marsh land coming up nearby. Industrialisation is a part of this process and we have already seen the effect of expanding industrial activities on ecology.

(b) *Activities emerging from political ambition of government*

Ecology is disturbed by the activities of the government for defence preparations and for accelerating the economic growth. The governments are often more interested in short term gains than working for real long term benefits to humanity. Political acumen would also suggest that they should show spectacular results so that people are impressed and they are re-elected. Many long-term plants regarding ecological preservation are likely to be sacrificed in favour of plans that endanger ecology if the latter would help them to retain their control. Pursuing this cause of action, the government engages in large scale development projects creating dams, reservoirs, hydroelectric or thermal power stations labour settlement colonies, various factories and mining projects, by devastating forest wealth, inundating large stretches of ecologically useful lands, filling up lakes and tanks, etc.

Military projects of the governments engaged in war (factual or cold) also add to the ecological degradation in a significant manner. Military activities are centered in uninhabited places along inaccessible boundaries where ecological sources survive in a peaceful manner. These sources are greatly disturbed and diminished on account of defence activities like road building, firing of explosives, laying mines etc.

(c) *Expanding development activities due to increasing population*

Expansion of population is one of the most important factors that directly affects ecological sources. Due to higher rate of population growth, natural resources are depleted at a much faster rate than they are regenerated. This reduces productivity and ultimately thwarts the rate of development itself. The sustainable development will depend on the quality and quantity of resources. Therefore, action for stabilising population should be taken on the basis of interrelationship between population, environment and natural resources development.

This is the responsibility of the present generation, which has in fact horrowed this planet from the future generation.

No doubt, the environmental problems have multiplied owing to the fast rate of growth of population, rising income and increasing per capita demand. The ecological balance of the habitat system is being disturbed by this factor as well as by growing urbanisation and industrialisation resulting from it.

Scientific and technological developments have been geared to serve the requirements of economic growth. Consequently, all these activities have imposed a great strain on the natural system, possibly beyond its carrying capacity. It means the size of the population the earth can support, without too much strain, is limited, the quantity of waste generated by man and his activities that the environment can absorb without damage to the quality of air, water and land is clearly limited. In case the carrying capacity is over loaded, or over strained, it is certain that the life support system of the earth may suffer irreversible damages. The situation existing now is the creation of the industrially advanced countries.

The population of India, is increasing at the rate of 12 million persons per year. It is adding to it the total population of the United States every fifteen years.

With the increase in population in larger cities, the number of class-I cities has risen up from 25 in 1901 to 300 in 1991. Due to heavy influx of people into metropolitan cities, the urban habitat is being crushed. Quality of life for majority of the urban residents has been continuously deteriorating. A larger population is now without shelters safewaters, sanitation and health care than what was a few decades before.

The cities have become major sources of pollution. Of the total sewage from major cities about 6 to 10 per cent originates from industrial waste, while 90 to 94 per cent consists of domestic sewage. Mumbai has more than 3,000 industries and as much as 65 per cent of the country's pesticide units are located there. The units pose a serious environmental threat.

Large number of industrial establishments and rapid growth of population have created a congested environment in Mumbai. Likewise, in Greater Calcutta, the average traffic noise level is about 75 decibels which is beyond the acceptable limit. The situation has been quite serious and alarming for the last ten years. Urban environment is being degraded day by day with the onslaught of population growth and migration from rural to urban areas. The crowded cities are facing more and more pressing problems with the arrival of new population. In urban areas, human life and society is being threatened by the occurance of slums, improper solid waste disposal, and water, air and noise pollutions.

Settlement and Colonisation

Human colonisation is responsible for environmental degradation, because housing is a basic need both in urban and rural areas, and due to nearly doubling of population from 1951 to 1981. The population density has increased from 77 to 221 persons/km^2. Therefore, at present there is shortage of 21 million houses, i.e. 16 million (76%) in rural areas and 5 million (24%) in urban areas. The population living in slums is likely to go up from 28 million to 55.2 million by 1990 (Khoskoo). There has been a tendency to put the blame on the rural people and also on the slum dwellers for environmental degradation, but it is actually the urban high energy and high consumptive life style which, in the ultimate analysis puts pressure on the rural and urban habitat alike.

According to National Building Organisation (NBO) the backlog in urban housing was to the tune of 4.9 million in 1981 and was estimated to jump to 7.2 million by the end of 1991. In Ahmedabad about 83 per cent and in Delhi nearly 93 per cent of the household leads in the slums are migrants from the countryside (Sinha, 1985). About 0.2 million people are migrating to Delhi every year and about 30,000 *Jhuggis* are added to already existing ones. In Mumbai between 1931 and 1981 there was a net migration of 1.06 million persons i.e. 290 persons/day. Near about 0.1 million people live on the pavements in Mumbai, whereas in Calcutta about 80,000 people take shelter on the streets.

In this way the metropolitan centres have recorded an alarming increase in the population of slum dwellers. Accoring to 1981 census about 40 per cent of the 1.68 million urban population in Kanpur lived in slums and the percentage was the highest in India. It was the lowest in Bangalore (10.20%). Numerically, Calcutta had the highest slum population of 3.24 million or 35.35 per cent in an urban population of 9.16 million. Mumbai was the next with a slum population of 3.15 million or 38.38 per cent of the urban population of 8.22 million. The percentage of slum population in other cities was similar or slightly lower, such as in Delhi (30.19), Chennai (31.87), Hyderabad (21.28), Ahmedabad (26.16), Lucknow (38.83) and Jaipur (15.62).

In the domestic sector, contribution to the air pollution arises because of the use of firewood, cowdung cake and coal which are the principal sources of fuel. Wood worth Rs. 11 crores is imported into Delhi alone. About 4 lakhs tonnes of dry wood is burnt per year by about one million people. Nearby 20,000 tonnes of smoke enters Delhi air per year and one is not surprised at the state of air pollution with the characteristic inversion layer in winter. Similar conditions exist in other metropolitan cities in our country. This is essentially a socioeconomic problem combined with environmental factors.

Another factor contributing to pollution is the high density of population and dwellings. It would be desirable to look at this problem at a micro-level in order to observe how the different patterns of human settlements (i.e. houses, schools, hospitals, business areas, industries, transportation etc.) can give rise to various types of environmental issues which may add to the welfare or inconvenience of habitants in general. The micro-density of population distribution is a very important parameter in determining the health of the environment as well as that of its human inhabitants.

Congested, unplanned and highly dense human settlement is responsible for water pollution which directly affects environment because the sewerage and sewage treatment systems in small settlements is the principal cause of water pollution in India. Although, sewerage system is inadequate in class-II towns, the pollution potential from class-II towns, is comparatively lower than class-I cities. The principal causes are the absence of infrastructural facilities such as inadequate collection, treatment and disposal of liquid and solid wastes, paucity of funds with the city civic authorities and assigning low priority to the sewerage and sewage treatment facilities.

Thus, considering the fact that extensive sewage treatment schemes may not be forthcoming in the major towns in future, the central and state boards for pollution control have contemplated to attack the problem in the reverse way. The terminal point of the sewered sector discharging the pollution load causes more damage to the environment as compared to the unsewered areas where pollution is more scattered. Such terminal locations should be treated first. In many major cities and towns, such a situation exists even with open Sutlage drain.

Environmental Management Strategy

The damage being done to the ecological system in the name of 'development' is quite alarming viz. industrial expansion, settlement colonies, defence and space research etc. It is also essential to preserve the ecological system and prevent future pollution and destruction before it is too late. We now better understand, the basis of ecological balances in nature, and its long-term advantages to man. The problem of solid waste and debries is creating environmental disorders threatening existence of human life as well as other organisms. Many countries do not even have laws to prevent pollution of environment by industries, automobiles and aeroplanes. The long-term effects of environmental pollution on human beings is not even properly understood.

In view of the serious nature of the problem of ecological preservation and the urgency surrounding it, the modern man has to take up and deal with the situation on a war footing. The strategy of environmental management is determined by the goals and objectives of development itself. The environmental considerations must enter as the basis in the process of economy of planning.

Protection to the Biosphere

There is need of environmental management for the stability of the biosphere in general. It is the biosphere that protects us from dangerous cosmic radiation from space and powerful heat rays of the sun. Any change in the composition of the biosphere may lead to a change in the atmospheric conditions of the earth which may result in total annihilation of the human race itself. Steps should be taken to prevent the release of harmful gases into the atmosphere that damage the biosphere. The chlorafluor carbons (CFCS) in aerosol sprays, exhaust fumes of missiles, rockets and jet planes are some of the factors that upset the equilibrium of the biosphere that needs to be controlled. The atmospheric test of atomic or nuclear explosives should be avoided. The composition of the biosphere should be monitored regularly in order to restore balance.

Forest Wealth

In order to ensure proper management of forest wealth the following measures should be adopted:

1. Restriction of the depletion of forest wealth through:
   (a) Checks or indiscriminate cutting of wood, and
   (b) The utilisation of low grade timber as substitute for high grade timber wherever possible.
2. Reducing the commercial cutting :
   (a) By inventing substitutes for wood.
   (b) By legal or government control on commercial cutting.
   (c) Enforcing planned commercial cutting so that needless cutting is avoided.
   (d) Cutting of huge or tall trees should not be allowed without permission from authorities.
3. Cooperation by industries and settlement development agencies through adoption of the following points :
   (a) Avoiding industries and colonies in forest areas.
   (b) In cases where it is necessary to acquire forest areas, as many trees as possible should be preserved by avoiding needless cutting.
4. Proper utilisation of forest areas in the following manner.

It is possible for human civilisation to co-exist with forest trees and plants. Settlements colonies and offices can be located within forest area by careful and ingenious planning so that least disturbance is created to the overall ecology of the forest and trees are saved from destruction to the maximum extent possible.

5. Encouragement of large scale tree plantation in urban planning.
6. Educating the public about the value of trees forest, wealth etc.

and creating a sympathy for trees.

7. Economising on fuel wood consumption by alternative fuels like biogas, liquid petroleum gas (LPG), electricity, solar power etc.
8. Maintenance of water supply for helping growth of trees and plantation by:
   (a) Ensuring that river beds do not dry up.
   (b) Increasing the number of lakes and tanks.

Protection of Aquatic Resources

Aquatic resources can be protected by adopting following measures:

1. Aquatic eco-systems as well as aquatic endangered species should be identified and a suitable strategy evolved for their conservation.
2. Much of the pollution in aquatic habitat is land-based and can be prevented by proper planning of the developmental work in the coastal regions so that the sea shores are not used as dumping ground for wastes.
3. Marsh lands and, swamps are often drained off under the tremendous pressures of human settlements. This should be prevented as they harbour habitats for a number of species beneficial to human beings.
4. Fresh water wet lands estuaries, mangroves and coral reefs are biologically very rich and ecologically important. They should be carefully preserved and maintained.
5. Research, development and education on aquatic habitats and their biota should be given considerable impetus in order to ensure their conservation on sound lines.

Reducing Pollution from Industrialisation and Urbanisation

Due to large scale industrialisation and urbanisation huge amounts of domestic and industrial waste (solid and liquid), poisonous fumes and gases, municipal wastes, etc. accumulate in a comparatively small areas. If not properly disposed they will find their way into the atmosphere water bodies and rivers around the place thus polluting the entire area.

In order to remedy the situation following control measures should be adopted :

1. Controlling industrial pollution by installation of effluent treatment plants.
2. Recycling waste products.
3. Converting municipal waste into bio-gas and fertilisers.
4. Legal provisions for prevention of discharging waste material into rivers, lakes etc.
5. Adopting necessary control measures against accidental leakage from factories etc.

Protection of Mineral Wealth and Mountains

1. Reducing mining activities:
   (a) Mineral wealth is a fund resource and non-renewable. Depletion of resource base may thwart of existence of the local economy. Therefore, the mineral wealth should be used in the most rationale manner.
   (b) A constant search for developing alternatives/substitutes of minerals that are presently being used.
   (c) Efficiency audit of minerals consumption, reduction of waste, and recycling of scraps.
   (d) Educating the public about the value and the need to save scarce resources and metals.
   (e) Alternative and non-conventional sources of energy should be developed and used to the maximum extent possible.
2. Mining should be carried out in a scientific manner, in order to reduce waste and to keep the damage to the ecology to a minimum.

Reducing Noise Pollution

The following measures should be useful in reducing noise pollution:

1. Finding out methods to produce machines and motors with least sound,
2. Legal prevention of noise pollution in selected areas, and
3. Educating public about dangers of noise pollution.

Saving Agricultural Land from Pollution through Pesticides and Fertilisers

1. Creating new type of fertilisers (bio-fertilisers) and pesticides which do not contribute to ecological damage.
2. Providing sufficient information to agriculturists about the harmful effects of pesticides etc. so that these may be used with utmost care without polluting land.

Preventing Pollution by Plastic and Polymers Materials

1. Discouraging manufacture of consumer items for the 'throw away culture'.
2. Efficient garbage collection system where by plastic items are separately collected and recycled.
3. Educating the public about bad consequences of plastics in polluting the soil.

Preventing Space Pollution

The dangers to ecological system from explosions, nuclear tests, wars, space pollution etc. are motivated by political manouvres. These can be controlled only by creating massive public opinion in the right direction at the global scale. One effective way is to create a public sympathy for preservation of ecological factors. This can be done by world wide campaigns managed by multinational and international organisations.

There is a greater urgency for environmental education including a value system which should help in conservation and management of resources and thus limiting the environmental degradation. We must create a general awareness about the dangers of environmental degradation and pollution by propagating environmental education on sound lines.

Protection of Marine Environment

Marine life, including animals, plants, insects etc. also contribute in a measure to the ecological balance. Realising the importance of marine resources great care must be taken to preserve them and their environment. The following are suggestions for protection of marine resources:

1. Limited licensing to fishermen, fishing vessels, etc. generally considered as a potential management tool for conservation of marine life. But this tool should be adopted in relation to the economic and social system prevalent in the region in such a manner as to assist in conservation efforts.
2. Protecting the mangroves for insuring judicious use of the fish and prawn resources of the ocean. The sea weeds, corals and turtles have also an interesting association and belong to an ecosystem, maintaining an ecological balance in environment. The interassociation of such flora and fauna is of special significance in the management of living as also certain mineral resources. An intelligent exploitation is important from the point of view of conservation. Special fishing farms in the ocean may be created to reduce the damage to ecology by commercial fishing.
3. The discharge of waste materials into the oceans should be a coordinated effort of industry, government and environmental experts. Great care should be taken in discharging large quantities of radio-active materials, hazardous chemicals or oil into the ocean.
4. Ocean science being inter-disciplinary in character, institutional machanism should be evolved to encourage and support collaborative work among scientists from different disciplines.

Reducing and Control of Population Pressure

1. Population explosion poses one of the greatest dangers to ecology. The only way this can be controlled is by educating people to adopt family planning methods and by making them understand the serious problem created by over population; governments, should institute incentives for planned families and education of women so that the future mothers could understand the problem of over population in correct perspective. These plans need to be area specific in order to achieve desired success.
2. Population and development are interrelated with environment and natural resources. Considering the age composition, India is a very youthful country. Three-fourth of its population consists

of children and youths. Such a situation is also very conducive to further population growth. It has a built-in demographic momentum. The crux of the problem is that the future population growth has to be related to the résources base. Since there is an upper limit to the availability of land (32,914 ha) and most of other resources, one needs to know the carrying capacity of the country. This will have to be taken in conjunction with the availability of food, water, land, energy etc. It means that the birth rate must come down from 35/1,000 to about 11/1,000 with death rate at 9/1,000. This has to be accomplished by the turn of the century or so. In turn, it takes us to the family welfare strategies in reducing both fertility and mortality of population. Zero population growth holds the key to our prosperity and to achieve this the government should even be prepared to adopt some unpopular measures. A population policy has to be enunciated not in isolation but in an integrated manner involving the relevant ministries and departments.

Improvement of Settlements/Slums

There has been a tendency to accuse the rural people as also the slum dwellers for environmental degradation but it is actually the urban high energy and high consumptive life styles which in the ultimate analysis puts pressure on the rural system. There is an urgent need to take stock of all the appropriate measures relevant to the different facets of human settlements and apply the same for prevention of population from human settlements.

There is also an immediate need for R&D initiatives in order to findout the best method of construction of houses, discharge of human waste and civic debris in such a manner as to create little disturbance in the ecological set up surrounding the settlements.

In the coming years, provision of environmentally compatible housing together with use of locally available building materials and basic environmental amenities and in particular slum improvement must engage the attention of the relevant agencies.

Conclusion

Here some of the most important issues that are posing threat to our environment as a whole and to the survival of human kind in particular have been mentioned. These require urgent attention, otherwise gains from the developmental activities will be ephemeral. The threats to our long term ecological security are real and are growing slowly and invisibly, but surely.

In order to formulate environmental policies the administrative machinery requires a lot more information and knowledge than it currently possesses. Therefore, one of the first priorities should be to broaden the base of our knowledge and information in the environmental field. It would be useful if India undertakes a survey of the present state of its environment and the major environmental hazards to which she is exposed. Centre for Science and Environment (New Delhi) is doing a commendable job in this field. The country should also undertake studies and research to define the kind of environmental problems that are likely to arise in the process of development during the next two or three decades. This accumulation of information and knowledge should enable the country to get a clear perspective of its environmental problems and the corrective actions that it requires at different stages of development. Thus, those who are actively engaged in environmental studies should always remember these words 'a healthy mind in a healthy body in a healthy environment'. Through better understanding of environment one can offer concrete suggestions and guidelines to the planners, the policy makers and the public for balanced and ecologically sustainable development.

References

Desai, Anjana (1985), *Environmental Perception: The Human Factor in Urban Planning,* Ashish Publication House, New Delhi.

Fedorov, E. (1980), *Man and Nature,* Progress Publishing, Moscow.

Haldar, A. (1984), *Pollution Rules over Calcutta,* Calcutta.

Hindustan Times (1985), October 20, New Delhi.

Khoskoo, T.N. (1986), *Environmental Priorities in India and Sustainable Development,* New Delhi.

Ryabchikov, A. (1975), *The Changing Face of the Earth,* Progress Publication.

Sinha, V.C. (1985), Resolving India's Urbanisation Problems: The Gandhian Way, *Khadi Gramodyog,* Vol. XXXI (9).

UNEP (1976), United Nations Environment Programme, Document No. UNEP/GC/58/4, Nairobi.

UN Conference on Human Environment (1972), *Identification and Control of Pollutants of Broad International Significance,* Document A/cone 48/8.

3

Cultural Basis for the Construction of the Environment

GOPAL KRISHNA CHAKRABARTI

"The deepest emotional investment of all is in the assumption of that there is a rule obeying universe, and that its rules are objective, independent of social validation."

(Douglas,1984, pp.243)

Now-a-days, the term environment has turned out to be a cliche. We talk of, every now and then, about environment—from environmental pollution to the deterioration of academic environment in the universities, for example. Desultory though it may be, we are actually encompassing a wide semantic range of environment—from natural to social environment. The anthropocentric view of the nature transposes the physical environment into the cognitive environment of the society and culture. But the construction and correspondence of natural environment into a social one is an important issue and being debated in the intellectual circles.

Scientific methods in their well known form are actually positivist approaches whose epistemology depend on the empirical experience. Far from being subjective, the experience must be firmly established on verifiable evidences and factual statements. The ontology of the positivistic interpretations of science is resting on the theory hoped to explain everything in the universe. Reasonably, one should remind here that the space-time reality in the *world out there* suffers a constraint for it depends on sense perception. Limitations are there in the capacity of the human senses which with the help of brains (keeping apart the existence and functioning of mind) *construct* the image of the *world out there*. These sense perceptible images nevertheless discontinuous come into continuity, as per the claim of positivists, aided by logical rigour. Literally science has been constructive of new facts. The earlier experiences (facts) must be verified in the light of new experiences and *vice versa* to formulate or conform the theory. Arguably, the positivists fail to appreciate that the facts or theories themselves are *value laden* and are expressed through *observation language.* Whatever experiments and observations are done with the nature, it can not be absolutely objective owing to the continuous presence of the human factor. *A priori* it is to reduce the human into the nature. It is not the nature who interprete itself, instead, it is the human who interpretes the nature.

The cognitive sciences themselves are the empirical studies of the epistemology. The term *construction* which is at the heart of cognitive sciences, means the acquisition of knowledge in human and in animal. Since the publication of *The Construction of Reality in the Child* by the pioneer psychoanalyst and constructivist Jean Piaget, constructivism gained importance in the cognitive sciences—describe Arbib and Hesse (1986). While they venture to address the problem of the artificial intelligence and God, we have attempted throughout the present article to describe the *construction of environment* by the human society.

The Arctic Eskimo and the Tropical Nuer

The eskimo lives in the hostile arctic environment of Asia and America covering large coastal areas of Greenland, Buffinland, Alaska and at the coasts of Hudson Bay and Labrador. They live in groups and the groups vary in dialects and customs. What is common among the groups, is the hunting and fishing—their way of livelihood. They are also acquinted with other primitive groups living in the same enviromental zone like Indians as well as with the advanced population from different enviroment through trading Walrus and Seal.

To the eskimo, as it has been revealed in the account of Mauss (1979), a year is divided into two halves the summer and winter when the climate changes visibly. Perception of this change is reflected into the eskimo social life—summer and winter. These two contrasted phases of social life stand in opposition to each other and is evidenced from dwelling house to family, communal and religious life.

The summer house of the eskimo is actually a conical tent covered by reindeer skin. No opening is there at the top for the ventilation and the bottom of the skins are held with large stones to withstand the strong wind. Each of the summer tent is occupied by a family consisting married couple, parents of either spouse, consanguineous and adopted children. The winter-house is made of wood, whale ribs, stone or snow (igloo) is composite in nature and is divided into several compartments. Each of the compartments is constituted of a platform for the use of the family and its guests and bench for the lamp. Each compartment covers equal space irrespective of the family size. The winter-houses are concentrated in a few selective places on the coastal line near the fiords. In the winter, a special type of communal house, particularly in Alaska, is built up. It is called *kashim* (literally *my place of assembly).* Practically it is the men's house where married and unmarried males assemble, take food from a central hearth, sleep apart from the women and children. The distinction between the summer tent and winter house points out that in the summer eskimo life is family centered and in the winter, the family boundary dissolves in the large winter house and particularly in the communal house. To them, the winter means intense collective social life. In the summer eskimos disperses throughout the extensive coastal line and at the end of the summer they return to congregate in the winter houses located in a few selective places.

Far from the eskimo, in the tropical environment of swamps and savannah, the Nuer lives on both sides of the Nile river and at the junction of its tributaries Sabat and Bahr el Ghazal in the south of Sudan. The following description of this pastoralist and horticulturist primitive society is from Evans-Pritchard's famous monograph On the Nuer (1940). The Nuer classifies the whole year into two seasons—the wet (tot) and dry (mai) season. The wet season ranges from March to the middle of the September and is characterised by heavy rainfall and flood. They take refuse in the villages located on the mounds as the whole Nuer land is flooded. But the actual village life starts from May and continue till the end of December. The dry season covers the time from the middle of September to the middle of March and is characterised by the cessation of rains excepting sporadic precipitation till December. In this season, Nuer moves to the dry camps located near tributaries, streams and water pools. This time, they maintain themselves hunting in the savannah grassland and fishing in the streams. The time actually spent in the dry camps ranges from the middle of November to the middle of March.

The Nuer youths begin to leave the village with the cattle herd during November and are joined by the elders at the camp site by the December. Similarly, the elderly people return to the village in the March to clear the dried grass for the millet horticulture. The youth returns to the village with the cattle herd just before the heavy rainfall in the June.

The Nuer's classification of the year into two seasons, standing opposite to each other, has been termed as ecological cycle by Evans-Pritchard. Careful reading of the description may end in ambiguity since the time spent in villages and in the camps overlaps the ecological time. But to the Nuer, it is not ambiguous for they produce another two minor classifications at the confluences of dry and wet seasons. *Jiom* covers the time ranges from mid-September to mid-December when the rain has not altogether been ceased. This time falls ecologically in the dry season and is also marked by the reaping of millet as well as movement of the youth with the cattle herd in search of water pools. *Rwil* covers the time ranges from mid-March to mid-June when it begins to precipitate. Ecologically it is in the wet season and is marked by the movement of elderly people for the preparation of millet horticulture. So, November to December and May to June are the critical times for the movement of the whole population. The Nuer maintains a strong bond with the fellow villagers though maintains contacts with the kinsmen from different villages. A homestead is occupied by a family and the cattle byre for its herd. Sometimes the total hamlet is formed with a composite homestead having a common cattle byre. In the dry camps, the Nuer of a particular village form a general camping area with their herds. Sometime kinsmen also congregate around a pool for the season.

Apparently, no similarity is there between these two ethnic groups who are historically disconnected and environmentally discrete having different economy. What is common among them is their classification of the year into two seasons which stands in clear opposition to each other. Accordingly, they move from one place to the other. This sort of movement (ecologists prefer to term it transhumance) is not peculiar to the eskimo or Nuer. The people inhabiting in extreme climate also shows extreme concentration in the winter and the reverse in the summer, as Mauss (1979) observes, "These are mainly the tribes of the north-west coast : Tlingit, Haida, Kwakiutl, Aht, Nootka, and a great number of Californian tribes such as the Hupa, and the Wintu".

From Culture Area to Eternal System

A section of the anthropologists has taken up culture as an inbuilt mechanism of the society to the environment. The concept of *culture area*[1] was developed by Wissler (1923) as a methodological device to show the relationship between culture traits according to their environment. The object of this concept was to classify culture traits according to their environment. The listing of those culture traits for universal comparison often leads to ambiguity. For example, the eskimo tent does not have any pore at the apex to emit the smoke out of lamps burning unlike their neighbour Indians. The eskimo burn seal fat in their lamps which do not produce any smoke at all. Once again, the Indians domesticate reindeers plenty available in that area while the eskimo hunt them for meat and skin.

Wissler (1923) views environment as the primary factor if not the only factor for shaping up of the culture traits. In the sphere of invention, environment limits the furnishing of material and can not claim sole authority as it emanates from human mind. Another type of environment he mentions, the ethnic environment which count mostly is nothing but cultural setting. The dualistic concept of environment forwarded by Wissler (1940) bears the similar analogy to the dualistic time concept of Evans–Pritchard–*ecological* and *structural time* comparing the culture areas. Wissler's intention to find out the universally discreet cultural patterns (types and sub-types) as a result of man's interaction with the physical environment, preaches *possibilism* if not *determinism.* This sort of approach along with *anthropogeography* presuppose the existence of environment before the culture.

Bennett's (1976) objection to possibilism is that the environment is totally subjected to the culture. The possibilist's over dependence on the choice of the community for the selection of environmental resources liquidates the critical steps for the selection which can not be in any way arbitrary. In order to remove the error embedded in determinism and possibilism, Steward proposes *cultural ecology* as an explanatory model. The model claims that the extra-subsistance social cultural factors can also be explained by techno-economic apparatus and its social factors. He has taken up one point seriously that had been missed by earlier anthropologists, i.e. time factor. His proposition to compare the different types of social organisation set in different ecological niche having no historical connection (thus minimizing the historical factors and avoiding diffusionists' view point) to find out *core cultural type* which change and arise out of human adaptation. Implicit in Steward's model is the feedback mechanism which influences the extra-subsistence social cultural factors, receives special attention in *cultural ecosystemicism* and *adaptive dynamics*. In the former one, Geertz claims that the technology selects the natural resources from the environment to produce energy and subsistence which further shapes up (feeds back) the culturally defined needs and institutions. Here, cultural factors and physical environment are postulated as mutually interdependent. Human efforts either distort, maintain or create a new balance in natural phenomena producing biotic or socio-cultural system. In the latter, technology's exploitation of physical enviroment is strategic to concerned community to create natural recourses for furnishing energy and goods that ultimately shapes up culturally defined needs (Bennett, 1976). In these two models, heavy emphasis has been given on technology as a means to harness potential energy from the natural environment.

If possibilism contains error in methodology having culture *a priori* to the enquiry, then Stewardian *cultural ecology, ecosystemicisms* and *adaptive dynamics* show some sort of technological determinism. Because, the achieved level of technology in a particular ecological setting can cause (or explain) the other aspects of social-cultural behaviour. Furthermore, it fails to address the problems of technological innovation. If technology (along with its social organisation) is to mean harnessing potential energy from the natural environment, at a level when produces sufficient or surplus energy, it is expected to maintain a *status quo* hindering further innovation. Often exaggerated *action* is redundant to be accommodated in *adaptive dynamics* since no group, unless otherwise coerced, would hardly undertake an experiment with the new being within a stable surplus system of food and energy.

The eskimo congregate in the winter camps located at a few places near the fiords, because, the hoards of seal and walrus assemble in those spots for the seals need large sheet of ice to protect their youngs and open area to breathe. Such requirement is not available excepting few places despite extensive coastline (Mauss, 1979). Hunting is easy in this season and the available food can support a large population. In the summer, the eskimo disperses to the summer tents following the seal and walrus. So far, the description conforms the analysis of ecosystematists. Yet, one might argue that plenty of food is not necessary pre-condition for disolution of family boundary in the winter long-house where many families along the kin line take shelter. Though it is the period of intensified collective life, the total food collected for the day is not distributed equally among the members excepting the large game. The family head is responsible to feed his family members. On the other hand, in the *kashim* (dormitory) the men stay sharing the food from the same hearth. Here, the cultural ecologists may confront an uneasy experience. The people endowed with the same technology meant for producing the food and energy is used in dichotomic fashions-one, the individual enterprise of the family heads and the other, the collective enterprise of the dormitory. This suggests that the technology is being used according to the cultural needs.

The Nuer practices millet horticulture adopted later along with pastoralism unlike their neighbour, Dinka who are fully pastoralist. Practice of horticulture extends the Nuer a sure stock of food along with acute problem of insects. Nuers are concious of it. Again they often raid Dinka groups for acquiring cattle since cattle is highly valued among the Nuer. They always love to talk about cattle and no bachelor can marry until he can provide the bride's family a number of cattle as bride-price. Evans-Pritchard (1940) admits the games in the Nuer land is available in abundance, ". . . . though the Nuer do not exploit this source of food extensively. Here, we find, not the ecological compulsion but the cultural selection is in operation.

"A system", as defined by Bennett and Chorley (1980), "is a *set of logical operations* acting upon or acted upon by, one or more inputs. These inputs lead to the production of outputs from the system and this process of through put is capable of sustaining the operational structure of the system, or of transforming it, perhaps catastrophically" (italics mine). Again, "It is hoped", by them, "to show that the systems approach provides a powerful vehicle for the statement of environmental situations of ever growing temporal and spatial magnitude, and for producing *the areas of uncertainty in our increasingly decision-making arenas*" (italics mine). This all pervading *eternal* system has been classified by them into *hard systems* and *soft systems* and the *interface* bridging the two. Hard system is "susceptible to rigorous specification, quantification and mathematical prediction in terms of their responses". On the other hand soft system is the cognitive system which is not amenable to mathematical methods. They also claim that the hard systems is *essential* to understand the *structural features* of the soft systems. It appears contradictory for the soft systems loops the inputs and outputs as well as the hard systems depends on soft systems (translation operator) for its sustenance and transformation. Hard systems is not the only part of the systems exposed to the state variables which are finite as derived from geographical environment and translatable in terms of energy but the soft system is also amenable to infinite cultural variables. Because cognitive systems can not compartmentalise environmental affairs, to the contrary, it judges other cultural affairs holistically. Every possibility is there for the soft systems to provide the structure of hard systems since the people who use the environment does not calculate the environmental inputs and outputs. They solely depend on *observation* not amenable to mathematics.

System interfacing has also been developed by Bennett and Chorley (1978) for the reason that ". . . . man is not apart from nature, but that he adjusts natural environmental behaviour within constraints of natural laws to his own goals and purposes". In their model "Environmental causal system" (combining geophysics, ecology, technology into *natures stratum)* is reciprocating *norms and goal system* (combining reciprocation of culture and values and human condition into *individual stratum)* via "organisational system" (combining reciprocation of socio-political and socio-economic organisations into *group stratum).* From the model it appears that the individual equipped with norm and value is scheming and struggling with fellow individuals to arrive at a decision in group stratum just to intervene into *nature's stratum* through technology for socio-economic reasons. Again, the intervention into the *nature's stratum* feeds back the individual norm and value via organisational system. The image of man presented here, is of *calculative man* who use the environment to his own utmost satisfaction without distortion of the environment. Nice to imagine it, but one has to be careful about one point. The individuals value (irrespective of individuality) does not belong outside the society (organisation) and its culture (world-view). The reception (cognition) of culture during the enculturation, (child training) process is vital to shape the view of the individual regarding the cultural and natural environment. So, the cognitive map (soft systems) embraces both the *organisational system* and *norms and goal system* and makes the idea of interface redundant. The soft systems along with innumerable variables remain space specific (despite the claim of Bennett and Chorley regarding the structural analysis of the universal cognitive system after Levi-Strauss) brushing aside the universality of systems analysis and encourages relativism.

In search of a universal theory for man's adaptive behaviour, different discipline proposes various models since the debate over environment gained central place. Different lifeforms exist interdependently in the natural environment recycling the energy. This process is given and is automatic. It never remains so when human being intervenes into the natural environment. He observes the nature and judges the environment which adds an extra dimension complicating the whole situation. Wissler's *culture area* is nothing but the selection of culture traits (invented therein as well as diffused from somewhere else) by the ecological zone. His concept of age area and its translation into *culture area* was a "......... borrowing from early plant ecologists who had found that other things being equal the older floral species were at the periphery of a geographical area, the newer in the central portions" (Benett, 1976). The *projection* of biological (automatic., natural) ecosystem approach into the human understanding fails to yield better resut for the reason that interaction between man and environment is culturally selected (among many possibilities) and not random like natural world. The ecological approach casts a shadow over the systems analysis. System is actually an explanatory model which combines its parts into a complex or unitary whole and not the reality itself. For a long time *system* or systemics used to refer to biology and later functioning of organism (Bennett, 1976) as digestive system, reproductive system etc. as an immutable natural order. This is almost self-regulative and *deterministic* in the sense that anything goes wrong in any part can be identified and corrected. Now, back to the definition of *system* which is defined as a set of logical operations. Whose logic is it anyway! It is the scientists model (and logic operation having no contradiction for if a part contradicts the other, the model can not operate) to explain the reality and particularly the environmental reality of different societies who use it. It is not unexpected that the *projection* of system scientist's model into the social-cultural system of human being might have headache to solve "the areas of uncertainty in our increasingly decision making arenas". It clearly reflects the authoritarian attitude' of the hard system scientists (or of politicians and policy makers at the cost of scientific knowledge and scientists) who rejoice explaining the other's reality in their own terms.

Classification: Scientific and Cognitive

Among the eskimo, the whole population is divided into two streams: those who born in the winter have a special collective name *Ptarmigans* (representing the land) and, those who born in the summer are *Aggirn* (representing the sea). In a part of the eskimo land, the *winter* child's first meal is consisted of a soup cooked from *sea animals* where as the summer child eat soup cooked from land animals or fish from a river. This sort of classification is not peripheral but central to the eskimo way of life. Some more examples may also be cited. The objects are also classified according to the season. The reindeer (summer animal) skin should not be brought into the contact of walrus (winter animal) skin. Eating the summer animal caribou is forbidden until the eskimo put off all of his winter clothes. The summer tent and dress must be buried under the rock before moving to the camps in the beginning of the winter. In this regard, one can not help but to quote Mauss (1979), "thus the way in which both men and objects are classified bears the imprint of this fundamental opposition between the two seasons".

In Sudan, the Nuer classifies the whole year (ecological cycle) into two : the wet and the dry in which one clearly stands in opposition to the other. The other minor classification of the year *jiom* and *rwil* which mark the movement, the first to the dry camp, and the second to the village, also shows opposition. This classification, according to Evans-Pritchard (1940), has been achieved by the Nuer through the observation of "seasonal and lunar changes" determined by "the movement of the heavenly bodies........ and such point of reference are paid attention to and selected as points only because they are significant for social activities". As if, the modern astronomers have changed place with the Nuer. Again, he writes that the political structure control actual behaviour between persons through values. These values appear contradictory. They are only seem to be consistent when the structure is viewed as sets of relations defined by reference to specific social situation. This will obviously appear contradictory since the study has not been made how the Nuer observes *their world* instead the observes world has been *projected* into Nuer's world. "When a Nuer is born he enters not only into a natural environment but also into a domesticated environment which is the product of human labour, and this inner world is constructed from the outer world, its form and content being strictly limited by natural resources". If it is not understood how the inner world (domesticated environment) correspond to the outer world (natural environment), the Nuer would appear, from ecological view point, as energy producing machine. No wonder, that their movements will be related to the "abundance of insect life which is an ever present menace" as one of the determinant factor. Ambiguous also relating the form and content of inner world to the limitation of natural resources since Evans-Pritchard agrees that "their country is more suitable for cattle husbandry than for horticulture". Again, "Nuer land is also rich in game though Nuer do not exploit this source of food exhustively. Nuer eat all animals except carnivora, monkeys, some of the smaller rodents and zebra". This example is enough to show that the interaction of the society with its surrounding environment poses certain unkind problem for universal generalisation.

To avoid the *projection* of observer's world (scientist's category) into the realm of *observed,* understanding of *Etic* and *Emic*– the twin approach from anthropology[2] may be considered. Evans Pritchard's description of the Nuer and eco-systemic analysis are etic type. The account of the eskimo given by Mauss implicitly bears emic essence. Describing the emic essence, one cannot help but to quote the French structuralist Levi-Strauss (1985), ". . . . the principle underlying a classification can never be postulated in advance; it can only be discovered a *posteriori* by ethnographic observation – that is by experience". The role of the observer is to mediate between the two worlds and communicate the observed world to the outsiders. He can not postulate the cultural classification (of environmental components) in advance because,—"Each culture settles on a few distinctive features of the environment, but no one can predict which these are or to what end they will be put. Furthermore, so great is the wealth and diversity of the raw material offered by the environment for observation and reflection that the mind is capable only of apprehending a faction of it. The mind can put it to use for elaborating one system among an affinity of other, equally conceivable ones : nothing predestines any one among them for a previledged fate". And again, "when confronted with technological and economic conditions linked with natural environment, the mind does not stay massive. It does not merely reflect upon these conditions; it reacts to them and works them into a logical system". This logical system tear apart (classifies) the continuity of natural environment into discrete phenomena and unite them into symbols to mediate the environment. "The coherence of each system of classification depends strictly on constraints specific to the functioning of human mind. These constraints determine how symbols are formed and explain their opposition and how they connect".

The functioning of mind is correlated to the *structure of ideas* through which the *world out there* is transformed into inner world. The age-old philosophical problem might appear again whether the mind has a separate entity out of the body. Avoiding the classical dualism, Levi-Strauss argues that in terms of anatomy, physiology, body is provided with an organic environment (an important point quite often ignored by eco-systemists and system analysts) tied with the physical environment which mediates the functioning of mind through the reception of sensory data processed in the brain the means of apprehending the world.

The natural phenomena which the scientists classify into categories are *continuous* and interconnected irrespective of the presence of the observer. Leach (1986) maintains that the sense image received by the observer from the external natural environmental forms the concept in the mind. The relation between sense-image and the concept of mind intrinsic and both way rendering the sense-image and the object in the world out—there is always arbitrary and symbolic. In cognitive classification, the observer intercepts the continuity of the natural phenomena creating artificial boundary in discrete fashion and forms different categories. These different categories are interconnected with each other and mediated through the arbitrary symbolic means.

The term *schema* was first used in neurology (body schema)[3] and then to experimental psychology although there is no formal definition of it (Arbib and Hesse, 1986). Piaget, the most influential user of the term defines schema as the structure of "the generalisable characteristics of this action, that is, those which allow the replication of the same action or its application to a new content" (Beth and Piaget 1966, quoted from *ibid.* To Arbib and Hesse, "schema is a unit of representation of the persons world. Schema theory is an attempt at an information processing theory of the mind". Schema activation is the dynamic processes of the brain to reduce all mental phenomena (1986). An individual's mind is contained of a collection of schemas that constitute the person's knowledge or long term memory. The schema changes, in course of time, to update itself with more information and form the perception of an individual through reflective abstraction. Piaget distinguishes four main stages of cognitive development—the sensori motor, the pre-operational, the concrete operational and the hypothetico-deductive or formal operational. But he is concerned with the cognitive development of the children. The process adaptation (environmental and social), for Piaget is extremely complex and interwined with the process of *assimilation* and *accommodation.* Among the children, the sensori-motor data of the world is *assimilated,* to currently available schema only for practical purposes, i.e. for success or utilisation. Thus, it is not passive representation. Since the children cannot translate and judge their observation and reflexive action, the sensori-motor data is registered into the schema. Accommodation is the process the repertoire of the schemas changes over time to reflect other aspects of the world beyond the currently available schemas. This process is linguistic and can be concious. After speech has been acquired by the children, the socialisation of thought starts through linguistic measures. The conceptual thought of a child being socialised leads to collective thought, i.e. the knowledge and therefore, yields to norms of truth

(Piaget, 1954; Arbib and Hesse, 1986).

Once the sensori-motor data (assimilated) of the child is socialised (accommodated) through the verbal means, what is the role of language (both verbal and non-verbal gestural) and for that matter symbol? Language put the schema into action (reflective abstraction) to express behaviour. It also stimulates the past experience (schematised memory) and act accordingly. Though perception stresses the construction of schema assemblages as representing the environment, but schemas may also be activated by the environment contained by the body (feeling the pain, for example) through the earlier (memorised) sensori-motor experience. So, the interaction between the mental schema and the object in the natural world is reciprocal. The meaning of the interacting schemas is out in the world and on the other hand the objects out in the world invoke the schemas via the senses and the language. "The process of reflective abstraction underlie the development of symbolisation and language. "(Arbib and Hesse, 1986).

The events (and phenomena as well) in the nature are perceived and communicated through signs and symbols. The sense perceptible events of the world out there carry message(s) to all living organism including human which trigger response. This is called signal in which the response is automatic, mechanical but dynamic and without human intervention. The static aspect of the signal is *index* where one event indicates the other. Such indication or reference may be by arbitrary human choice called *signum*, or by natural association but selected by human being called *natural index*. Signum gives rise to symbol and sign. Symbol is an object, property or action which stands for something else by arbitrary choice. The meaning is always external to the object, property or action it represents where there is no prior intrinsic relationship. On the other hand, the meaning of the sign which is also an object, property or action, has got a prior and intrinsic relationship with the thing it represents and taken as *part for the whole* relationship. Both *natural index* and *sign* form the *metonymy*. Symbols are again divided into finer categories. They are *standardised symbol* (arbitrary but habitual) which is of two types – *conventional but wholly arbitrary* and *icon* (shows planned resemblances); and 'nonce (private/individual) symbol. All the types of symbol form *metaphor.* The above description has been given from Leach (1976). In the context of communication, both the metonymy and metaphor are culturally created and the whole *world out there* is *represented* (or projected) in terms of them. So, when the sense-perceptible world is schematised, the schema assimilation accommodates the semantic categories through socialisation process in terms of metonymy and metaphor.

A particular community can not and does not use all of the natural and environmental resources for the components of the world are classified into binany semantic categories in opposition to each other, for example, useful/useless; benevolent/malevolent. Eskimo's construction of the society on the basis of summer and winter seasons as opposite points of the calendrical year, their metaphorical use of the environment in naming the children, and prohibition to bring the summer and winter dress and implements in close contact unveils their conception of the environment. The movement of the Nuer appears to be in harmony with the ecological time because they construct and transform the ecological time into *structural time* (as Evans Pritchard terms it, 1940) which invokes the interpersonal relationships during dry and rainy seasons. The minor phases at the junctions of the two opposite points in an ecological year determines the movement of the youngs from village to camp and back.

The present article is not so ambitious to disregard the efficacy of environment for the society's function for survival but to stress the importance of people's classification and cognitive categories of the environment in which they live in. In the context of the Ilongot[4] metaphor, Rosaldo and Atkinson (1975) write, "Semantic conceptions and environmental categories operate in a similar way in the selection of plants for agriculture and hunting". Tilc communitics who livc in the same environment will not necessarily conceive the environment in the same way. The Lele and the Bushong live in the opposite banks of the Kasai river in Africa. Lele in the left and Bushong on the right of the river in Belgian Congo. Thcy classify the year in two opposite points of the calendar—the rainy and the dry seasons. Ecologically the spell of the dry seasons between the two habitats differ only in 5 to 10 days. Yet the Lele consider the dry season unbearably hot and the Bushong pleasant (Douglas, 1984).

Ritual: Primitive and Scientific

Innate to the human individual as well as society is the symbolic classificatory schema which makes and remakes the world of their own. Environmental semantic categories are invoked in day to day activities and in rituals performed by the individual or the specialist in the dire consequences. Among the Ilongot, the magical spells contain plants name which grow in different part of their own environment, for example, forest, secondary forest, river banks, grass lands, garden etc. Rosaldo and Atkinson (1975) have provided a detail list of the plants, their botanical identification, interpretation as well as the environments where the plants are found. Metaphors of the sexes was the said authors intention to discuss, although, the magical spells are good example to show the metaphorical use of the environment. "... *an equation between plants and people* is common in Ilongot metaphor (as it may be universally) plants, specially wild species, are associated with *intangible forces which give life.* So, in oratory, plants are used to speak of humans; *shoots* can mean children, and *young leaf* describes a child's growing up (*Ibid*, italics mine). Metaphorical magical spells are known to every adult Ilongot and they use it profusely accompanied by the plants before starting their occupational activity—men the hunting and woman the gardening (swidden cultivation). Since the gardening require heavy intervention in the nature and natural environment and success is uncertain, magical spells are cast to ensure the good yield. Ilongot exercises a cultural choice for the sustenance of natural forests for it supplies the wild plants which enliven the magical spell and impart the 'tangible life force' to the cultivated plants. The same is in the case of hunting. The association of wild plants with the all pervading forces of nature sustains the enterprises of life.

Metaphorical is also the eskimo use of environmental classification into the social classification. Abstinence rules are there to avoid contagion. Even the contagion may occur, as they believe, in the stomach of the individual and so, the salmon (a summer fish) must not have a contact with the winter animal. If a girl eat caribou during the winter she is banished in the mid-winter which is equivalent to capital punishment. Breach of abstinence rule put the environment into risk. Rightly mentions Douglas, "we are far from being the first civilisation to realise that our environment is at risk. Most tribal environments are held to be in danger in much the same way as ours. The dangers are not naturally identical. Here and now we are concerned with over population. Often, they are worried by under population. But we pin the responsibility in quite the same way as they do. Always and everywhere it is the human folly, hate and greed which puts the environment at risk" (1984). Ilongot considers some of their magical spells sensitive to the human greed, carelessness and other human abuse. The practitioner of these spells undergo certain abstinence rule for the successful use of the spells. Otherwise they would fell severely ill (Rosaldo and Atkinson, 1975).

Prohibitory rules are inherent in the system of classification in the primitive societies. Pollution is generated when there arises ambiguity in the system of classification. The offenders who break the prohibitory rules are coerced to save the system of classification i.e. the concept of the world from being threatened. By inflicting punishment to the offenders, the primitive people control and manipulate the natural environment.

The schemas with which the scientific communities are accommodated are complex since as human individuals they have got the imprint of their own social tradition on the one hand, and those schemas further accommodate their training in the respective academic field. There communication about the world out there in terms of symbols when projected into the world others become metaphorical. *Technology*, and *energy* are the two metaphors of the ecologists. Actually, this is the reflection of the ecologists' own technology oriented society. Technology is becoming the key metaphor of the day. Advanced societies are striving for the more and more advanced technology. Since, the day in the human history, the primitive man who perfected a stone implement of hunting, technology never stopped its pace. The quest for the knowledge can never be stopped because of the inquisitiveness of the human mind to know the nature of the universe. As the days pass by, technology is breaking the barrier of human limitation. Earlier classification of the world life enterprise is threatened, the pollution is being generated in the natural environment. We are provided with abstinence rules, for example, not to destroy (interfere) the reserve of the natural resources (e.g. forest) further. What are about the minerals? We are *eating* energy. Industries are dependent on natural resources to feed us. Magical spells are also casted to revitalise the nature in the form of planting fast growing plants (*e.g.* Eucalyptus in India). Is it not the replacement of the nature instead of naturalising (revitalising the nature) of the environment? Will this magical spells survive out of human greed—to save the world from the total environmental disaster and serving the industry at the same time? Where are we going to banish the *cultural meaning* of the plants? If technology is taken to be responsible for environmental pollution, why the environmental catastrophy occurred millions of years back during glaciation and pluviation period?

The systematists are checking their metaphorical spells at the time of performing scientific rituals. In the alters of the University laboratory they are correcting their magical spells for better environmental management not knowing when the nature will comply with them. The environmentalists attempt to cleanse the world out of pollution must take a pause to think where the problem is rooted.

Man has created a mask for himself by attaching *meaning* to the objects of environment and nature. Torn apart the meaning, where rests the Man and his mirror nature. This is a separate reality indeed.

Notes

1. Culture area is actually a geographical area where culture traits tend cluster together. The centre of the, geographical area is responsible for the origination of culture traits and known as culture centre. The human communities living in a particular culture area shares the majority of the culture traits. At the junction of the two or more culture areas i.e. the frontier (marginal) area, the inhabiting community or communities will share the culture traits from two or more culture areas. Related to it is the concept of age-area. It is given in this concept that the culture traits diffuse from culture centre to the periphery. So, the traits found in the periphery are relatively older than the traits near the centre.
2. K. Pike derives Etic and Emic models from phonetic and phonemic analysis in linguistics to observe the non-verbal behaviour of human groups (Language in Relation to a Unified Theory of the Structure of Human behaviour, Vol. I., Glendale California Summer Institutes of Linguistics, 1954). Etic is generally known as *out siders'* view of culture in which the categories used for the study are meaningful to the scientists. On the other hand, Emic is the *in siders'* view of culture where the cognitive categories, taxonomy etc. used for the study are meaningful to the culture participants.
3. The use of the tem schema (body schema) can be traced back to Head and Holmes (1911) who introduced the notion into neurology. "People with parietal lobe lesion may ignore half of their body— not-feeling pain in it, not even dressing in it, since they do not see or feel it as part of the body (their

own) they are dressing. Conversely., an amputee may still have a phantom limb and speak of feeling of the movements of the missing limb. The representation in the brain, the body schema constructs the peoples reality, making them extra-ordinary resistant to discordant sensory data about the state of their bodies (Arbib and Hesse, 1976, pp. 43).

4. Ilongot is a Philippine group of Malayo-Polynesian speakers who are referred in the ethnographic literature since the eighteenth century. But they refer themselves and their language as *bugkalut* or *qiru Nut* ('from the forest') (Rosaldo & Atkinson, 1975, pp. 71, notes 1).

References

Arbib, M.A., Hesse, M.B. (1986) *The Construction of Reality;* Cambridge University Press; Cambridge London, New York New Rochelle Mellbourne Sydney.

Bennett, J.W. (1976) *The Ecological Transition : Cultural Anthropology and Human Adaptation;* Pergamon Press Inc., New York Toronto Oxford Sydney Frankfurt.

Bemett, R.J., Chorley, R.J. (1978) *Environmental Systems: Philosophy, Analysis and Control,* Methuen & Co. Ltd., London.

Douglas, M. (1984) (1975) *Implicit Meanings : Essays in Anthropoloy;* Routledge & Kegan Paul, London Boston Mellbourne and Henley.

Evans-Pritchard, E.E. (1940) *The Nuer : A Description of the Modes of Livelyhood and Political Institution of a Nilotic People;* Oxford at the Clarendon Press.

Leach, E. (1986) (1976) *Culture and Communication : The Logic by which the Symbols are connected - An Introduction to the use of Structuralist Analysis in Social Anthropology;* Cambridge University Press, Cambridge London New York, New Rochelle Mellbourne Sydney.

Levi Strauss, C. (1985) *The view from Afar;* Translated by J. Neugroschel and P. Hoss; Penguine Books.

Mauss, M. (1979) (1950) *Seasonal Variations of the Eskimo: A Study in Social Morphology,* in collaboration with Henry Beuchat, translated with a forward by James J, Fox; Routledge & Kegan Paul, London Boston and Henley.

Rosaldo, M.Z., Atkinson, J.M. (1975) Man the Hunter and woman: Metaphors for the sexes in Ilongot Magical spells, in Roy Willis (ed.) *Interpretation of Symbolism;* ASA Studies No. 3; Malaby Press, London.

Wissler, C. (1923) *Man and Culture,* Thomas Y. Crowell Company, New York, 5th Printing.

4

ECOPSYCHOLOGY

An Emerging Frontier in Environmental Psychology

SUKUMAR BOSE

Echopsychology may be defined as the study of interrelationships between a person and his psychological environment, at any given span or situation or in the midst of companions of his immediate social milieu. It is mainly concerned with the *psychological environment* or *phenomenal world* of man, as the product of his individuated experiences. It remains within a person as a covert, dynamic psychological field to control and guide his social life. It admits the influence of ecosystem and exerts selfish efforts to dominate that for effective survival of the self and progeny by a variety of purposive behaviour. It functions to build up a life-space, befitting and psychologically indispensable, to express individuality and a quality of group-life in different spheres of competition for survival.

An attempt has been made in this paper to present its basic concepts, major hypotheses, supportive research evidences, and implications in the domain of purposive behaviour of man for survival with competency.

Basic Concepts

"*Oekologie* is a Greek word with two components, viz., *Oikos* and *logy*. In English it may be translated as the knowledge or study of family life and environment. Being first used by Haeckel in the late 1860s, it became much known when Warming used the word as title of a book in 1895 and more public when British Ecological Society was founded in 1913. Presently, the term is frequently used in the issues related to human life and environment. It is defined as the study of the relations "between organisms and the totality of the physical and biological factors affecting them or influenced by them" (Pianka,

1988). In connection with the human life and environment the above definition bears two important indications, *viz.*, (a) a reciprocal relationship or transactions, and (b) a befitting environment for that, i.e., social environment. Here the term *social environment* refers to a suitably built-up environment for comfortable life of self and progeny, for creation, recreation, and procreation. The inevitable role of mind is admitted here through the major issues involved, e.g. intention, purposeful behaviour, a group spirit and feelings, and a perceived comfortable space. Knowledge and study of the above psychological variables, practically, have built the infrastructure of *ecopsychology*.

Ecopsychology is solely concerned with the dynamics of human behaviour with reference to several covert psychological systems—how they function harmoniously, cope with, and restore a balanced condition in the cognitive field. In a sense, it follows the principle of *homeostasis,* the maintenance of a relatively stable internal state under a much wider range of reality conditions. It is achieved by psychological means and through behavioural responses. It allows the person(s) to stick to actively in achieving his goal by overcoming uncongenialities in and around him. It considers certain issues of *ergonomics* as well as almost all issues of mental economics (Singleton & Dirkx, 1991; Cohen, 1970).

To survive in the socio-cultural milieu, to become happy in family life, and to develop healthy interpersonal relationships in the social world, a person has to cope with frustrations and concomitant stress reaction. It refers to a typical dynamics of social behaviour, aiming at maximum protection of self and its prime interest by self-improvised strategies. It may be described as his effort to conform with social and cultural norm, to avoid confrontations and earn social recognition and accomplishments. A person can conform to and cope with informal or primary group life more easily than with formal or secondary group life because of the variations in the nature and intensity of conflicting demands in the said two group lives—along the dimension of rigidity-flexibility. Any unexpected deterioration in the *psychosocial* field of any kind of group life when tax upon the personality *limit* and perceived as distressful then, by way of directional selection, the person concerned seeks accommodation to a new, suitable group life and social milieu. The more the person can maintain his rationality in the critical moments of deterioration and distress the more he can enjoy the benefits of directional selection and the more he gets dominated by emotionality the more he has to depend on risk and anticipations, on the laws of probability or chance factors.

In this connection a simple model of adaptation and undirected environmental deterioration, developed by Fisher (1930) may be utilised to explain certain highly emotional behaviour of man, for overcoming self and family distress and his coping strategies. After Fisher (1930), it may be inferred also that perfect coping is an ideal state-men fail to cope with distressful social environment in some ways and to differing degrees.

From ecopsychological viewpoint, affinity between individuals is the sole binding force in primary group life or in any kind of psychological group relationship. The tie of affinity, its strength or weakness is commonly explained by the presence of altruistic spirit in man. This perceived relationship may develop in between two individuals, irrespective of genetic affinity. The said phenomenon bear high semblance with the phenomenon of *reciprocal altruism* (Trivers, 1971). "In *reciprocal altruism,* some behavioural act incurs a relatively minor loss to a donor but provides a recepient with a large gain; thus, two entirely unrelated beings can both benefit from mutual assistance" (Pianka, 1988). Many current affairs of modern social life like marital relationship, divorce, dessertion, increasing trends of single affinal family, etc. can be explained in the light of a modified *reciprocal altruism* rationale.

Modern organisational life and behaviour rest basically on the *espirit de corpse,* aim at a highly cohesive group life, and cherish to develop high morale amongst its members and in organisational behaviour. Destiny of man and civilisation yet may be found progressing and retrogressing along the dimension of animosity-humanity, of war peace. Ecopsychological explanation for the said bipolar movement, in lieu of a steady progress, has identified here only one genus of intervening variable as the causal factor viz., the competition. Its nature and mode of expression may be explained after the speculations of Odum (1959) and Haskell (1947) in regard to various sorts of direct pairwise interactions that may occur between two populations. For example:

— Competition (where one population inhibits the other) as observed in any pursuit to dominate or rule over by force revolution, battle, war.

- Predation (the predator group kills or exploits others) as observed in any pursuit for cognisable crime or offences.
- Neutralism (neither population affects the other) as observed in any pursuit for co-existence, in any treatise amongst corporate body.
- Mutualism (interaction is favourable to both, obligatory or facilitative) as observed in any pursuit of intra and inter-institutional relationships.
- Commensalism (one gains advantage while others not affected) as observed in any pursuit of benevolence.
- Amensalism (one is inhibited but the other is unaffected) as observed in any pursuit of bureaucratic organisational life and behaviour. Unhealthy and distressfull amensalism for a long duration generally indulge and intensify urge for coercion, as stated under the first point.

Besides, several other authors have mentioned about four types of indirect interactions, viz., (a) apparent competition (Holt, 1977); (b) competitive mutualism (Pianka, 1981); (c) facilitation (Vandermeer *et al.*, 1985); and (d) exploitative competition to cause resource depression (Pianka, 1988).

Ecopsychology in the Matrix of Environmental Psychology

The study of environmental psychology was initiated in earlier decades with the main aim to understand the implications of human behaviour and well being in relation to the sociophysical environment, as the applied social psychologists observed in the phenomenon of overcrowding, shrinkage of natural resources, deterioration of environmental quality, and some of the constraints of ecological environment. Its focus was the *self-contained psychological processes* and locus was *person-environment relationship.* But researchers became more interested in exploring causal relationships centering around the *locus*, without doing equal justice to the *focus*. Emphasis was gradually shifted to sociological views and affairs while the *focal* theme remain attached as an adjunct. The interactional approaches in environmental psychology thus became more acceptable to the researchers concerned. At this stage, the general trend of researches concentrated on studying post-fact influences of mainly the *context* (environment), as independent predictor variables, on the response of persons (psychological functioning), as dependent outcome variables, along the dimension of time. At this juncture the researchers became conscious of giving equal importance to both *focus* and *locus* in a dynamic process of transaction. Thus, person-in-environment became considered as unit of analysis as-well-as a system (cognitive-affective-conative) to achieve goal. An important characteristic of the said system, as stated by Wapner (1981), seem akin to mechanism of the autonomic nervous system in protecting *homeostasis* in the internal-environment within human body. Later, it was admitted by Stokol and his collaborators that psychological functioning may involve unique configuration of actors, settings, and cultures" (Stokols & Altman, 1987). In their admission it remains implied that all those may not be objects in the external world in as is condition but their existence in the *phenomenal reality* are undeniable. They remain alive as *dramatis personae* in the subjective reality of the person.

Behaviour settings has been described as *psychological ecology* and where it has been qualified as *psychological habitat* (Barkar & Wright, 1950). Later Barker has used (1968, 1978) ecological psychology and ecobehavioural science. His settings are superordinate, self-regulating, dynamic entities which manipulate the behaviour of their human components and which function also to protect the equilibrium within the *settings*. (Barker, 1987). In spite of a lack in the clarity of emphasis on the distinct prime role of *phenomenal reality* to control and guide social behaviour of man, Barker has expressed an undercurrent of his firm support to the regulatory function of a person's *cognitive frame of reference* on his behaviour. The said *frame of reference*, a product of individual experiences of a person, has been given prime importance in *ecopsychology*.

"One task of the psychologist is to illumine the world of meaningful experience by resolving subjective-objective antithesis. Such an achievement requires the two-edged sword constituted by our two basic questions. The psychological would thus be rendered *objective* and the so-called *objective* would be recognised as, in the last resort, psychological" (Cohen, 1970).

Further, almost the entire task of obtaining a good description of the perceived social worlds of different people remains yet to be done.

For the student interested in obtaining insight into the social behaviour of people, the whole area offers an exciting and rewarding field of research (Krech & Crutchfield, 1948).

When an experience (of any person) *reverberates* within and finds expresion in a social behaviour it provides then an evidence of a covert, inner reality-subjective but undeniably—the phenomenal world (reality). It is a subjectively built in environment in the midst of which self and everything incorporated as extended self live comfortably and act to defend self from pain and misery, for survival in a manner which seems most desirable to self. The study of the said *comfortable environment for self* has formed the subject matter of *ecopsychology.* It is mainly interested to explore the peculiarities of the perceived or *phenomenal* world (reality) which governs social behaviour of man, his motives, his loyalty, and his faith on humanity.

Major Hypotheses

According to Sir Charles Sherrington, a radical distinction exists between life and mind. The former is an affair of chemistry and physics; the latter escapes chemistry and physics (Sherrington, 1975). Neuroscientist Sir John Eccles agrees that the events in the neuronal machinary are a necessary condition for experience, though they are not in a sufficient condition (Eccles, 1970). "Sense perception is real yet it is not matter, it is not a quality of matter, and it cannot be explained by matter. Hence Sherrington concludes, that our being should consist of two fundamental elements offers no greater inherent improbability than it should rest on one only. The new story *posits* two *irreducible* elements in man: body and mind" (Augros & Stanciu, 1984). As presented by the authors (*ibid,* 1984), the following assumptions deserve due consideration:

(a) There must be in us an internal sense that can perceive and distinguish all the qualities apprehended by the external senses.
(b) Memory not only recalls the previous experience but recalls it *as past*, and can catalogue it chronologically in relation to other past experiences. Man can recall *something forgotton.*
(c) Man can picture not only things perceived but even things not perceived by them (imagination) which works freely and

creatively.

(d) Emotion connects an individual with the world uniquely, different from perception (known as feeling) and directs him either toward or away from the external object concerned (around which emotion is felt).

(e) Animals perceive where they are going but not *why.* Man has the power to understand that by reason—the mind or intellect. Mind can understand, by investigation, the *what* and *why* of things; and man can help others to understand that, similarly.

"It is the mind (not the brain) that watches and at the same time directs" (Penfield, 1975), integrates the world of experiences into a whole, involving thoughts, memory, imaginations, emotions and actions of past. "The unity of conscious experience is provided by the conscious mind and not by the neuronal machinery" (Eccles, 1979). *Ecopsychology* considers the said "unity of conscious experience" as the *frame of reference* in the *cognitive field* or *psychological field* (Krech & Crutchfield, 1948). It can reverberates individuated experiences to form social *attitudes,* as pro-action tendencies, to guide human behaviour. After Wigner (1967), it may be assumed that "there are two kinds of reality or existence: the existence of my consciousness (the phenomenal reality), and the reality or existence of everything else" (the external reality).

Biological systems, in any organism, refuse; or avoid irritations (unfavourable stimulus) while accept and accommodate congenial stimulation (pleasant stimulus) to maintain stability and harmony in the cellular environment (to avoid any kind of protoplasmic distress). In the cellular environments biological substances follow the principle of complementary nature for attractions and unision (positive and negative force, presence and absence of structural peculiarities). The component systems in the *cognitive field* of a person follow the same procedure for acceptance or non-acceptance of any experience in course of individuation. *Cognitive field* generates needs and drives (purposeful behaviour) to achieve that object which is absent in him but present in that object following directly or indirectly the principle of complementation. The said object (man wants to possess) becomes the target or goal of any purpose of the man concerned (in whose *cognitive field* the purpose is generated). By way of possession the goal is achieved, purpose is solved, pleasure or satisfaction is attained and consequently the said object is incorporated in the *cognitive field* structure as *love object*—an important component of extended self. The exact nature of any *love object* remains known to *cognitive field* as the treasure of *phenomenal reality.* The above rationale helps *ecopsychology* to explain the *why* of any preference, choice, and intense urge to get a particular object of any particular individual.

"What decides the purpose of life is simply the programme of the pleasure principle. This principle dominates the operation of the mental apparatus from the start. There can be no doubt about its efficacy, and yet its programme is at loggerheads with the whole world, with the macrocosm as much as with the microcosm. There is no possibility at all of its being carried through; all the regulations of the universe run counter to it" (Freud, 1955). Thus, frustration becomes unavoidable and thus human propensity for pleasure pursue in search of a suitable substitute.

To explain the *why* of a deviant behaviour of some persons, Maslow has remarked: "What he wants to do may be bad for him; even if he does it he may not enjoy it; even if he enjoys it, he may simultaneously disapprove it, so that the enjoyment is itself poisoned or may disappear quickly. What he enjoys at first he may not enjoy later. His impulses, desires and enjoyments then become a poor guide to living. He must accordingly mistrust and fear the impulses and the enjoyments which lead him astray, and so he is caught in conflict, dissociation, indecision; in a word, he is caught in civil war" (Maslow, 1968). When any pursuit becomes acceptable to any particular *cognitive field,* it is not repelled as undesirable (deviancy) but when it is found not acceptable it usually generates conflicts. Under acute conflict, *cognitive field* becomes disorganised. The said disorganisation may be temporary or may last for any long duration. Till reorganisation of the *cognitive field* the person concerned continues to behave deviantly (to others), with strong argument to justify or defend his behaviour as *not deviancy.*

Ecopsychology considers that any person can undertake any pursuit if his *cognitive field* generates favourable attitude to do that (irrespective of what would be its interpretation to others), due to the fact that in his private world or phenomenal reality the said pursuit has a distinct meaning and purpose (pleasurable and gainful). Till his individuated experience speaks otherwise, he would stick to his pursuit (ignoring others around him). Without sufficient understanding about his cognitive *frame of reference* the *why* of his behaviour in social life cannot be justly explained and dealt with.

Cognitive field or *phenomenal reality* is distinct from external world or the reality around a person. It is formed of the individuated experiences which have subjective importance to a person—the perceiver of the environment around him. From the pool of external world some subjectively significant experiences are individuated or they are *taken away,* removed and stored in the memory repertoire to form cognitive frame or *phenomenal reality.* The whole affair is akin to alienation to cause a separation to occur (Klein, 1966). *Ecopsychology* considers that *phenomenal reality,* with its alienated existence, has a reference to the person's affective experience associated with the state separation (self-love). The person remains more at the command of his own *cognitive field, his own phenomenal reality* with the belief that the world around him remains alien to him. So for self-interest, he efforts to adjust, to conform with, to keep pace, as far as it is permissible to his *phenomenal reality.* It is a typical subjective state of existence, where self reserves his right, autonomy, and control, and fight his best to maintain them at any cost. When his social behaviour is visibly predominated by his *phenomenal reality* he is then perceived by others (around him in the external world) as an *estranged man.* In almost all cases the said *estranged man* fails to consider his behaviour as estranged behaviour—he thinks that he is on the right track while others are at fault. He is not obssessed, he has no incoherency, while others are behaving with him peculiarly without any reason.

Ecopsychology considers that *phenomenal reality* is built up by a person solely by his own psychological effort as his the best comfortable place to live. Environment around is external to him. He will try to oblige that world strategically his best but not at its good cost or denying that. He therefore, always prefer to enjoy its charm by obeying its directives, by setting aside even the directives of societal norms. Socialisation has taught him that he has no social freedom to move under the directive of his *phenomenal reality* and conformity with others around him is the only way for social survival. When the *phenomenal reality* suffer from acute psychological distress or apprehend a severe disorganisation then the person concerned either tries to protect his self-built environment by any extreme deviant behaviour with others or destroys that by self-destruction (suicide is not, here, an escape from external world but truly an escape from *phenomenal reality*). In a good many suicidal note, it is clearly written that "not making responsible anybody for the act, going to a place where eternal peace and happiness exist"—an extension of phenomenal reality, build up by his own psychological effort for self-comfort. In daily life to overcome psychological distress, normally, an individual badly needs to live in that self-built *phenomenal reality* for a good stretch of time and he feels a compulsion to do it, without being interfered. So he sleeps and after a good sleep he, with freshness, starts his sophisticated role-playing in social life. Sleep is his conscious effort and he can alter or modify his sleep-rhythm under emergency. That is why, one of the main objectives of *ecopsychology* is to explore and explain that *phenomenal reality.* In the external social world, "man does not live in a world of hard facts, or according to immediate needs and desires. He lives rather in the midst of imaginary emotions, in hopes and fears, in illusions and delusions, in his fantasies and dreams, what disturbs and alarms man (said Epictetus) are not the things, but his opinions and fancies about the things" (Cassirer, 1944).

Ecopsychology has selected, rather borrowed, the concept *life space* from Lewinian Theory. Accordingly, to understand the mysteries of *phenomenal reality* it is necessary to explain what is meant by the term *phase space,* which characteristically different from physical space. A phase space is something definitely different from that three-dimensional physical space within which physical objects are moving. In the same way the *life-space* (psychological space) or psychological field *(phenomenal reality),* in which psycholgical locomotion or structural changes, organisation or disorganisation take place is not three dimensional; where dimensions mean merely gradations of properties like hierarchy of needs and which is intended to represent the concrete situation whose interdependent properties determine a specific psychological event.

"A phase space may be constructed to express the laws governing the properties of life spaces" (Cartwright, 1959). Further, "asked to account for why an individual does something at a particular time, the field theorist describes the situation in which the individual exists at that time. He does not describe some state of affairs in the individual's past" *(Ibid,* 1959). Because any given present bears the residues of his past social life. According to Lewin, one's "cognitive structure has a time dimension, extending from the psychological past into the psychological future. An individual's behaviour is oriented to both the future and the past as they exist for him at any given time" *(Ibid,* 1959).

Ecopsychology considers the *behaviour* of any person at any given time span as the function of his *life-space,* where *life-space* includes both the person and his *phenomenal reality* (psychological space or environment). The task of explaining *behaviour* then becomes identical with (a) finding a scientific representation of the life-space and (b) detemining the function which links the behaviour to the life-space. The function is what one usually calls a *law* (Lewin, 1951). Here, the behaviour might be a change of location (in the *phenomenal reality*) and/or a cognitive reorganisation in the *phenomenal reality* or its restructuring. There are two basic properties of the *life-space,* viz.,(a) degree of psychological reality-irreality (levels), and (b) psychological time-location (view of the past-present-future) or the dimensions of life-space.

Empirical Support

A psychological foundation for environmental aesthetics was grounded through a series of empirical studies on psychological aesthetics by Berlyne during the period 1958 to 1971. Zube conducted a series of investigation on the land use perception during the period 1968 to 1975. Persons conducted indepth studies on the perceptions of employees about their work environments in the' 70s, and Verhaegen and his collaborators during 1981-1985. In the area of behavioural ecology and individual health several interesting papers were contributed by Willems in the '70s. All these works mainly strengthened the empirical infrastructure of *ecopsychology.*

In Calcutta University a study was conducted on the modes of reverberated behaviour potentials of handicapped children was reported (Bhattacharyya & Bose, 1979) and its appreciation encouraged the authors to formulate the structure of *ecopsycholgy* and its methodology. During 1985-87 another study was conducted at the Department of Applied Psychology of the University by Kannanpuzha on the psychological determinants of crowding behaviour in residents of single-room housing flats of Bombay and Calcutta (doctoral dissertation) and its appreciation spoke in favour of formulating ecopsychological research. A very interesting work has been done recently by Ghosh at the Department of Geography of the University (1988) by studying the ecological aspects of anti-social behaviour of the townships in a West Bengal district and whose appreciation confirmed the future possibilities of *ecopsychological* research.

In this connection the author considers it necessary to mention the work of Judi Marshall (1986) to understand "how system characteristics contribute to individual's experience of stress" and the work of Rudolf H. Moos (1985), a progress report of some 20 years' hard labour of him and his colleagues to identify the major coping and social resources that help individuals avoid and alleviate stress and through what processes do these resources fulfill their function in the light of his *integrative conceptual framework.*

References

Augros, R.M. & Stanciu, G.N. (1984) *The New Story of Science: Mind and the Universe.* Regnery Gateway. Illinois.

Barker, R.G. (1968). *Ecological Psychology: Concepts and Methods for studying the Environment of Human Behaviour,* Stanford Univ. Press. Stanford.

—— (ed.) (1978) *Habitat, Environment and Human Behaviour: Studies in Ecological Psychology and Eco-behavioural Sciences, from the Midwest Psychological Field Station. 1947-1972,* Jossey-Bass. San Francisco.

——(1987). Prospecting in Environmental Psychology: Oskalooga Revisited, In Stokols & Altman (eds.) *Handbook of Environmental Psychology,* Vol. 1. John Wiley, N.Y.

Barker, R.G. & Wright, H.F. (1950). *Methods in Psychological Ecology: A Progress Report.* Ray's Printing Services. Topeka, K.S.

Bhattacharyya, N.K. & Bose, Sukumar (1979). Modes of reverberated behaviour potentials of handicapped children. *Japanese Psychological Research,* vol. 21(2), pp. 70-77.

Cartwright, D. (1959). Lewininan theory as a systematic framework, in S. Koch (ed.) *Psychology: A Study of Science,* Vol. 2, McGrawhill, N.Y., pp. 7-91.

Cassirer, E. (1944). *An Essay on Man.* republished by Doubleday, N.Y., in 1953.

Cohen, John. (1970). *Homopsychologious.* Allen & Unwin. London.

Eccles, John. (1970). *The Human Mystery.* Springer-Verlag, N.Y.

Fisher, R.A. (1930). *The Genetical Theory of Natural Selection.* Clarendon Press,

Oxford.

Freud, S. (1955). Civilizations and its discontentments. In J. Strachey (ed.), *Complete Psychological Works of Sigmund Freud.* Hogarth Press. London.

Ghosh, S. (1988). A Study on the ecological aspects of antisocial behaviour in the township of South 24 Parganas. Unpub. *Doctoral Dissertation* (Geography), Calcutta University.

Haskell, E.F. (1947). A natural classification of societies. *Academy of Sciences,* Translation Series 2, 9: 186-96.

Holt, R.D. (1977). Predation, apparent competition, and the structure of prey communities. *Theoretical Population Biology,* vol. 12, pp. 197-229.

Kannampuzha, George J., (1987). A study on the determinants of crowding behaviour in residents of single-room housing units of Bombay and Calcutta cities. Unpub. *Doctoral Dissertation* (Applied Psychology), Calcutta University.

Klein, E. (1966) *A Comprehensive Etymological Dictionary of the English Language.* Elsevier, N.Y.

Krech, D. & Crutchfield, R.S. (1948). *Theory and Problems of Social Psychology.* McGrawhill-Kogakusha (Asian Students Edn.), New Delhi.

Lewin, K. (1951). *Field Theory in Social Sçiences.* Harper, New York.

Marshall, Judi (1986). Towards ecological understanding of occupational stress. *International Rev. of Applied Psychology,* Vol. 35, pp. 271-86.

Maslow, Abraham (1968). *Towards a psychology of Being.* Van Nostrand Reinholt, New York.

Moos, R.H. (1985). Creations healthy human contexts—environmental and individual strategies. In Rosen & Solomn (eds.), *Prevention in Health Psychology.* Univ. Press of New England, Hanover. Ch. 14, pp. 366-89.

Odum, E.P. (1959). *Fundqmentals of Ecology* (2nd Edn.), Saunders, Philadelphia.

Parsons, H.M. (1974). Human factors in the constructed environment, and what happened in Hawthorne? In Altman & Wohlwill (eds.) *Ibid,* pp. 163-209.

Penfield, W. (1975). *The Mystery of the Mind.* Princeton Univ. Press, Princeton.

Pianka, Eric R. (1988). *Evolutionary Ecology* (4th Edn.). Harper & Row, N.Y.

—(1981). Competition and Niche theory. In R.M. May (ed.), *Theoretical Ecology: Principles and Applications.* Blackwell, Oxford.

Sherrington, Charles (1975). *Man on His Nature.* Cambridge Univ. Press.

Singleton & Dirkx (eds.) *Economics, Health, and Safety,* Leuven Univ. Press. Leuven.

Stokols, Daniel & Altman Irwin (1987). Introduction in Stokols & Altman (eds.) *Handbook of Environmental Psychology* John Wiley, N.Y. vol. I, pp. 1-5.

Trivers, R.L. (1971). The evolution of reciprocal altruism, *Quart. Rev. Biol.,* vol. 46, pp. 35-7.

Vandermeer, J., Hazlett, B., and Ratheke, B. (1985). Indirect facilitation and Mutualism, In D.H. Boncher (ed.) *The Biology of Mutualism.* Oxford University Press, N.Y., pp. 326-43.

Wapner, S. (1981). Transaction of persons in-environment: Some critical transition. *J. Env. Psychol.,* Vol. 1, pp. 223-39.

Willems, E.P. (1965-1976). *Behavioural Ecology and Health status, and Care,* in Altman & Wohlwill (eds.) *Ibid,* pp. 211-63.

Wohlwill, Joachim F. Environmental Aesthetics, in Altman & Wohlwill (eds.) *Human Behaviour and Environment,* Vol. 1, 37-86. Plenum Press, London.

Zube, E.H., (1968-1975). Perception and Landuse, in Altman & Wohlwill (Eds.), *Ibid.,* pp. 7-121.

5

Nature of Mass Poverty in Tribal Regions:

A Case Study of Babhani Block of Sonbhadra District of U.P.

ANAND PRASAD MISHRA

Since independence the problem of poverty and malnutrition has been a cause for great concern to the government as well as to social scientists in India. Various development and poverty alleviation programmes have been launched in the country, particularly with both in long term perspective and short term objectives in last four decades. Inspite of these efforts, the extent and magnitude of poverty has remained more or less unchanged. Expression of the magnitude of absolute poverty in the most underdeveloped primitive agrarian societies can be seen in the form of either severe malnutrition as a consequence of unemployment or employment with high dependency ratio, which has only helped to regulate and perpetuate poverty over a long period of time.

Several formulations have been presented regarding the regional characteristics of poverty in a spatio temporal framework. Among these, it is pertinent to mention here that there is a close empirical association between material poverty (whether of assets or income) and malnutrition.

The dynamics of poverty can be best understood by analysing this single aspect. In an underdeveloped tribal region the severeness of poverty is manifested by scarcity of both working days and development activities.

A number of reasons have been elaborated to understand the casual relationship between poverty and malnutrition. The basic causes for perpetuation of poverty in tribal regions may be conceptualised as under:

(a) In tribal egalitarian societies poverty is considered as a natural phenomenon, which has been ordained by God.

(b) Malnutrition is a function of deficient agrarian economy characterised by an underdeveloped cropping pattern that generates insufficient food production.

(c) Calculation of the indices of malnutrition on the one hand and income as well as expenditure on the other has to be done. Malnutrition may vary seasonally with variations in aggregate indicator of socio-economic status, such as temporal imbalances in income, consumption, work effort, wage-price relation, availability of working days, diseases and important events such as birth, marriage and death ceremonies

(d) In underdeveloped tribal areas where poverty is quite common, implementation of scheme has helped to reduce the severeness of poverty to some extent.

Consequently, there is a close association between the working calendar and the malnutritional status of tribals.

Considering the above aspects in view, the present study attempts to synthesize the dynamics of poverty in an underdeveloped tribal region. The objective of this study is to explore the casual relationship between malnutrition among the tribals on one hand, and socio-economic and cultural elements of the tribal societies on the other. It seems that existing socio-economic system has helped to develop a conceptual framework. Through this study an attempt has been made to monitor the impact of poverty eradication programmes, particularly in tribal regions.

The Study Area

The Babhani block which is tribal dominating region comprises of the most backward poverty stricken area of newly constituted district Sonbhadra of Uttar Pradesh. It lies south of River Son and touches the boundary of Bihar and Madhya Pradesh. The area is dominated by tribal population with a total area coverage of 39,71,490 ha. Geographically, the study area is called 'South Mirzapur Region' because previously it was a part of Mirzapur district. The area has its special significance. It comprises of hills, streams, forests, various mineral resources and tribal population. (Figure 5.1).

The study area (with 70 inhabited villages) reflects most

contradictory features of socio-economic conditions. The Babhani block has been experiencing the various aspect of industrial developments. In fact, it has emerged as major industrial-cum-power region on the map of India. After independence, most of the developmental results have been enjoyed by bureaucrates, contractors, businessmen and other social parasits who do not belong to this region. But the local people which are mostly scheduled castes and scheduled tribes are having very low socio-economic status. They even partly live on forest leaves, jungle roots, and flowers for three to four months in a year for survival. The analysis of this region reveals different facets of the problem of poverty in an underdeveloped Indian society.

Structural Analysis of Poverty

Poverty in underdeveloped Babhani block of Sonbhadra district of U.P., has distinct expression in the form of severe malnutrition among the tribals. It not only shows areal variations according to level of development but also variations among the different strata of tribal society. A salient feature of this egalitarian society, is that it is mostly with minimum gap between rich and poor. The underdeveloped tribal regions having input of modern development measures have no doubt reduced the level of poverty to a certain extent, but it is not necessary that these measures have uplifted the tribals.

Ahluwalia (1978) has observed that increase in per capita production of foodgrains has not led to a relative decline in the degree of rural poverty in the Punjab. In fact, Punjab is the only state where high yielding varieties of wheat have led to unprecedented growth in output; in other states, only small areas had benefited from the adoption of high yielding varieties of seed, and hence interstate differences in yield and output appear to small, if one excludes Punjab (Krishnaji, 1975).

Considering the above emerging view, it is all the more important to examine, in detail, the dynamics of rural poverty in the primitive agrarian society of Babhani block. To achieve this objective a detailed diagnostic survey has been conducted in 10 representative villages of the block with a view to generate data on the various aspects of poverty, with participatory observation, from official records and other sources. Though there may be some error in the data processing and the

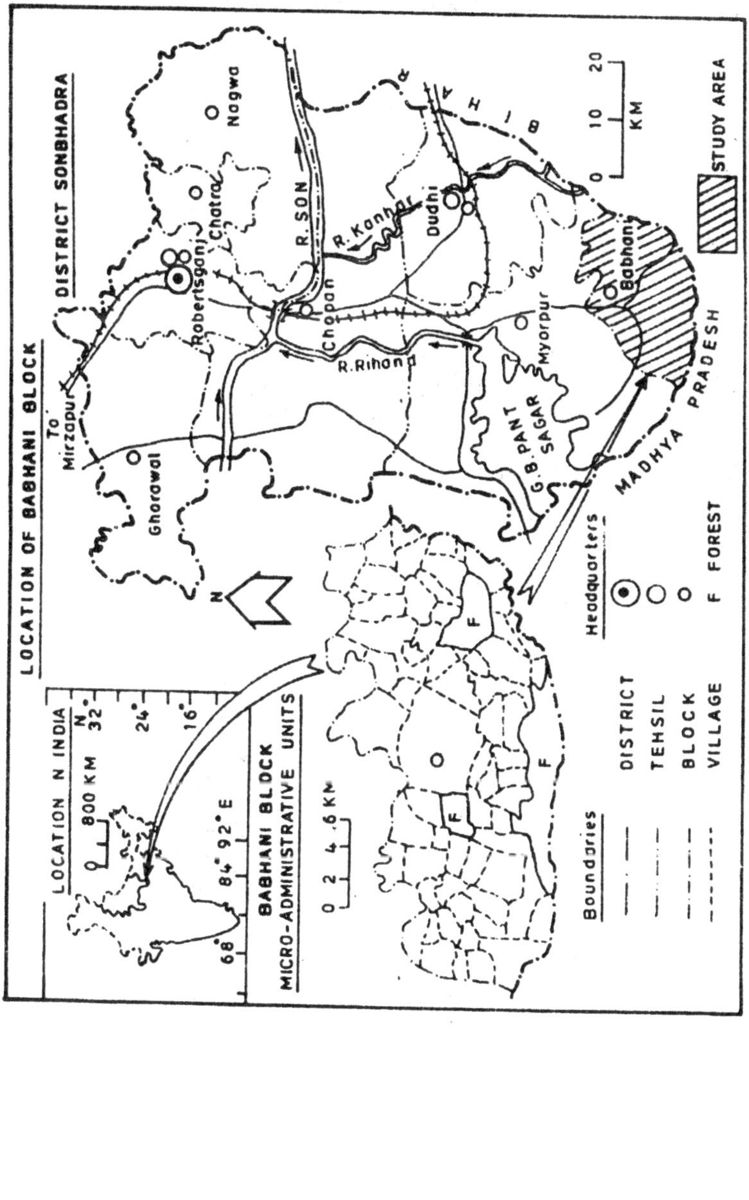

Fig. 5.1 : Location of Babhani Block

consequent observations, it is expected that the formulations derived from the study, reflects to true nature of the socio-economic dynamics of poverty among the tribals in general. The following table 5.1 reveals the level of poverty structure in sample villages.

Employment and Income Analysis

In a primitive tribal region, people live under the condition of severe malnutrition; as the agrarian production is generally not able to fulfil the requirerment of foodgrain for local population. People find gainful employment only at the time of sowing and harvesting of crops. Lack of modern infrastructural facilities in the predominantly agrarian tribal economy of the study area, has hindered the growth of the other occupational activities (Table 5.1). In view of low per capita agricultural production and limited employment opportunities, an important measure of poverty can be incorporated through the availability of foodgrains and the number of working days of gainful employment. Thus, it has been observed that both employment and income pattern in this tribal society are highly seasonal in character. However, association between employment and malnutrition and per capita income has been found to be the more dominant component in

Table 5.1: Monthwise number of working days

S.No.	Village	Months (from January to December)											
		J	F	M	A	M	J	Ju	A	S	O	N	D
1.	Ranah	10	9	15	20	25	20	8	5	7	10	11	10
2.	Hathiyar	8	7	15	21	19	15	5	4	6	9	12	13
3.	Chapki	10	7	14	24	26	14	7	6	5	10	11	15
4.	Jiganhawa	6	10	11	15	17	7	3	8	5	11	15	11
5.	Duba	10	17	11	14	10	4	7	2	4	9	7	7
6.	Babhani	12	15	25	17	18	·9	6	3	5	10	12	11
7.	Pokhara	10	18	19	20	17	8	9	4	5	7	10	11
8.	Chauna	9	13	12	17	18	8	4	3	5	6	13	15
9.	Machban-dhawa	9	14	16	15	18	8	5	3	6	8	15	10
10.	Asandih	7	10	17	16	13	5	5	6	5	15	16	10

Source: Based on personal field survey, 1991.

the delimitation of poverty level in the study area (Table 5.2).

Anand and Harriss (1989) have observed that income criterion may be justified on other ground as well because, in practice, it is proxy both for aggregate agricultural, non-agricultural and domestic

Table 5.2: Monthwise per capita income of working population (1991)

S.No.	Village	Months (from January to December)											
		J	F	M	A	M	J	Ju	A	S	O	N	D
1.	Ranah	50	40	75	100	175	40	50	50	35	50	155	150
2.	Hathiyar	100	75	150	200	200	100	50	40	60	100	145	150
3.	Chapki	150	105	160	250	300	170	50	35	50	150	122	175
4.	Jiganhawa	90	150	165	175	225	110	45	40	35	110	180	190
5.	Duba	150	200	190	150	190	60	105	30	50	170	150	140
6.	Babhani	130	160	300	300	200	135	90	45	60	135	150	165
7.	Pokhara	150	190	225	300	300	96	90	50	50	115	150	150
8.	Chauna	135	150	180	300	350	115	60	57	90	100	160	180
9.	Machbandhawa	140	200	250	290	295	115	75	45	60	119	200	190
10.	Asandih	105	190	300	280	251	90	75	60	75	180	170	200

Source: Based on personal field observation, 1991.

assets and for certain types of this is known to behave less reliably in statistical terms than does expenditure. However, the local market and surplus stocks of consumer items, especially food, has affected the nutritional status of poor masses. Poverty is also related with increased dependence upon purchased staples and nutritionally inadequate control over foodgrain (Table 5.3). The field survey has been carried on the basis of the above mentioned hypothesis and expression of the various component of poverty have been noted and tabulated.

Field Observation on Poverty

A large chunk of the populaiton, in the study area is engaged in primary occupation, i.e. agriculture. The prevailing agricultural pursuits are highly primitive in nature, yielding very low per capita production. Employment of agricultural labour is highly seasonal with low

Table 5.3: Monthly Catalory Status

S.No.	Village	Months (from January to December)											
		J	F	M	A	M	J	Ju	A	S	O	N	D
1.	Ranah	2	2	1	1	1	2	3	3	3	2	1	1
2.	Hathiyar	2	2	1	1	1	2	3	3	3	2	1	1
3.	Chapki	2	2	1	1	1	1	3	3	3	2	1	2
4.	Jiganhawa	3	2	1	1	1	1	3	3	3	2	1	1
5.	Duba	2	1	1	1	1	3	2	3	3	1	1	2
6.	Babhani	2	1	1	1	1	2	3	3	3	2	1	1
7.	Pokhara	2	1	1	1	1	2	3	3	3	2	2	2
8.	Chauna	2	2	1	1	1	2	3	3	3	2	1	1
9.	Machban-dhawa	2	1	1	1	1	2	3	3	3	2	1	1
10.	Asandih	2	1	1	1	1	2	3	3	3	1	1	1

Note: 1. Represents normal calory status and represented by 2400 calories

2. Represents testitude calory status represented by 1500-2000 calories intake.

3. Represents severe testitude conditions of calory status and represented by below 1500 calories intake.

Source: Based on personal field observation, 1991.

earnings. During the summer months (March to May), gathering of *Mahua* fruits and other minor agricultural produce are the only local source of income for the poor people. Both employment and wage obtain their peaks in summer (March to May), are practically non-existent during the rains (July to September). Consequently, it explains the periodic character of the poverty status. The spatial distribution of different poverty component, i.e. illiteracy, malnutrition, unemployment, etc. shows less inequality, however, than have been observed elsewhere in developed society, it shows the absolute or 'ultrapoor' (as defined by Lipton 1983) nature of poverty in egalitarian tribal society (Fig. 5.2).

It is observed that as a result of various employment gaurantee schemes and other development as well as poverty eradication programmes, the problem of malnutrition has been somewhat reduced recently. These schemes, do not operate on a continuous basis but

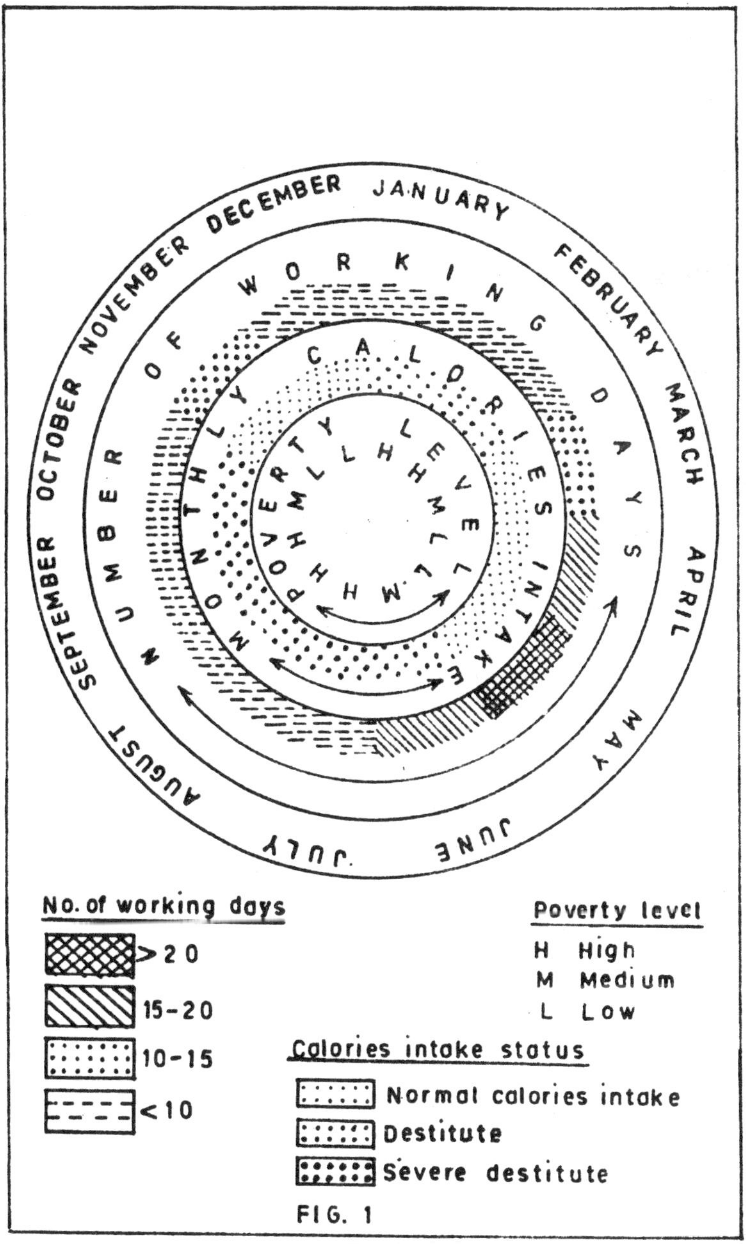

Fig. 5.2 : Poverty Calendar (In Tribal Society)

have a seasonal duration. As the number of working days, under these schemes, increase during the periods of October to Feburary, this has benefited the poor masses in these months.

The physiographic appearance of the local people is presently bad. Their houses are of bamboo, mud, and their sanitation standards are not high. The settlements are infrastructurally undeveloped. There is no electricity and lack of other modern facilities. Water is drawn from the public well with great difficulties. Facilities of education and primary health care are not sufficient. Consequently, the study-area is characterised by severe destitution. The empirical observation regarding mass poverty in the area reflects the severeness of hunger among all tribal community. It is observed that the very severe nature of destitution could be minimised through certain injection of modern development programmes (Table 5.3). On the basis of above mentioned observations a conceptual framework may be presented for eradication of poverty from the backward tribal regions of the country,

Dynamics of Tribal Poverty

The tribal community in Babhani block of Sonbhadra district due to lack of modern infrastructural facilities, is characterised by primitive stage of socio-economic development. Consequently, the scenario of mass poverty has synthesised differently in the tribal economy as compared to other developed regions. The investigation has also established the fact that existing network of self-supporting tribal economy has helped to regulate the gap between the 'haves'and 'have nots'. But low productivity and resource scarcity have accentuated the problems of hunger and severe destitution in the area.

Empirical evidences suggest that expression of the poverty in tribal area are somewhat different than in other developed area. For example, in agriculturally more advanced areas, there has been a tendency to replace animal and human power by mechanical devices, which are by their nature labour displacing (W.H. Bartsch, 1973, Martin and Singh, 1972).

But the implementation of certain development works in tribal society has resulted in change in the calender of income and employment spectrum. In fact, the problem of tribal poverty is endangered by the prevailing dominance of primitive social base and the penetration. of state-sponsored capitalism. These emerging development models have characterised the poor tribal masses in the following ways:

1. The local tribal resources have been exploited by the new emerging traders and government agencies.
2. The fertile land resource has been generally utilised by outsiders and other government power projects. Consequently, the traditional rights of the tribals in respect of forest have been drastically curtailed, creating increasing poverty among the tribals.
3. Inadequate irrigation facility has stagnated agricultural productivity, which has helped to perpetuate poverty. In contrast, wherever irrigation facilities have been increased, the occupational intensity has also increased.
4. The breakdown of traditional rights over local tribal resources and change in ownership pattern, without compensatory development has destroyed the old economic security and has compelled the tribals to move out of their home at the time of draught or other hardships.

Issues for Poverty Eradication

The foregoing discussion has highlighted the several aspects of poverty in a predominantly tribal area. These need to be examined in detail for tackling the problem of poverty and deprivation among the tribals. Several eminent social scientists, have studied the impact of poverty eradication programmes in various parts of India. Their findings are however, different. Consequently, the planners and researchers, should consider the issue of poverty eradication in tribal area, in a fresh way.

Rath (1985) has argued in favour of expanding wage employment programmes and has criticised the usefullness of I.R.D.P. as anti-poverty strategy. On other hand, Dantwala, M.L. (1985) and Indira Hirway (1985) have observed that in the existing socio-political system distortions have emerged irrespective of the nature of the programmes. Hence, the strategy of poverty alleviation needs to be evaluated on the basis of its suitability in the existing environments of particular area (Dantwala, 1985).

Considering the above emerging theme, a local anti-poverty strategy can be formulated on the basis of the prevailing dimensions of poverty components in Babhani tribal area. Table 5.1, 5.2, 5.3 and calender Fig. 5.2 reveal more details about the nature of poverty in the study area. The following points need be considered in the formulation of anti-poverty strategy for this tribal area:

1. Due to destruction of local forest and other resources, people are mostly dependent on non-traditional occupations.
2. The people of the area are engaged in agricultural pursuits only during March to May, when, the number of working days are maximum and both income and calories intake are high.
3. The Government sponsored road and other construction activities have also generated additional number of working days, which has helped to minimise the severeness of malnutrition among the masses (Tables 5.1 & 5.3).
4. Any package scheme through I.R.D.P. has not influenced the poverty level and its impact is more or less unsatisfactory.

Conclusion

It is concluded that poverty mitigation programmes, especially in underdeveloped regions, should be based on employment generation. This study also supports the earlier findings of Rath who has advocated for the expansion of wage employment for ameliorating the problem of poverty. But, the parastic nexus of contractors- bureaucrats and liquour traders, should be broken to ensure proper wages to the tribals with system for suitable utilisation of such wages. High illiteracy among the tribals has been a poverty generating factor. This evil has to be eradicated through intensive literacy programmes. Thus, absolute poverty among the tribals of Babhani block can be ameliorated on the basis of above mentioned anti-poverty strategy.

References

Ahluwalia, M.S. (1978), 'Rural Poverty and Agricultural Performance in India; *Journal of Development Studies.*

Bartsch, W.H. (1973). *Employment Effect of Alternative Technology and Techniques in Asian Crop Production,* I.I.O. Geneva.

Dantwala, M.L. (1985), "Garibi Hatao", Strategy options, *Economic and Political Weekly.*

Indira, Hirway (1985), "Garibi Hatao. Can I.R.D.P. Do it?" *Economic and Political Weekly.*

Krishnaji, N. (1975). 'Inter-Regional Disparities in Per Capita production and productivity of Foodgrains:, *Economic and Political Weekly.*

Lipton, M., "Poverty, undernutrition and hunger", world bank staff working paper, No. 597, Washington, DC.

Martin, Billings and Singh, Arjan (1972), *Farm Mechanisation and the Green Revolution The Punjab Care,* USAID, New Delhi.

6

Himalayan Environment and Development

S.C. KHARKWAL

Mountains and uplands constitute about 20 per cent of the surface of the earth supporting about 10 per cent of the global population and because of their orographic character, they control not only the climate of the surrounding region but also forms a storehouse of water in the form of ice which provide the necessary meltflows into the rivers during the hot and dry seasons. In the recent decades, the ecological system in the Himalayan region has gone under great stress due to growing pressure of human as well as livestock population, overuse and rapid exploitation of rich economic resources, extension of cultivated land in the forest areas, construction of heavy roads, tunnels, dams, etc., have greatly weakened the recuperative power of the mountain system, thereby introducing events counterproductive to development.

Degradation of the Himalayan environment has now become a matter of concern as several of its ill effects have become increasingly evident, both in the Ganga plain and within the region itself. The magnitude of the problem is still not clear as there are not enough studies to show its totality. But, there are plenty of disturbing evidences to show that the onslaught of modern technologies is leading to a large number of environmental problems. Apart from the depletion of resources, degradation of the ecosystem is manifested in a number of ways. The most conspicuous types of degradation in the Himalayan system are bare surfaces, damaged soils, accelerated erosion, landslides, land subsidence, alteration of water flow regimes, siltation of floods, outrephication of water bodies, climatic changes, and pollution of air and water. Consequences of these types of degradation are spilling over to the adjoining Ganga plain specially in the form of siltation of reservoirs, raising of river beds and frequent floods. Increasing contact of local population with the outside people has now changed the attitude and requirements of hill people and has brought in a major transformation of man-environment relationships. Out-migration of the labour force is another fallout of degradation specially from pockets where soils do not yield enough for sustenance.

In order to understand the causes of large scale degradation, one needs to have an idea of the ecological vulnerability of the region as well as the modalities of resource uses. A mountain ecosystem is not a stable system, for it represents a stage in the cycle of development of land tending towards a balanced state. In this cycle, the Himalayas are still very young which implies that their erosive energies are at their peak. Being a young system, they are prone to several geological weaknesses like unconsolidated deposits, abundance of soluble rock like limestone, highly crushed and faulted regions and seismic zones, all being highly erosion prone. The height of the Himalayas, the copious monsoonal rainfall and the sub-tropical temperates, etc., all favour weathering and active erosion. In many places, specially in the Central and Eastern Himalayas, these highly weathered materials are hold together by the massive root systems of the tropical and sub-tropical forests. The present mode of human use of the Himalaya, which necessitates the removal of the natural cover consciously or unconsciously has led to the acceleration of the erosional processes and creation of large scale degradation. Both the higher Himalayas and the Siwaliks are ecologically more vulnerable than the middle Himalayan zone, the former because of altitude, lower temperatures and drier climate; and the latter because of hot, wet conditions, loose deposits and seismicity.

The human population has lived in and used this region for at least 7,000-8,000 years, but they have lived on a sustenance basis and utilised the least vulnerable lands for year round utilisation like the broad mature valleys, alluvial terraces and the lower slopes of the middle Himalayan zone. Their trade consisted of sharing of surpluses only, both within and outside the region. For a centuary and a half, and particularly in the last three to four decades, commercial economy has made inroads whereby the region has to sustain itself for the basic needs and produce commercial products as well. The measure commercial products being timber and forest produce, fruits and vegetables, and livestock products supplied to the growing urban areas, and tourist industry in the region as well as the plains below. This practice has led to undue pressure on the use of fragile areas, i.e. steeper slopes, and the highly vulnerable Siwaliks and higher Himalayas.

In the Himalayan region, particularly in the Central Himalaya (U.P. Himalaya) (Fig. 6.1), the post independence resource development scenario got under way from the 1960s onwards being, based on three major parameters: (a) need for accessibility to the northern border because of the political conflict, (b) economic development of a backward area as part of the overall planning process, and (c) resource development in terms of raw materials and electric power for industrialisation of the country as a whole. Policy measures to raise the living standards of the Himalayan people were largly translated into schemes for generating cash and employment in the region by boosting agricultural and horticultural products, industrialisation, large scale tourism and so on. The objective of accessibility and resource development led to building of infrasturctural facilities like roads and hydroelectric power projects. Both the above activities have led to enhanced population growth as well as urbanisation largely because of the incoming of labour and other service groups. In the Himalayan region, the decadal population growth during 1981-1991 was 29.3 per cent, in contrast to that of 49.3 per cent in the eight decades of 1901—1981. The number of urban centres in this region has become more than triple in two decades of 1971—91.

Review of Literature

Considerable number of studies on the problems of this issue have been conducted by various authors. All have tried to put their views directly or indirectly in various ways and methods, but a sound solution of this problem is still awaited. The economic status and progress of the region is in fact measured by the way in which the land resource is used and maintained. There have been multiple schemes, without taking care of grassroot factors and non-coordination in different levels of developmental functionaries, etc., have made these schemes welfare oriented rather than development (Misra, *et. al.* 1974). Since 1962,

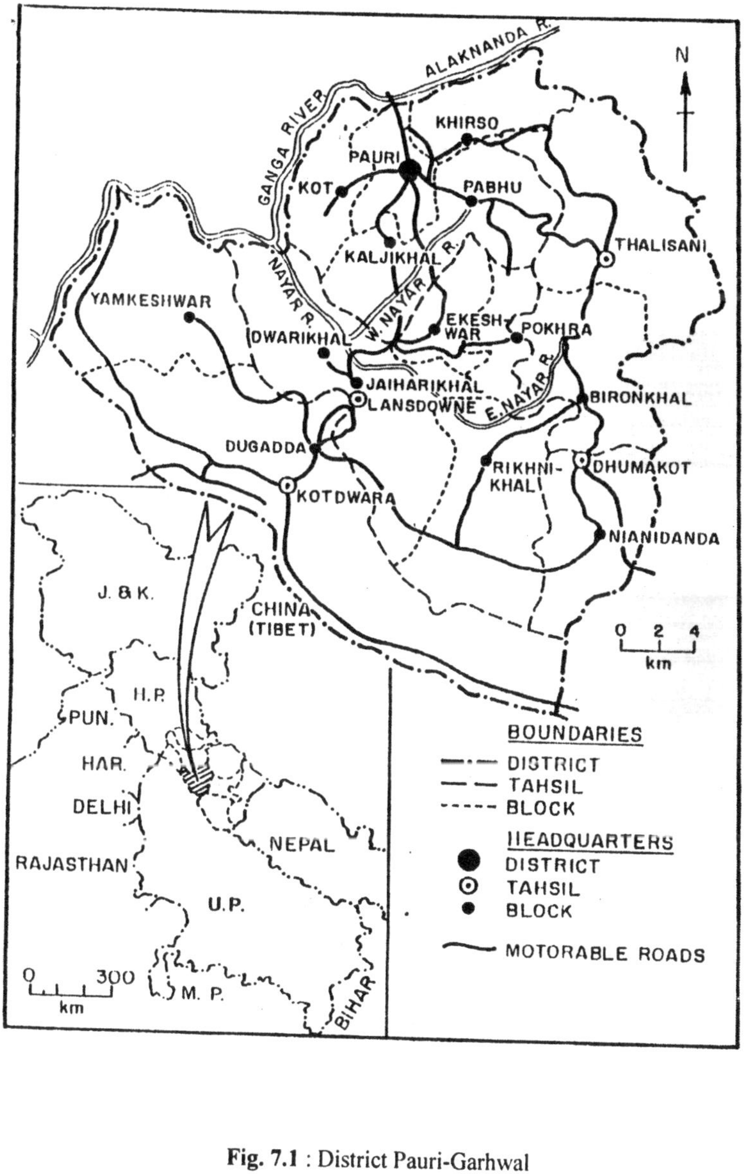

Fig. 7.1 : District Pauri-Garhwal

major tributary of the Ganga on the one side i.e the northwest and the Ganga for a short stretch and river.Hinyul in the southwest. The Gumkhal-Dwarikhal-Chelusain -Silogi ridge rising upto the height of 6000 ft. above sea level stands as the watershed divider between the Nayar and Hinyul, both flowing northwesterly direction to meet the Ganga between Devprayag and Rishikesh, at Vyasghat and a little above Laxmanjhula, respectively.

From this description it can be easily imagined that the block area is heavily drained by these major streams in the region. The land area for the most part is dominated by steep slopes hardly with any open valley worth the name. Even undulating gentler slopes are scarcely found. Thus, the land endowment is generally poor for agricultural activities. In fact, Dhangu is almost synonimous with a rugged mountainous and rather dry geographic personality. As an index it has the lowest density of population among the 15 blocks in Pauri-Garhwal district, though climatically it is in the more temperate zone and in closer proximity, with the more populous plain region.

Area-wise the largest block unit in the district, it has the lowest population density. It recorded a population of 44.09 thousand in 1981. It also recorded a relatively moderate population growth rate. The population of the district had grown by 16.0 per cent during the census decade 1971–81, as compared to 15.3 per cent of the block. Surprisingly, in the following decade (1981–91) the population growth rate of the district drastically decreased to 6.9 per cent, and maintaining the earlier trend the population growth rate of the block decreased even more steeply to 5.2 per cent only.

The nature of the terrain and poor land endowment of the block area are also reflected in the settlement pattern. As stated above, there are no open valleys or gently undulating slopes in the block. Accordingly the settlements are relatively tiny or small and scattered. Most of the villages are situated on steep slopes, in two-tier and rarely three-tier arrangement and are designated frequently with the suffix *malla* (upper) or *talla* (lower). There are in all 225 villages (*mauzas*) in the block which are organised into 111 *Gram Sabhas* and 10 *nyay panchayat* (Fig. 7.2) and their size distribution is given in Table 7.1.

The indicators even though not uniformly available for the district and the block and may only afford rather partial view of the ground level realities in side the study block , one or two factors stand out,

Table 7.1: Size of Villages

Size-Class	Number of villages	Per cent
200 or less	150	58.8
200-499	68	26.7
500-999	6	2.4
1000-1999	1	0.4
Total	225	100

A list of selected demographic and socio-economic indicators for the block and the disrict is given in Table 7.2.

Table 7.2: Demographic Indices

S.No.	Indices		Block	District
1	Population Density	1991	49 km^2	117 km^2
2	Population Growth	1971-81	15.30%	16.00%
		1981-91	6.90%	5.25%
3	Rural Population	1991	100.00%	90.20%
4	Sex Ratio	1981	1154	1092
		1991	1103	1112
5	SC/ST population	1991	7.50%	11.70%
6	Literacy	1991	66.50%	65.30%
	Male	1991	84.50%	82.50%
	Female	1991	50.60%	49.40%
7	Main Workers	1991	36.40%	26.70%
8	Cultivators	1991	92.80%	70.40%
9	Agricultural Labour	1991	3.60%	0.80%
10	Animal Husbandry, etc.	1991	1.70%	1.70%
11	Small & Marginal Farmers	1991	77.90%	78.60%
12	Area occupied by small and marginal farms	1991	21.2%	37.20%
13	Area Cultivated	1991	9.70%	13.70%
14	Cultivated Area Irrigated	1991	5.20%	8.80%
15	Village Provided Drinking Water Facility	1991	78.00%	54.90%

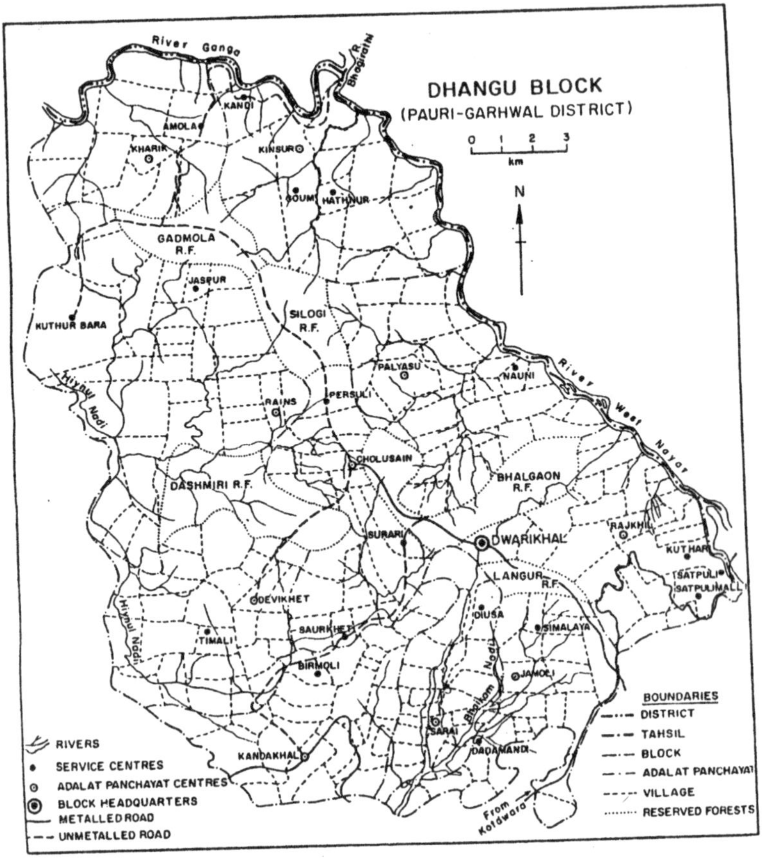

Fig. 7.2: Dhangu Block (Pauri-Garhwal District)

rather strikingly. Firstly, about the poorer land endowment of the block it will be seen that the area available for cultivation and proportion of the area cultivated under irrigation are significantly lower for the block as compared to the district norms. At the sam e time the proportion of the "workers" (main) is remarkably higher in the block, as compared to the district average. Add to this the preponderance of small and marginal farmers and the picture of the subsistence crisis becomes complete. Thus, migration has emerged as an important coming-up mechanism. The significant increase in the sex-ratio in 1991, over 1981 rather strongly reflects the heightened trend of single male migration in search of employment. Needless, this appears to be the major explanation for the low probably the lowest in the country-rate of population growth in the block and in the district for that matter. Indeed to say, that such large scale migration and the transferred incomes (through remittances) is the mainstay of family subsistence, could be environmentally beneficial in sheer terms of reduced load on the carrying capacity. But, the depletion of the active labour force also has several adverse implications for the maintenance and regeneration of the environment.

It is important to recall here that historically speaking the sub-region of the block has also experienced the longest and most intensive pressure on its ecological or environmental resources. Being most proximate to the Ganga plain below from where the modernisation/technological pressures rise, the area was the earliest to be exposed to such pressures. It is pertinent to note that the first out-post of the British colonial rule at Lansdowne, which was developed as a major army recruitment and training centre much before the world war I is situated right at the edge of the block area. The major forest resources of the area were easily accessible for commercial exploitation with the opening of the area and it was the first to face the onslaught. Thus, the area is also in greater need for ecological reconstruction and regeneration.

Issues in Thematic Mapping

The development of geography has been associated with cartography and most maps are accumulation of experience. They serve as the most powerful tools of technology for investigation and record of the present (Roy, 1992). Yet, thematic cartography has been of late development in the history of geography and the literature available on thematic cartography is still rather scare; more so in the Indian context. Most of the geographer think that if thematic cartography has received any special emphasis, it was from the 1960s only with the establishment of a specialised agency by the Indian Government; the National Atlas and Thematic Mapping Organisation (NATMO). Simultaneously, some other government organisation notably the census organisation as well as certain research institutions, and university departments have also contributed to the development of thematic cartography in this phase. Resultantly, by this time we have many a fine specimen of thematic maps/atlases on various issues such as resources planning forest resources, tourism, irrigation systems, land use and land forms, and so on. *The National Atlas of India,* which is a collection of thematic maps on different issues is regarded as another landmark in development of thematic mapping in our country. On this basis, it can be said that thematic mapping has taken roots in India and with continuous experimentation a better representation of data is being incorporated in new atlases.

Thematic mapping has no rigid defination. It is an art to project through maps of various scale in such an integrated manner that the status of resources can be seen correlatively in its whole (Roy, 1992). Likewise, the issues in thematic mapping have been continuously changing. Developing countries of Asia and Africa are trying to project their problems, such as population, planning and resource utilisation and the like in many dimensions through thematic mapping.

On the whole, the thematic mapping tasks are difficult, complex and time-consuming. Consequently the rate of progress in this specialisation has been rather slow, though its utility was always greatly acknowledged. Moreover, previously the contents and issues of the thematic maps were generally simple and direct more often in the form of diagrammatic representations. However, gradually with the rising trend of quantification in geography through application of statistical methods econometric and stochastic processes the process becomes complex. In the recent years specially the use of computer and digital inputs has given further impetus to cartography and geography. According to purpose, the nationally important goal of development planning has enhanced the need for thematic mapping. Resource identification, regionalisation planning and utilisation remain to be the most emphasised issues in thematic mapping.

Quantitatively the application of computers and digital technology has made it possible to manage large variations of data in one single cartographic representation. However, development of Geographic Information System (GIS) has started gaining importance as part of some multilateral and bilateral development project, specially in the area of integrated watershed management projects. In the early 1970s we have seen interesting application of thematic cartography in the area of micro-regional planning as the part of the growth centres pilot project that were undertaken with technical support of the Ford

Foundation. The methodology has heavily drawn from geographers like Christaller, Loesch, Brain Berry and others.

Any kind of mapping, and no less thematic mapping, involves specific guidelines with regard to its purpose (including the background of the users in view) and perspective beside the technical requirements of scale symbols etc. The basic issue has been raised and discussed in this paper, in the context of eco-development planning for a decentralised unit (roughly coinciding with a development block) relate to these critical elements of purpose and prespective. Admitted that resource identification and planning have become key themes in thematic mapping, but the guiding perspective is still propelled by the need to raise the level of resource utilisation, whereas, the eco-development philosophy and approach would seem to impose a different goal and perspective: resource conservation and accelerated utilisation being the important one. The spatial organisation and integration models generally used in micro-regional planning heavily rely on the center place theory and vertical integration models. These therefore, tend to strengthen the processes of centralisation. These issues could be posed as the major challenges in thematic mapping in the perspective, and for the purpose of eco-development.

Given this background, we are confronted with many problems, both of theoretical and practical nature, in thematic mapping for the application of the eco-development approach to micro-regional planning for the lowest decentralised unit, that is, a development block popularly known as *prakhand* in the state. These issues could be summarised as under:

(1) The cartographic models or illustrations available are strongly influenced by the two theoretical orientations, namely the Growth Centre theory and the Central Place theory both with a strong emphasis on hierarchical spatial organisation. The underlying assumption in both is that "development viewed as a process of innovation as well as growth does not appear everywhere at the same time, but manifests itself at favoured points from it tends to propagate outwards. Growth Centre policy will try to favour both concentration and outward propagation". The aproach essentially involves delineation of a micro-region around a growth centre and the service centres of lower order and the rural settlements dependent

on them for various economic and service functions. Most of the thematic mapping exercises at that level also proceed with this framework. Needless to say that the policy is mainly geared to the development of infrastructural facilities and promoting rural-urban integration as a vehicle of 'development' of the rural sector.

The methodological problem involved is that these concepts and the models derived from them are heavily based on projection of the past trends of 'development' and see future development in the same image. The models may not foreclose any introduction of new elements or conditions of growth, but it is needless to say that the demands of the eco-development approach are radically different. It calls for reversal of the dominant pattern of development in the past. There can be no denying that these models may be useful for certain purposes, for example optimal utilisation analysis for development of marketing industrial location, transportation facilities, and so on, but in the ultimate analysis development of vertical linkages with the larger economy, that is the urban-industrial centres may only accelerate the pace of exploitation of the ecological/environmental resources, as generally witnessed in the past and become counter-productive to the eco-development approach and strategy. Concentration/centralisation of growth process is something very different from decentralisation, which is emphasised not only a strategy but as a 'value' in eco-development. Thus, the basic task in thematic mapping becomes locating and strengthening the horizontal linkages between and among the rural communities in the given micro-region.

(2) At the practical level the basic most problem faced is that the block boundaries are drawn rather aribitrarily on administrative basis, i.e. the old *patties (patwari* or smallest revenue circle) drawn during the colonial rule. It is neither an economic unit nor a natural geographic unit. Thus it is totally inappropriate to eco-development planning. The map of the blocks which represent very rough sketches and the official statistics/information are available for the present administrative unit or sub-units and are practically of very

little help for any exercise in eco-development planning. Given the geographic personality of the region, land, soil, water and vegetation have to be developed as the basic 'themes' in eco-development planning and the watershed pattern (main, sub, micro, mini-micro) is the most natural framework for this kind of planning. Every geographer has to begin with delineation of the watershed pattern and redraw the planning units on this basis. The survey of India topo-sheet is the only 'resource' to fall back upon.

(An excellent exposition of the approach could be found in Central Place Studies; a bibliography by Brian Berry and Allen Pred and Regional Development and Planning, edited by John Friedman and William Alonso.)

Reference

Roy, B.K., Statistical Thematic Mapping in Prithvish Nag (ed.), *Thematic Cartography and Remote Sensing*, Concept Publishing Company, New Delhi, 1992, pp. 51-68.

8

Landuse Pattern and its Environmental Consequences

V.K. KUMRA AND C.S. KUMAR

The landuse pattern and its consequences on natural ecosystem, although being a matter of global concern, has become inevitably most alarming in developing countries in general and India in particular. At present, roughly 3.6 billion ha of world's dryland, about one third of the total, has been lying in stage of degradation. The world would earn roughly US $ 42 billion every year as an extra income if these lands were nursed and reclaimed properly. Most of this income loss (estimated as US $ 30 billion) is shared by poor countries of Asia and Africa. According to UNEP estimates, approximately US $ 8.6 to 18.2 billions of exchange is needed every year for 20 years ahead so as to rehabilitate these degraded lands.

Agriculture has intruded into marginal lands where the structure of the soil and the climate are particularly unfavourable. There are three types of lands where farm expansion is likely to cause problems for (a) land along steep hillsides, (b) land in the humid tropics that supports forests, and (c) the land close to the edge of the desert. On steep slopes, removal of an established vegetation cover for agriculture often leads to severe erosion of the top soil, as rains wash away soil particles formerly bounded by roots. Consequent upon increasing population pressure there is an excessive demand of more land both for agriculture as well as non-agricultural pursuits. This has resulted creation of vast stretches of wastelands including degraded land, saline and alkaline soils, waterlogged areas, desertified, soil eroded areas, etc. and a decrease in per capita cultivable land, besides ecological imbalances. Keeping these facts in view, an attempt has been made to analyse problem of land degradation and its consequence on ecological status of India. In addition, remedial measures are suggested for an integrated landuse management for the country.

Landuse

India, with a total area of 2.4 per cent of the world, supports about 15 per cent of the total population of the globe. The per capita land availability is about 0.4 ha as against 8.43 ha in formerly USSR and 0.98 ha in China and Pakistan. The landuse classification laid down according to the Directorate of Economics, Ministry of Agriculture also records the change in landuse pattern of the country (Table 8.1). It reflects that net sown area has increased from 41.8 per cent to 46.5 per cent during 1950-51 to 1988-89. Further there is also much disparity in rainfed and net irrigated area in the country. Table 8.2 highlights the problem of Indian agriculture and needs to develop strategies for sustainable development, so that further increase in area under wasteland may be restricted to significant level.

The arable and permanent croplands increased from 55 per cent to 57 during 1965-67 to 1981-83 period while meadows and permanent pastures decreased from 5 per cent to 4 per cent in the same period. Forests and woodlands also increased from 20 to 23 per cent while other lands decreased from 20 to 17 per cent. Since most of the population are concentrated in agricultural plains, the per capita usable land in India has been reduced from 0.33 ha in 1951 to 0.29 ha in 1971 and further, decreasing on account of population pressure. The projection for per capita usable land is 0.10 ha or less by the end of 2000 AD. During 1951–83 an additional 24 million ha land was brought under intensive cultivation. About 4.3 m ha of forested area is degraded on account of deforestation. The loss of top soil is maximum in India, being 18.53 per cent of the total global soil loss. Considering 5 tonnes/ha soil loss as tolerable, about 58.08 million ha of arable land and 6.31 m ha of forest area in the country require immediate attention. In the lower Ganga alluvial plains alone, total annual soil loss is roughly 132.5 MT from agricultural lands and 1.4 MT from forest.

The destruction of the stability of agro-ecosystems on account of utilisation of soil and water resources is one the most relevant factors for low crop yields. Practices like deforestation, overgrazing, increasing cropping on undulating lands, bundling without vegetation

Table 8.1: India: Land Use Classification

(Area in million ha)

Classification	1950-51	1960-61	1970-71	1980-81	1988-89
Geographical Area : 328.73					
1. Forests	40.48	54.05	63.92	67.47	67.08
	(14.2)	(18.2)	(21.0)	(22.2)	(22.0)
2. Area put to non-agricultural uses	9.36	14.84	16.48	19.66	21.25
	(3.3)	(5.0)	(5.4)	(6.4)	(7.0)
3. Barren and un-culturable land	28.16	36.91	28.16	19.96	19.99
	(13.4)	(12.0)	(9.3)	(6.6)	(6.6)
4. Permanant pastures and other grazing lands	6.88	13.97	13.26	11.97	11.80
	(2.3)	(4.7)	(4.4)	(3.9)	(3.9)
5. Land under orchards and groves	19.83	4.46	4.30	3.60	3.45
	(7.0)	(1,5)	(1.4)	(1.2)	(1.1)
6. Culturable waste	22.94	19.21	17.50	16.74	15.23
	(8.1)	(6.4)	(5.8)	(5.5)	(5.0)
7. Old fallows	17.44	11.18	8.76	9.92	10.46
	(6.1)	(3.8)	(2.9)	(3.3)	(3.4)
8. Current fallows	10.68	11.64	11.12	14.83	13.86
	(3.8)	(3.9)	(3.5)	(4.9)	(4.5)
9. Net sown area	118.75	133.20	140.27	140.00	141.73
	(41.8)	(44.6)	(46.3)	(46.0)	(46.5)
10. Report area	284.52	298.46	303.77	304.15	304.83
	100.00	100.00	100.00	100.00	100.00
11. Gross cropped area	131.89	152.77	165.79	172.63	180.11
12. Area sown more than once	13.14	19.57	25.52	32.63	38.38

Source : Indian Agriculture in Brief 24th Edition, Directorate of Economics and Statistics, Department of Agriculture and Co-operation, Ministry of Agriculture, Government of India, pp. 18-19.

Note: Figures in brackets show area in percentage of total geographical area.

Table 8.2: India Net Rainfed and Net Irrigated Area (in 000 ha) 1990-91

Sl. No	State	Net rainfed area	% of rainfed area to all Indian rainfed area	Net irrigated area	% of Net irrigated area to net sown area
1.	Andhra Pradesh	6,964	7.04	4,305	39.1
2.	Arunachal Pradesh	96	0.10	31	20.8
3.	Assam	2,124	2.15	572	21.1
4.	Bihar	4,848	4.90	3,347	43.5
5.	Goa	110	0.11	20	14.8
6.	Gujarat	7,343	7.43	2,493	26.8
7.	Haryana	1,427	1.45	2,599	72.7
8.	Himachal Pradesh	495	0.50	99	17.0
9.	Jammu & Kashmir	426	0.43	298	40.8
10.	Karnataka	8,856	9.0	2,113	20.4
11.	Kerala	1,913	1.93	333	14.8
12.	Madhya Pradesh	16,195	16.38	4,314	22.1
13.	Maharashtra	16,083	16.28	2,036	11.3
14.	Manipur	75	0.07	65	46.4
15.	Meghalaya	143	0.14	46	22.8
16.	Mizoram	57	0.06	8	12.3
17.	Nagaland	133	0.13	59	31.1
18.	Orissa	4,822	4.88	1,934	30.7
19.	Punjab	568	0.57	3,910	92.7
20.	Rajasthan	12,011	12.14	3,904	23.8
21.	Sikkim	62	0.06	16	16.8
22.	Tamil Nadu	3,148	3.19	2,373	42.5
23.	Tripura	217	0.22	41	15.2
24.	Uttar Pradesh	7,369	7.46	10,542	60.2
25.	West Bengal	3,361	3.40	1,911	35.8
	India	98,946	100.00	47,434	33.3

Source : (a) CMIE Basic Statistics Relating to States of India, Sept. 1994.
(b) Govt. of India, Ministry of Agriculture, National Watershed Development Project for Rainfed Areas: Guidelines, New Delhi, December, 1990.

cover, shifting cultivation, cropping pattern and unscientific management, overirrigation and similar other programmes without a holistic approach are causing increased runoff, reduced groundwater recharge and severe erosion. The ultimate results are degradation of soil, salt infestation, nutrient stress and soil acidity. Such degraded lands are categorised as wastelands and regarded as ecologically unstable with almost complete loss of top soils or have developed toxicity to life forms. Landuse practices and soil conservation measures have much more effect on soil loss than on runoff. Again, it has to be emphasised that forest removal as such as not so important with regard to runoff and soil erosion as the other use to which the land is converted. In the humid tropics, on more or less level ground, the effects of clearing the established vegetation are different. Decomposition of waste matter is so rapid that the nutrients on which life in the forest depends are largely content in the living plants and animals themselves. When these are removed there is little reserve in the soil to sustain crops and attempts at conventional farming usually fails after a few seasons. Besides, slash and burn type of primitive farming (*jhum* cultivation) in hilly and tribal areas has caused considerable damage to the natural ecosystem. In the context, various consequences of landuse change are discussed in sequel.

Deforestation

With increasing human and livestock population, forests face the onslaught of increasing agricultural and industrial demands with consequent adverse effects such as inadequate extent, ill distribution and low productivity. Forests are to be cared to fulfill their protective, productive and bio-aesthetic role with the cooperation of the government and the people.

Changing landuse pattern under forest in the country has brought attention of planners, policy and decision makers as it has caused considerable damage to the environment. According to an estimate about 2.5 million people in India live by shifting agriculture, utilising annually about 5.5 m ha of land and thus the total area involved on account of shifting process is estimated as 81 m ha. Considerable area is subjected to excessive erosion as a result of destruction of forests.

The National Forest Policy of 1962 felt concern about the indiscriminate extension of agriculture. It entitles forest to an adequate share of land (33% of the total) to maintain balance of natural ecosystem and thereby the economy. Table 8.3 shows the area under forest for 1972-75, 1980-82 and 1990-92 periods. It reveals that not only forest area is inadequate but their distribution is not regionally balanced. It gives rise to scarcity of forest products and concerned about environmental quality and well-being. Low percentage of area under forests in various states has also been well marked. The per capita forest in India is only 0.14 ha against the world's average of 1.19 ha. The averages are very low .in some of the states in India, (Punjab 0.01 ha, West Bengal and Haryana 0.03 ha, Kerala and U.P. 0.05 ha, Bihar and Tamil Nadu 0.06 ha, Gujarat 0.08 ha and Karnataka 0.03 ha). According to state of Forest Report 1991, based on satellite imagery, area under dense forest (with a crown density of 40% or more) is 11.70 per cent, open forest (crown density < 40% - > 10%) is 7.59 per cent, scrub (crown density < 10%) is 1.82 per cent, marginal forest 0.18 per cent and other lands 78.76 per cent. Area under forest by all measures is very much lower than that required ecologically and economically and must be increased both quantitatively and qualitatively. Forest area has decreased from 55.20 m ha (16.83%) to 45.70 m ha (13.94%) i.e. by 2.89 per cent during 1972-75 to 1980-82 period. The maximum damage has been suffered by closed forests (3.25%) which is a cause for concern as these types of forests are more valuable from ecological and economic point of view. The maximum decline of 11.37 per cent has been reported in Tripura followed by Himachal Pradesh (10.68%), Dadar and Nagar Haveli (10.61%), Mizoram (8.95%), Meghalaya (8.59%) and Manipur (6.79%). The forest cover in the Himalaya has come down from 60 per cent of the lands to less than 20 per cent (Kayastha, 1982). With the present rate of environmental assault, India is steadily losing more than 1.5 m ha of good forest land each year. In other words, every five years forest cover, the size of Haryana, vanishes leaving behind constantly eroding soil that hold neither water nor nutrients (*India Today*, 1986).

Deforestation has not only caused changes in landuse pattern and severe damage to environment but has adversely affected the whole ecosystem. Most people in India including small landholders, landless

Table 8.3 : India: Actual Area Under Forest (per cent of total geographical area)

	1972-75	1980-82	1990-92
Uttar Pradesh	08.8	07.1	12.0
Jammu & Kashmir	10.0	06.5	9.0
Bihar	13.0	11.6	15.0
Karnataka	15.4	13.4	17.0
Kerala	22.1	19.0	26.0
Madhya Pradesh	24.5	20.4	30.0
Himachal	27.1	16.4	24.0
Goa, Daman & Diu	32.0	29.9	35.0
Andaman & Nicobar Islands	39.7	77.5	92.0
Tripura	60.4	49.5	51.0
Meghalaya	64.0	55.4	70.0
Manipur	67.5	60.7	80.0
Mizoram	65.7	56.8	86.0
Arunachal Pradesh	61.5	62.3	82.0
Dadra & Nagar Haveli	36.1	25.5	31.0
Orissa	31.1	25.3	30.0
Assam	26.8	25.2	33.0
Sikkim	24.1	39.5	46.0
Andhra Pradesh	17.7	14.6	17.0
Tamil Nadu	12.8	10.1	14.0
West Bengal	9.5	7.8	10.0
Gujarat	4.8	3.3	2.0
Haryana	0.5	0.6	1.0
Punjab	0.4	0.5	2.0
Rajasthan	1.0	1.5	3.0

agricultural labourers, rural artisians, tribals and herders, depend on biomass gathered from surroundings to meet their basic needs. Their quality of life declines as the availability of the biomass diminishes. In addition, it would result extension of cultivable land and thus changing the landuse pattern of the area.

As a consequence of deforestation, the entire Himalayan ecosystem is threatened under severe imbalance as snow-lines have thinned and perennial springs have dried up. Rajasthan's already poor land is getting further degraded, turning large area into wasteland. The water levels in the River Ganga have decreased due to siltation and sedimentation on acount of discriminate deforestation, large scale soil erosion and siltation in watersheds of river basins and dams, have resulted reduced flow of water. Sub-surface water has gone down and agricultural productivity has decreased to considerable level. Deforestation has made Himalaya an "Unsteady Region" and can be termed as ecocide (Kayastha, 1982).

Flood

The growing man-made activities, including habitation, unscientific cultivation and industrialisation in the flood plain zone resulted higher flood damage in terms of loss to human and cattle life as well as property. Denudation of the catchment areas of the rivers owing to unabated deforestation is the largest single factor responsible for increase in density as well as frequency of floods (Sud, 1995). As assessed by Rashtriya Barh Ayog (RBA) set up by the Government of India during 1976, an area of about 40 m ha in the country is prone to flood (Fig. 8.1). The flood-prone area has increased from 6.48 m ha in 1950 to 9.0 m ha in 1970 and 1980. According to the report of Centre for Science and Environment (CSE), flood-prone area has risen even further from nearly 19 m ha in 1959 to 23 m ha in 1969, 43 m ha in 1979 and in 1984 it was 58.5 m ha. Since Rashtriya Barh Ayog (RBA) underestimates flood-prone area as large proportion of flood affected area is liable to get flooded are not counted, thus the total flood-prone area in the country will be over 60 m ha.

According to Central Water Commission (CWC), the percentage of crop area affected has increased from 29 per cent in 1950 to 51 per cent in 1980 and the average annual damage from Rs. 600 million in

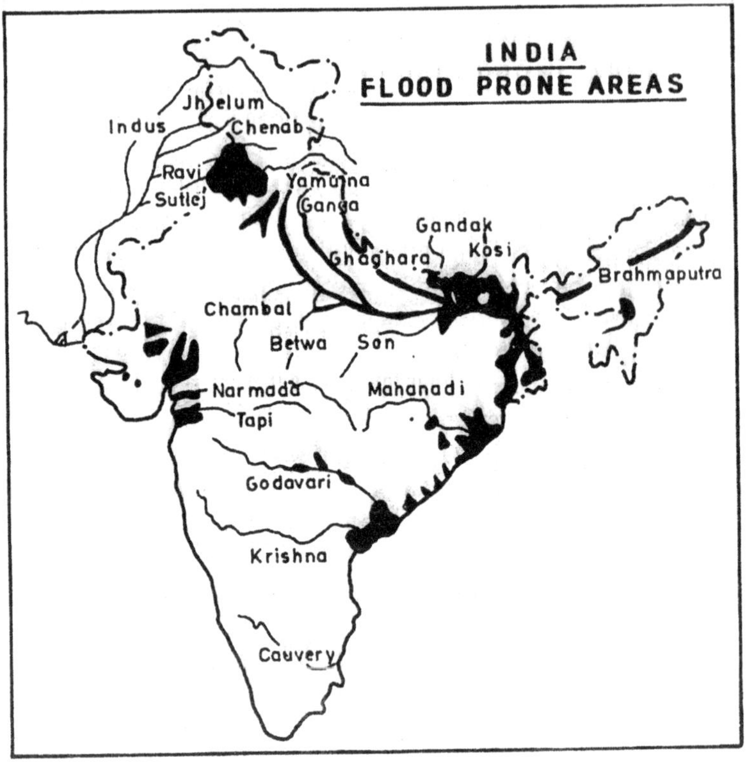

Fig. 8.1: India : Flood Prone Areas

1950 to Rs. 3,070 million in 1980. On an average about 7.75 m ha land gets inundated by gushing flood water every year, resulting a loss of 518 human lives and 1,00,706 heads of cattles.

Of the total flood-prone area (41 m ha), nearly 32 m ha is the protectable through various judiciously conceived measures. Despite the efforts made in successive plans, not more than 14 m ha area has actually been provided reasonable protection.

Flood forecasting and early warning to flood affected areas are among the most important and cost-effective measures for flood management. A country wide flood forecasting and warning system has successfully being established by Central Water Commission (CWC) to cover most of the flood-prone interstate rivers. The concept of flood plain zoning or risk zoning certainly regulate landuse in the flood plains in order to restrict damage through floods which are bound to occur from time to time. Satellite remote sensing is providing timely information on drainage congested areas, extent of infrastructure damage and watershed characteristics enabling identification of flood-risk zones which is important for flood-plain management.

Drought

On account of large changes in landuse pattern—expansion of crop lands onto grazing and forest lands, reduction in area under grazing and their over exploitation, heavy grazing of forest areas, decline in tree cover, adverse effect on marginal crop land and increase in wasteland, the area under droughts has increased considerably. About 260 m ha (79%) of total geographical area of India is subject to different degrees of drought—20 m ha falling victim to disastrous drought, 117 m ha experiencing severe drought, 76 m ha having moderate drought, and 47 m ha subjected to milder drought (Kaul and Das, 1982). One-third of the country's area comprising 76 districts have been identified as drought prone (Fig. 8.2). Out of these, about 40 per cent of land mainly from western parts of Rajasthan and Kutch district of Gujarat is chronically drought prone (Shukla, 1995). Two-third of the country's crops including 42 per cent food grains and industrial raw materials like cotton and groundnuts depend upon rains. During 1979-80, the period of severe drought, production of foodgrains pulses and oilseeds fell by 28 million tonnes (Valdia, 1987). Besides these problems, during drought years drinking water problem and the number of villages affected are quite considerable (Table 8.4).

Table 8.4: Number of Drought Years (1876–1990)

Period	Slight	Moderate	Severe	Calamitous	Total
1876-1900	4	2	0	2	8
1901-1920	4	5	0	1	10
1921-1940	3	0	0	0	3
1941-1960	1	2	0	0	3
1961-1980	3	4	1	0	8
1981-1990	1	2	1	0	4
Total	16	15	2	3	36

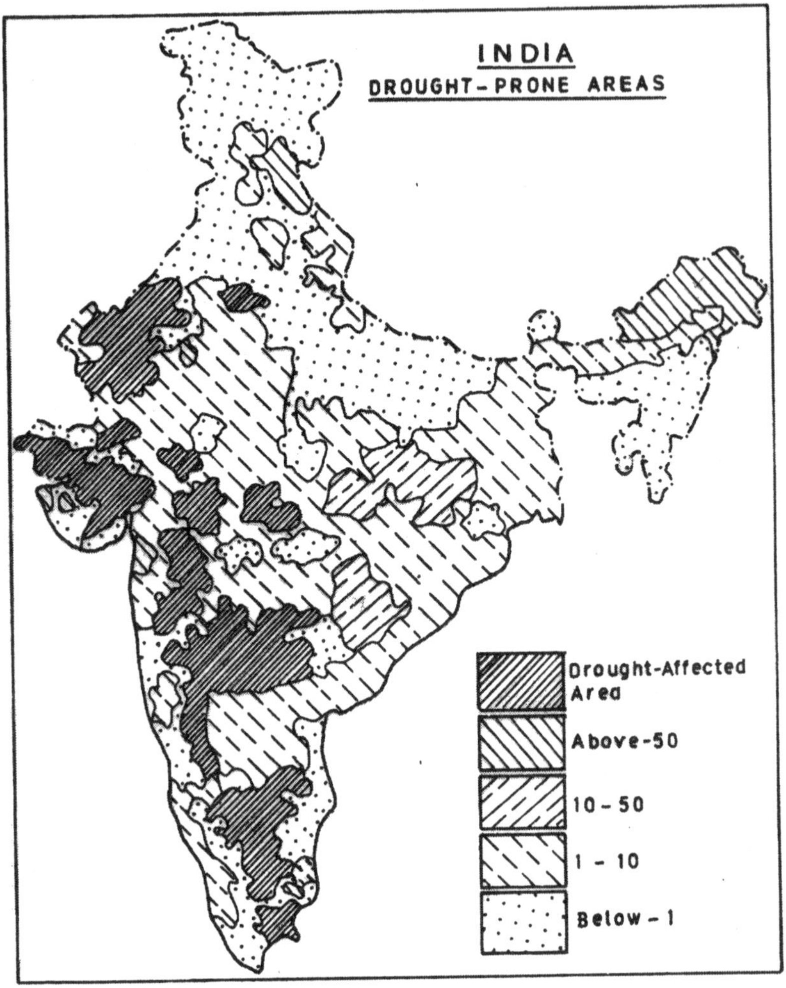

Fig. 8.2: India : Drought Prone Areas

The Government of India has given due emphasis on development of drought prone areas. During 1991, Rs. 1220.49 million were spent which involved afforestation and development of pastures on 1452.08 thousand ha of land. Large areas in semi-arid regions need to be rescued from improvident agriculture and overgrazing and directed into forestry and controlled grazing not the other way round. The main policy for drought areas should be the conjunctive use of water from rainfall, surface and underground, and the introduction of a crop pattern which would provide optimum protection from drought and at the same time ensures a reasonable income per hectare.

Desertification

On account of changing landuse pattern, more and more land is being converted to deserts. About 80 per cent of the total productive land in the arid and semi-arid areas of the globe is believed to the affected by desertification (Hillary, 1984). The UNEP estimates that world's deserts (approximately 8.0 million km^2) could eventually treble in size.

The significant factors for desertification are identified as illegal cutting and felling of trees than are planted; more food production from marginal land in semi-arid areas, increase in number of people and animals. Besides, modernisation of agriculture—planting single crop instead of more traditional and varied crop-cycle and cutting down fallow periods—leave the soil washed. away and also at risk of soil erosion; the over exploitation of futile land of low rainfall and irrigation schemes in areas without adequate and proper drainage facilities contribute to desertification.

In the tropics alone roughly 1.2 million km^2 area is under saline and alkaline deserts and turning into unproductive every year on account of waterlogging and salinity or alkalinity of soil. Besides these, much of the world's prime agricultural land is being concerted over.

The major forces of desertification in Indian desert region have been overgrazing, over cultivation and over irrigation (Mensching Sharma, 1983; Singh, 1987). About 12 per cent of the country's land suffers from threat of desertification in the arid regions of western Rajasthan and in the broad zone of the central territory classified as semi arid from Punjab to Tamil Nadu. The Thar desert of Rajasthan has advanced 1,500 km eastward during the last 8,000 years. In the period 1927-77 desert has encroached upon 1,300 ha of land in Haryana and Uttar Pradesh – the rate of increase being 0.8 ha/year (Das, 1977). According to an estimate roughly 3,059.70 km^2 area is water-logged which requires immediate attention. Various initiatives were taken by geographers to evaluate desertification control programme i.e. D.D.P., D.P.A.P., Social Forestry under W.F.P. and Indira Canal Project.

Wasteland

Wasteland is defined as a degraded land which may be brought under vegetative cover with reasonable effort and is currently underutilised and land which is deteriorating owing to lack of appropriate soil and water management or on account of natural causes. It is formed due to inherent or imposed constraints such as by location, environment, chemical and physical properties of soil or financial and management constraints (N.W.D.B., 1987), Table 8.5 provides detailed account of degraded land in India.

Table 8.5: Land Degradation Problem in India

S. N.	Particulars HA (1984-85)	Area in million (km^2)
1.	Geographical Area	329.00
2.	Area subject to water and wind erosion	144.43
3.	Total drought prone area	260.00
4.	Typical land degradation problem	
	(a) Water logged area	8.53
	(b) Alkaline soils	3.58
	(c) Saline soils	5.50
	(d) Ravines	3.97
	(e) Shifting cultivation	4.91
	(f) Average annual area affected by shifting cultivation	1.00
	(g) Rivers and torrents by ravines	2.73
	(i) Total of typical problem	29.22
5.	Total problem area	173.65
6.	Average annual loss of soil per ha (ICAR)	16 tonnes
7.	Average annual loss of plant nutrients	5.7 - 8.4 mt.

Source : Ministry of Agriculture, Soil and Water Conservation Problem (1988).

As per the waste land statistics generated from the maps prepared by (NRSA); about 16 per cent of the country's geographical area is under wastelands. Districts having more than 15 per cent of the geographical area under wasteland have been considered as critically affected and selected for mapping in phase I and phase II. Further

181 districts were selected for mapping based on above criteria. Table 8.6, and Fig 8.3 show category-wise wasteland area of 146 districts in India and statewise wasteland area respectively.

According to NRSA, very high percentage of area under wasteland is reported in Jammu & Kashmir (47.67%) and Himachal Pradesh (37.72%) due to snow cover and degraded forest; whereas in Nagaland (53.46%), Assam (57.02%) and Manipur (54.21 %) is owing to shifting cultivation and in Rajasthan (34.01%) is due to sandy area.

Table 8.6 : Categorywise Wasteland Area of 146 critically affected districts in India (*Area in km²*)

Sl. No.	Category	Total area	% to total geographical area covered
1.	Gullied and/or Ravinous land	12,976.94	1.09
2.	Upland with or without scrub	78,166.45	6.56
3.	Water logged and marshy land	3,059.79	0.26
4.	Land affected by salinity/alkalinity—coastal/inland	10,484.41	0.88
5.	Shifting cultivation area	16,302.14	1.37
6.	Under utilised/degraded notified forest land	60,661.42	5.11
7.	Degraded pastures/grazing land	11,029.2916	0.93
8.	Degraded landunder plantation crop	4,620.96	0.39
9.	Sands – Desertic/Coastal	13,590. 90	1.14
10.	Mining/industrial wastelands	418.25	0.04
11.	Barren rocky/stony waste/sheet rock area	21,585.56	1.82
12.	Steep sloping area	3,421.16	0.29
13.	Snow cover and/or glacial area	2,830.73	0.24
	Total wasteland area	2,39,093.88	20.12

Source : Wasteland Atlas of India, Vol. 1. 1993, NRSA, Hyderabad.

Among all the districts covered Punjab has minimum (2.5%) and Assam has maximum (57.02%) area under wasteland. Table 8.6 shows that high percentage (6.56%) of total geographical area is occupied by uplands with or without scrub followed by underutilised notified forest land. Out of 146 districts, seven districts are distributed in the northeastern states, Rajasthan and Himachal Pradesh have got more than 50 per cent of the geographical area under wasteland. Four districts have got 41-50 per cent range, 13 districts 31-40 per cent range, 28 districts have got 21-30 per cent range, 52 districts 11-20 per cent range and 42 districts less than 10 per cent of wasteland (Fig. 8.3). The uplands, with or without scrub cover maximum area accounting to 7.81 m ha (6.56%) followed by underutilised or degraded notified forest land covering 6.0 m ha area (5.11%). Further, gullied. or ravine lands accounted for 1.29 m ha (1.01%) and degraded pasture or grazing land 1.10 m ha (0.93%). This study provides an elusive account of desertification of wasteland in India. It may be quite useful for planners and authorities, for suggesting remedial measures to reclaim wasteland.

Concluding Remarks

The foregoing discussion reveals that a significant proportion of our land has been degraded and cannot be used for cultivation. There are a number of competing demands on land like agriculture, forestry, grassland, urban and industrial development and transport. Our land statistics are indeed very confusing. For sustainable development of land resources, we need a set of figures agreed upon by the concerned departments (agriculture, forestry, revenue records etc.) utilising modern tools like GIS, and remote sensing. A time bound nationwide survey programme of micro level landuse planning may be taken up, starting with the village. Such a survey should also highlight country's long term requirements for competing landuse and landcapability. Obviously, landuse will have to be related with soil characteristics and its uses. This will help in delineating landuse classes. The existing legislations are weak and require, to be reviewed and updated with reference to changing landuse in the country. A strategy should be prepared to renew and restore past damages as also insulate the country from future damage. It will lead to dynamic landuse policy. India has a National Landuse and Wasteland Development Council with two

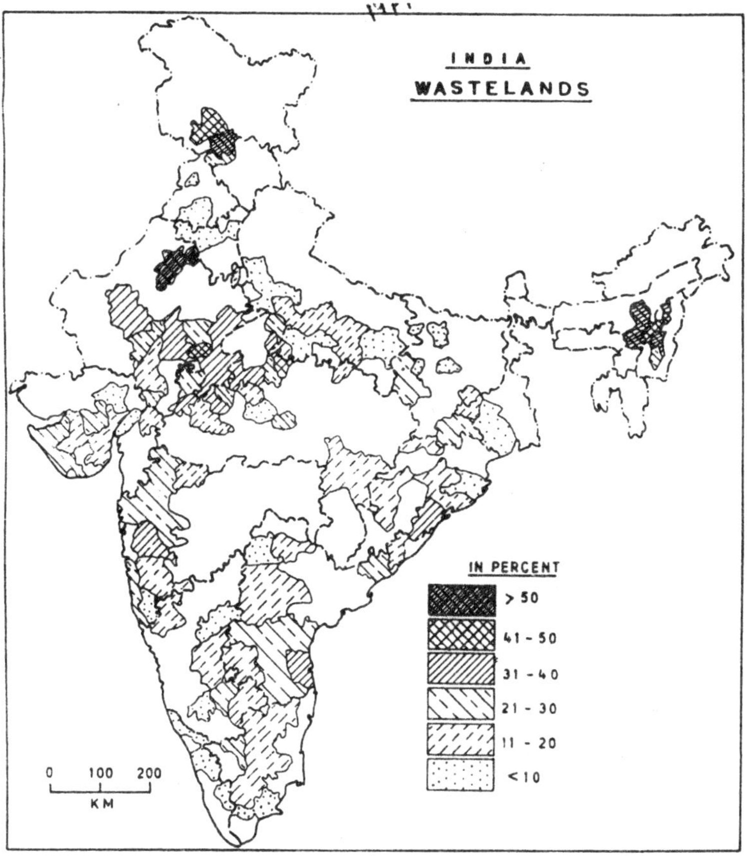

FIG. 8.3: India : Wastelands

boards—National Wasteland Development Board and National Landuse and Conservation Board. These organisations should work together for the sustainable development of landuse in the country.

References

Das, D.C. (1977), Soil Conservation Practices and Erosion Control in India A Case Study, *Soil Bulletin of F. A. O.*, vol. 33, pp. 11 -50.

Hillary, E. (eds) (1984), *Ecology 2000*, Multimedia Publications, U.K. pp. 58-81.

India Today (1986), Deforestation An Environmental Holocaust, August, 1986, pp. 77-88.

Kayastha, S. L. (1982), Forest in India and the impact on economic developments, Lecture delivered in I.G.U. Latin American Regional Conference Brazil, in the Commission on Environmental Problems, Sao Paulo, pp. 1-12.

Kaul, R.N. and Das, D.C. (1982), Soil and Water Conservation, *State of Environment—Some Aspects,* Dept. of Environment, New Delhi. pp. 1-16.

Mensching, H.G. and Sharma, R.C. (1983), *Resource Management in Arid Lands,* Rajesh Publications, New Delhi.

Messerli, B. and Hofer, Th. (1995), Assessing the impact of anthropogenic landuse change in the Himalayas, in G.P. Chapman & M. Thompson (eds), *Water and the Quest for Sustainable Development in the Ganga Valley.* Mansell, p. 71.

Singh, S. (1987) Ecological Impact on irrigation on biophysical potentials of landforms of the Rajasthan desert, *The Geographer,* vol. 34(2), pp. 39-45.

Shukla, V. (1995), Harnessing water resources effectively, *Yojana*, vol. 39, No. 1&2, p. 7.

Sud, S. (1995), Tackling the flood menance, *Yojana,* Vol. 39, No. 182, p. 59.

Valdiya, K.S. (1987), *Environmental Geology: Indian Context,* Tata McGraw Hill Pub. Co. New Delhi p. 217.

9

Developmental Planning in Coastal Environment

ANJANA DESAI and GOUTAM BHATTACHARYA

The coastal environment is a special ecosystem where interaction takes place between land and sea. As a result, it is seen in many cases that people have progressed economically through trade and exchange of commodities by taking the advantage of the coastal location. However, coastal people have to face many environmental hazards like marine erosion or deposition, cyclones, storm surge, marine ingression and soil salinity. Hence, specific planning is necessary for the development of coastal environment. The coastal area of the Bhal region, in the Gulf of Khambhat in Gujarat, suffers both from drought and flood, salinity, marine ingression and deposition. The region is not sustainable environmentally and economically, but can be made so, if resources are managed properly.

The main objectives of this study are:

(a) to find the land capability of the coastal unit with the help of Geographical Information System or in short GIS.
(b) to propose utilisation of land, based on land capability for the purpose of sustainable developmental planning.

Methodology

The whole Bhal region has been divided into three units, based on fourteen variables. Those divisions are coastal unit, sub-coastal unit and interior unit. In this study, only coastal unit has been analysed. A series of physical variables has been superimposed by using GIS, to find a map of land capability. Some of the variables are: flooded and marine ingressed area, distribution of soil salinity, quality of ground water, geomorphic features and drainage system, rainfall distribution, type of soil and wasteland, change in landuse, levels of land degradation, and change in density of population. Landuse and economic activities have been proposed on the basis of land capability. Data have been used from various sources, e.g., remote sensing data from Space Applications Centre, topographical sheets from Survey of India, soil data from soil testing laboratories, department of agriculture and irrigation and census reports. For a micro-level analysis field work has been done in the village of Khalatalav with the help of a questionnaire.

The Study area

The coastal unit of the Bhal region, covers 48 villages, without any town having a total area of 2,142.23 km^2 in the districts of Bhavnagar, Ahmedabad and Kheda, boarding the Gulf of Khambhat (Fig. 9.1) Physically the unit is a fragile ecosystem, being a part of the sea in the tertiary period. Geomorphologically the area is a flat mud plain with very little slope. It has both high and low tide mud flats, flood plain and submerged shoals. It is only five meters above the sea level with high ground water table (2m) which is saline. Soil is dark and clayey loam, being formed on the fissure erupted basalt in cretaceous period. The inner section has calcareous red and brown soil. The coast is exposed to marine ingression which makes high pH value in soil, mostly above 8.6. It is lower (7.6-8.0) in Kheda district. Due to lack of slope, high water table and clayey soil, the area gets flooded during the monsoon, but it suffers from drought in the pre-monsoon period when capillary action brings salt to the surface of the soil. Consequently, agricultural activities are limited both in time and space in the coastal area. The mean annual rainfall in this area varies between 500-600 mm.

It is observed that, there has been a lot of land degradation in the coastal area mainly due to marine ingression (Fig. 9.2). This has affected the general landuse, cropped area and the cropping pattern and the density of population, leading to seasonal or permanent out-migration of labour. Nearly 40 per cent of the coastal unit was degraded due to marine ingression before 1967, as indicated in the topographical

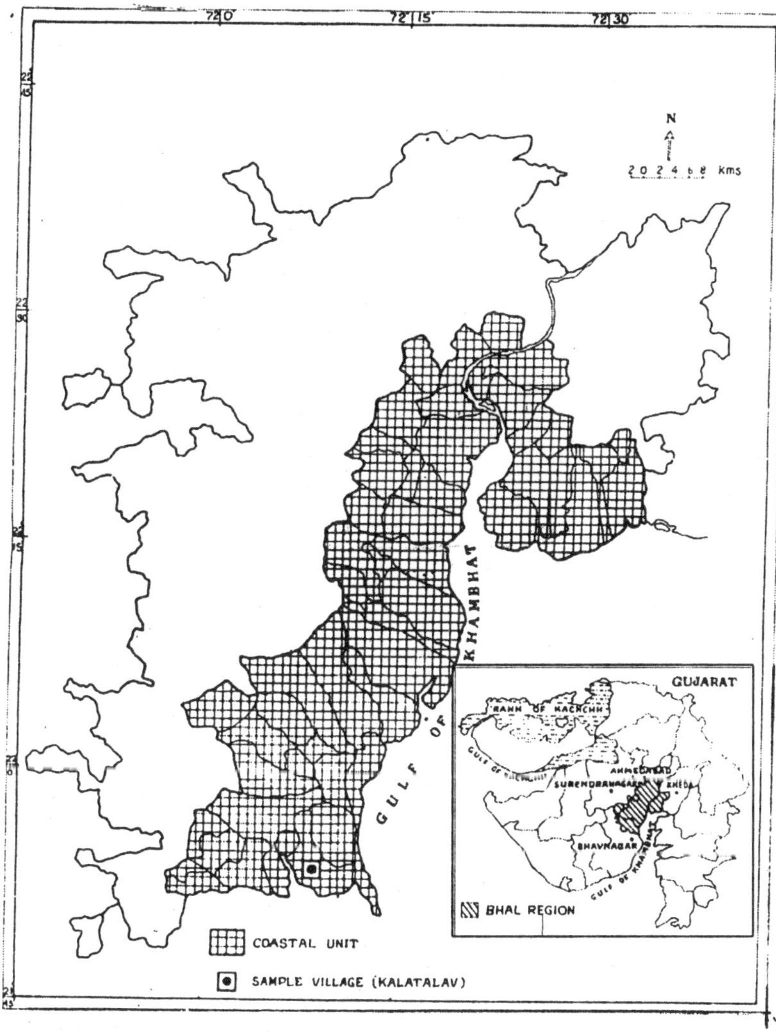

Fig. 9.1: Location of Coastal Unit in the BHAL Region

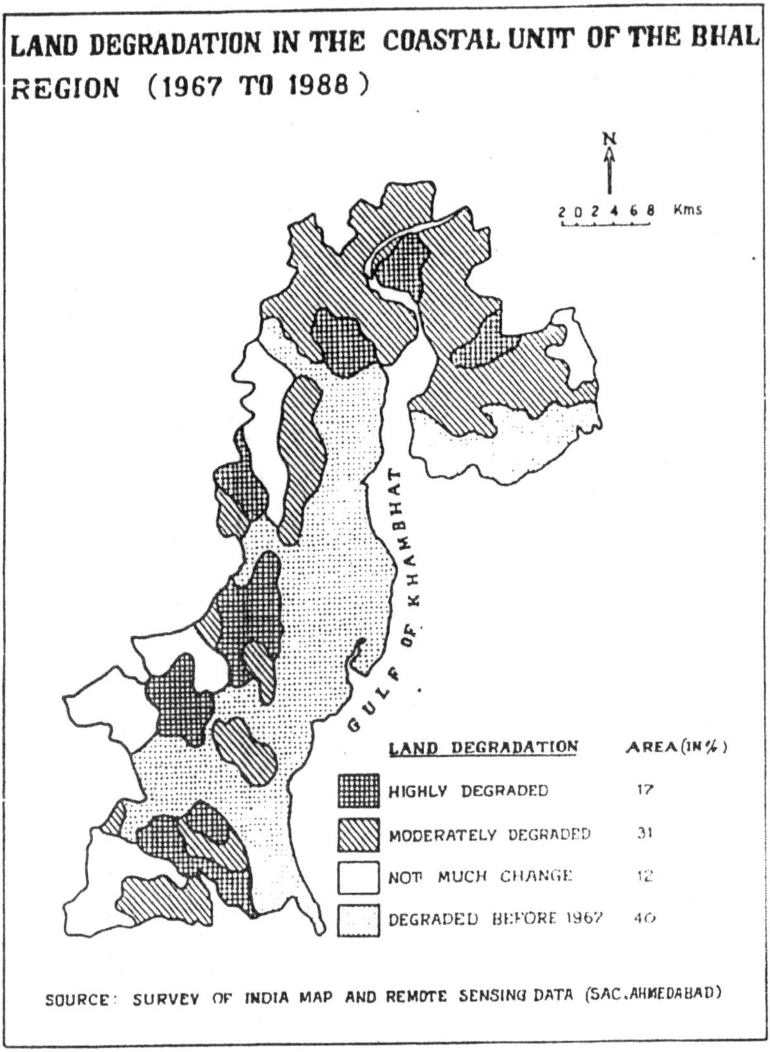

Fig. 9.2: Land Degradation in the Coastal Unit of the BHAL Region (1967 to 1988)

sheets. This process of land degradation has been spreading towards the interior, which has been detected in the satellite imagery of 1988. During the period of 1967—1988, 48 per cent of the coastal unit has been either highly degraded of moderately degraded. Only 12 per cent of the area has not been affected and has remained without much changes in the western side. In short, the coastal unit of the Bhal region has been badly affected by marine ingression. Unless it is checked immediately the whole Bhal region will be affected. The interior section of the Bhal region produces the best quality of wheat in Gujarat. Both land and production of wheat will be lost forever if marine ingression is not checked in time.

As a result of land degradation there have been changes in the landuse. In Kheda and Ahmedabad districts, agricultural land has remained fallow. Moreover, in Kheda there has been an increase in soil salinity due to over irrigation of saline ground water. Here human interference and ignorance have resulted in land degradation. In Bhavnagar and Ahmedabad areas, agriculture land is gradually changing into salt waste. Demographically the coastal unit has been affected also due to land degradation (Table 9.1). The growth rate of population during 1971–1981 was 25 per cent in the coastal unit, compared to 27 per cent in total Gujarat. But the growth rate of population has decreased very much in the coastal unit during 1981–91 (i.e. 5 %), compared to that of Gujarat in the same period. Regarding the density of population during 1971–91, it is very low in the coastal unit, i.e. 20 persons per square kilometers in 1971 and 26 in 1991 as compared to 136 for Gujarat in 1971 and 210 in 1991. In fact, more than 50 per cent of villages in the coastal unit has shown negative change in population density during 1971–91.

The occupational structure of the coastal unit has also been affected by land degradation. In 1971, 54 per cent of the total workers were cultivators, but the figure is reduced to 30 per cent in 1991. It has become very difficult to do farming in land which has turned salt waste. Hence, the cultivators have become agricultural labour. The percentage of agricultural labour was 33 per cent in 1971, which has increased to 54 per cent in 1991. Similarly, percentage of other primary workers, like fishing, and household industry, like diamond polishing, have become increased.

A close view of a village, *Kalatalav,* in the coastal unit, gives a

Table 9.1: Coastal Unit of Bhal Region (48 villages)

1. Demographic components

	a. Coastal unit	b. Gujarat
Total population		
1971	43,460	26,697,475
1981	54,439	33,960,905
1991	57,043	41,174,060
Density Km^2		
1971	20	136
1981	25	173
1991	26	210
Growth rate		
1971-81	25	27
1981-91	5	21

2. Occupational structure

Type of workers	Year	
(in %)	1971	1991
Workers to total population	30	37
Cultivators to total workers	54	30
Agricultural labour	33	54
Other primary workers	3	5
House hold industry	8	10
Services	2	1

3. Landuse (in%)

Type of landuse	Year	
	1971	1981
Irrigated	0.02	0.5
Un-irrigated	24	25
Total cultivated land	24	26
Culturable waste	9	4
Area not available for cultivation	67	68
Forest	—	1.5

Source: Census of India: 1971, 1981 & 1991.

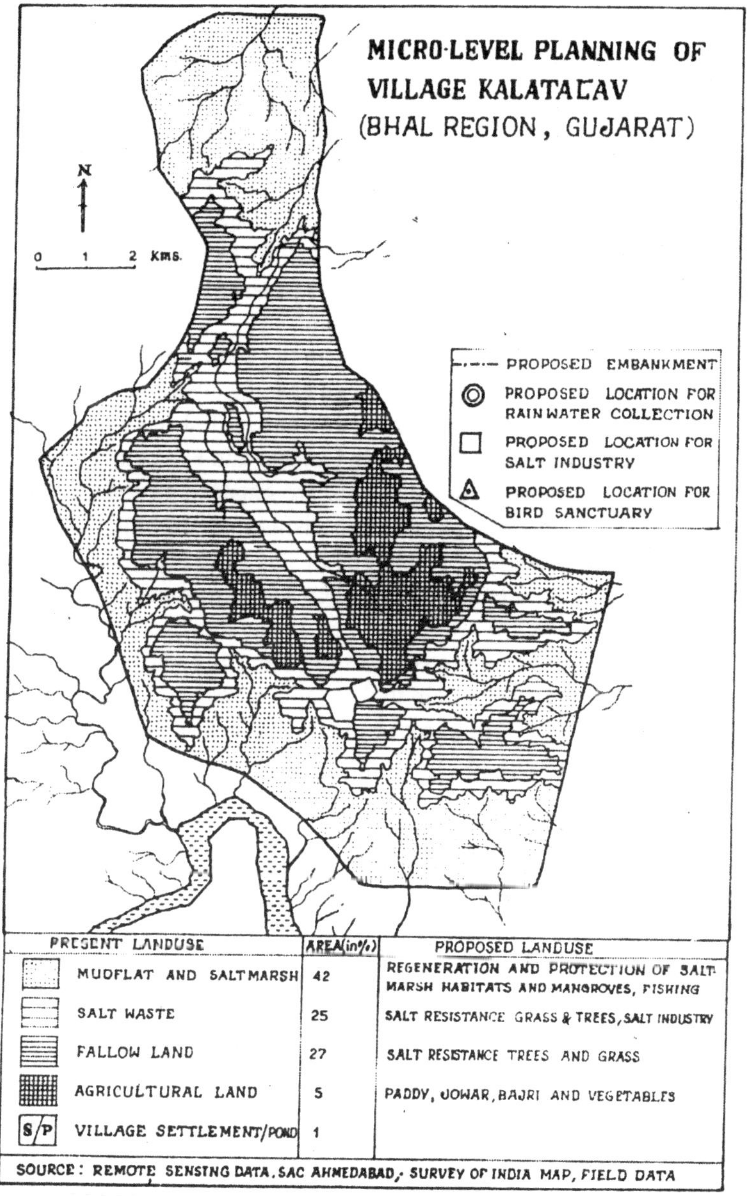

PRESENT LANDUSE	AREA (in %)	PROPOSED LANDUSE
MUDFLAT AND SALTMARSH	42	REGENERATION AND PROTECTION OF SALT MARSH HABITATS AND MANGROVES, FISHING
SALT WASTE	25	SALT RESISTANCE GRASS & TREES, SALT INDUSTRY
FALLOW LAND	27	SALT RESISTANCE TREES AND GRASS
AGRICULTURAL LAND	5	PADDY, JOWAR, BAJRI AND VEGETABLES
S/P VILLAGE SETTLEMENT/POND	1	

SOURCE : REMOTE SENSING DATA, SAC AHMEDABAD, SURVEY OF INDIA MAP, FIELD DATA

Fig. 9.3: Micro-Level Planning of Village Kalatalav (Bhal Region, Gujarat)

better idea for a micro-level planning (Fig. 9.3). Field work shows that nearly 95 per cent of the land in Kalatalav is unavailable for agriculture, of which 42 per cent is mudflat and salt marsh, 25 per cent is salt waste, 27 per cent is fallow land and 1 per cent is the village settlement. During 1981–1991, the cropping pattern has also changed. Earlier wheat, cotton and oil seeds were cultivated alongwith *jawar*. Recently, farmer can only cultivate *jawar* for fodder as other crops are not suitable for growth due to land degradation. Population has decreased in this village during 1971-1991 from 972 to 428. The density of population is very low. It was 2 persons per km^2 in 1971 and now in 1991 is only 1 person per km^2.

The profit from migration have been the main means of survival for residents of Kalatalav. Being located near the city of Bhavnagar, a large portion of population is away for nearly 300 days in a year. They return for a couple of months in October and November to get a winter crop. At destination, migrants are engaged in loading and unloading type of work, diamond polishing, household work or services. They also work as labourer in the ship breaking yard of Alang in Bhavnagar. Migrants can sustain themselves in a meagre way as the mean wage is Rs. 20 per day per person and they work nearly 230 days per year. Individual sustainability can be achieved through migration, but, it does not help to develop the area of origin. In fact, the coastal area is suffering from a vicious cycle of land degradation and rural outmigration. Moreover, the process of land degradation is engulfing the better land in a fast manner to the extent that, rural sustainability will totally be lost in near future. As more and more land is kept uncared, more degradation takes place. Perhaps outmigration for survival has done much harm to conservation and protection of land in this area.

Regional and Micro Level Planning

It has been suggested in this study, that a micro level planning both with measures of checking land degradation on one hand, and implementing economic activities, as per the land capability on the other hand. With the help of various physical factors related to soil, water, slope, flooded area, level of land degradation etc., and by using the Geographic Information System (GIS), the coastal area has been further classified into sub-regions of land capability (Fig. 9.4). Out of

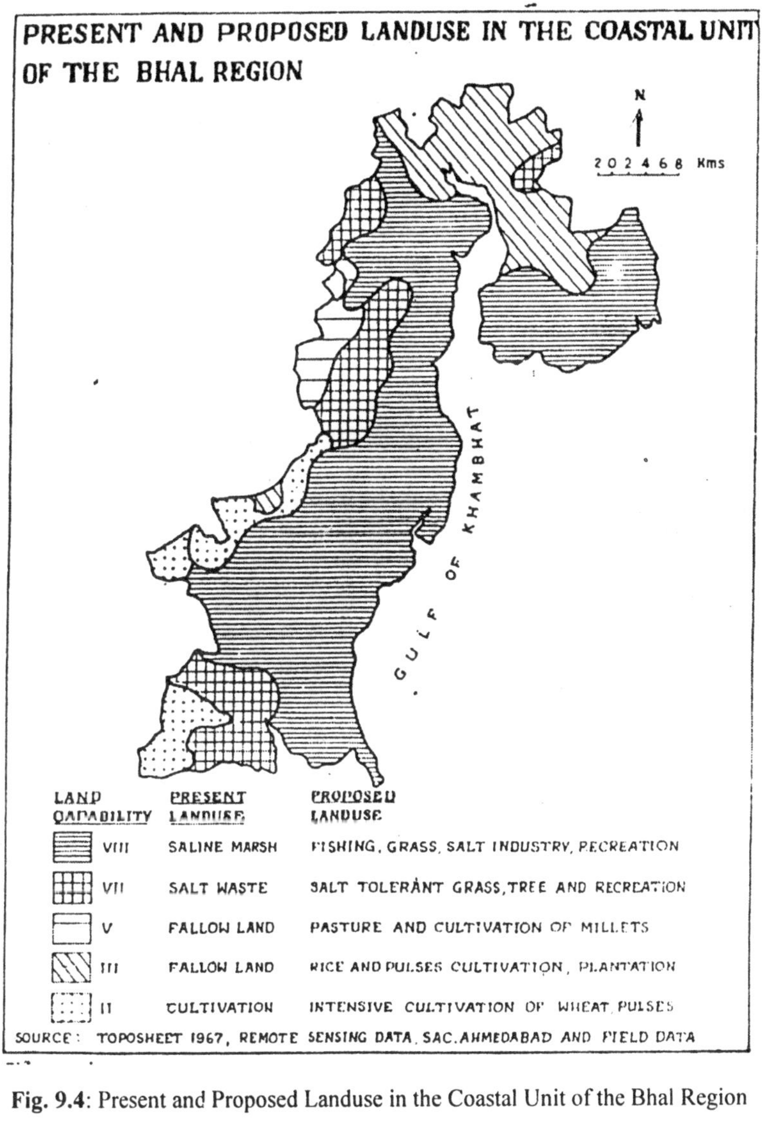

Fig. 9.4: Present and Proposed Landuse in the Coastal Unit of the Bhal Region

eight groups of land capability, categories of I to IV are suitable of agriculture activities, of which class I is the best suited. Classes between V to VIII are not suitable for farming. The major part of the coastal unit is having land capability of class VIII which is at present a saline and or marshy area. The proposed landuse and economic activities which are not farming, for this area are fishing, grass growing for fodder, salt industry, and recreation and tourism. To the west of this section, the land capability comes in the category VII which is located in western part of Bhavnagar district. It can be developed into a pasture with salt tolerant grass, bordered with salt tolerant trees. Recreational activities can be promoted. Further, west is fallow land of category V. This patch can be developed into a pasture and millets can be cultivated here as well. In Kheda and Ahmedabad districts, fallow land of class III can be developed through the cultivation of rice and pulses. Intensive cultivation of wheat and pulses has been proposed in the western margin where the land capability falls in class II.

Conservation of land can be done also through a proper management of land and water along with land use planning based on land capability. Marine ingression has to be checked with proper embankment. Local people, tend to move out for employment instead of working for their own area. Relief work should be extended by the government in villages rather than in cities. Rain water can be harvested in man-made deep ponds having plastic lining at the bottom so that fresh water does not get mixed with saline ground water. Coastal embankment should be fully protected with a line of thick mangrove forests for checking marine ingression. Estuaries and gullies should be plugged for the same purpose. Along with development of the embankment, aquaculture bird sanctuary, and tourism can be introduced for economic sustainability as well. Salt industry and diamond polishing are suitable activities which will be able to hold rural population. Salt resistant grass can restore the land into a rich pastural land on which dairy co-operatives can thrive and can raise the level of income of residents. Similar sustainable planning in the villages of Kalatalav is needed, which is environmentally and economically viable so that outmigration of people can be checked. Local resources will be then utilised in such a way that sustainability can be achieved through a combination of farm and non-farm activities, fully organised by local people and are based on local ecosystem. Such a planning will increase man-days of labour. However, in order to increase the efficiency of rural labour, organisation and management are absolutely necessary in term of production and marketing. There has to be a proper linkage from research and information through government and non-government organisations so that the people use local, appropriate technology for sustainable planning. Developmental planning should have a comprehensive approach of conservation of resources, utilisation of land in a proper way, diversification of economic activities for full employment, training and organisation of labours for efficiency and better management both at micro and regional levels.

Acknowledgement

This paper is based on the ongoing project entitled "Management of Land Resources Through Geographic Information System", funded by the Department of Science and Technology, New Delhi. The authors are grateful to the funding agency.

References

Davidson, Donald A (1992), *The Evaluation of Land Resources*, Longman Scientific and Technical, England.

Rammaiah, Y.V. and Reddy, N B K (1983), Carrying capacity of land in Andhra Pradesh, *Indian Geographical Journal*, Vol. 58.

10

Response to Environmental Hazards in Coastal Villages of Sagar Island: An Opinion Survey

SUNANDO BANDYOPADHYAY

Sagar is a 580 km^2 estuarine island located at the south-western corner of South 24-Parganas district, West Bengal (21°37′ to 21°57′ N and 88°02′ E to 88°11′ E). The island has an 11 km stretch of bay-facing coastline that is occupied by four villages.

A sample opinion survey was conducted among the families of three coastal hamlets of the Sagar island in order to ascertain the local peoples' response to the environmental problems they are facing. These problems include coastal erosion, tide-water incursion and dunal encroachment. While in Beguakhali village all of these problems have reached a critical level, in Raspur-Gokulnagar and Shibpur the major problems are sand-encroachment and coastal erosion respectively. In general, there is some dissatisfaction among the villagers about implementation and maintenance of the management project.

The Environmental Problems

The littoral tract of Sagar island is mostly eroding except the central in-filling section (Bandyopadhyay & Chakraborty, 1990), where the coastal dunes are currently prograding landwards (Bandyopadhyay, 1990). The changes in the island's coastline between 1942 and 1975 has been shown in Figure 10.1. It may be noted from this figure, how an entire village, Bishalakshipur, has been destroyed by coastal erosion between 1942 and 1969. During the same period, however, some area was also acereted where the present Raspur-Gokulnagar village is situated. From west to east, the four hamlets, Beguakhali, Gangasagar,

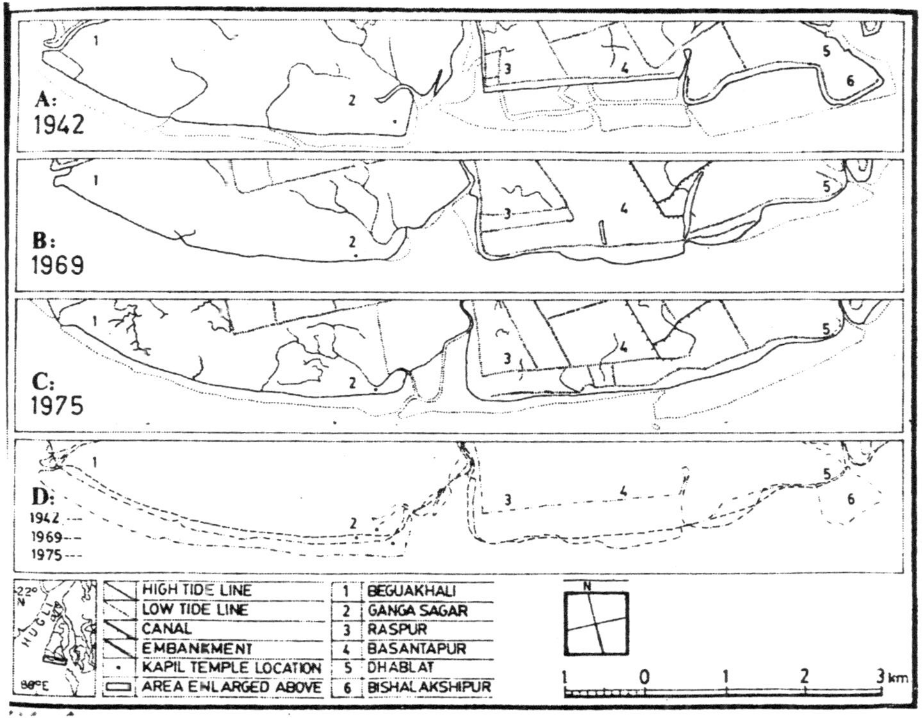

Fig. 10.1: Changes in the Sagar island coast : 1942—1975. The low tide line is not fully represented in 1B. Raspur-Gokulnagar village is situated to the south of Raspur. Basantpur is not a coastal village at present.

Source : 1:63,360 and 1:50,000 Survey of India topographical sheets, 1:25,000 aerial photos of years 1942, 1969 and 1975 respectively. All these materials were brought to a similar scale before comparing.

Raspur-Gokulnagar and Shibpur are spread along the southern coast of Sagar. The post independence pattern of population growth of the mouzas in which they belong has been shown in Table 10.1. The main environmental problems being faced by these villages are, coastal erosion, dunal encroachment and salt-water incursion through breached marginal embankments—all causing temporary or permanent loss of agricultural area. The preventive measures taken to combat these hazards include construction of coastal embankments and afforestation using *Casuarina equisetifolia.*

Table 10.1: Growth of Population in four coastal *mauzas* of Sagar island (1951—1991)

Village (*Mauza*)	1951 (actual population	1951 (Index)	1961 (Index)	1971 (Index)	1981 (Index)	1991 (estimated population)	1991 (estimated index)
Beguakhali	1,542	100	134.3	159.3	193.6	3,425	224.7
Gangasagar	1,560	100	148.3	199.2	243.9	4,553	291.9
Bishalakshipur	33	100	75.8	*Delluviated*			
Dhablat	1,513	100	174.2	193.2	236.9	4,275	259.3

Note: 1. Base of all the index figures is 1951 (=100).
2. Data source is the District Census Handbooks of 24-Parganas, 1951, 1961, 1971 and 1981.

Methodology of the Survey

Among the four coastal villages of the island, Gangasagar was excluded from the sample opinion survey. This was done because, being an important place for pilgrimage from ancient times, the settlement has grown an urban affinity into its identity. The place has a large non-resident population related to tourism and religious practices. With a bus terminus, several government houses and diesel generated electricity, Gangasagar is greatly different from the interior villages of the island, where one can reach only by cycle-rickshaws on un-paved roads or by foot.

The opinion survey was conducted by interviewing randomly selected families on the basis of a pre-determined questionnaire. The summary of its findings has been shown in Table 10.2.

It should be mentioned here that the number of the families (synonymous to the definition of *households* in the Indian census), of the surveyed hamlets does not necessarily match the census figures for the villages *(mauzas)* bearing an indentical name. The reason for this is, in Sagar like elsewhere, a *census-village* or *mauzas* bounded by revenue boundaries, is formed of a number of smaller hamlets – each disconnected from the other. For example, the census-village of Dhablat, comprising of 586 households (1981), is constituted by nine small hamlets. Of these, only two, Raspur-Gokulnagar and Shibpur, are coastal (the Shibpur hamlet of Dhablat should not be confused with the census-village of Shibpur – situated to the north of Shibpur hamlet).

Discussion and Analysis

From the survey it is seen that, coastal erosion and dunal encroachment have been defined as the major problems of Shibpur and Raspur-Gokulnagar villages respectively. Beguakhali has a mixed character.

The intensity of these problems are best reflected by the villagers' attitude towards outward migration. While as high as 76.9 per cent of the surveyed families intend to leave Shibpur because of the problems they face, none of the families thinks of migration in Raspur-Gokulnagar. However, among all the families surveyed, only one in Shibpur has any kind of alternative arrangement for leaving (in this case, an alternative housing in another part of Sagar). Thus, the intention for migration is unlikely to be undertake in reality. Another contradictory finding of the survey concerns the village of Raspur-Gokulnagar. Here, 77.8 per cent of the families (the sample size of this small village covered all of the existing number of families) has lost all or part of their land due to coastal erosion, but they have rated the problem as less important than sand encroachment and none of them wishes to migrate. This is because, in the Raspur area coastal erosion is currently not quite evident (Bandyopadhyay & Chakraborty, 1990). The loss of land had occurred in the past-and at present the nature of the problem has changed.

During the survey, it has been seen that there is an apparent lack of knowledge among the villagers about which agency or body is

Table 10.2: Summary of the sample opinion survey of three coastal villages in Sagar Island (February, 1991)

Questions	Choice of answers	Response (in per cent of surveyed families)		
		Beguakhali	Raspur-Gokulnagar	Shibpur
What is the most critical problem being faced by the village?	(a) Coastal erosion	64.7	0	96.2
	(b) Dunal encroachment	0	83.4	0
	(c) Salt-water incursion	14.7	0	3.8
	(a) and (b)	0	16.6	0
	(a) and (c)	0	0	0
	(b) and (c)	0	0	0
	(a), (b) and (c)	20.6	0	0
Do you intend to leave your village because of the problem it faces?	Yes	29.4	0	76.9
	No	70.6	100	23.1.
Have you lost all or part of your agricultural land due to coastal erosion?	Yes	35.3	77.8	42.1
	No	64.7	22.2	57.9
Have you changed your profession (e.g.: from farming to fishing) because of coastal erosion	Yes, all of the family members have changed	14.7	0	34.6
	Yes, a part of the family members have changed	0	0	0
	No, none has changed	85.3	100	65.4
Does coastal erosion endanger you?	Yes	85.3	22.2	96.2
	No	14.7	77.8	
Are you satisfied with the embankments constructed to prevent coastal erosion	Yes	14.7	77.8	3.8
	No	85.3	22.2	96.2

(Contd.)

Table 10.2: *(Contd.)*

Questions	Choice of answers		Response (in per cent of surveyed families)		
			Beguakhali	Raspur-Gokulnagar	Shibpur
Which agency is best maintaining the embankments	(a)	The Panchayat	55.9	66.7	79.9
	(b)	The State Irrigation Dept.	44.1	38.9	23.1
	(c)	Not Sure	0	0	0
Have done progradation encroached all or part of your agricultural land?	Yes		50	78.8	3.8
	No		50	22.2	96.2
Are you satisfied with the afforestation programmes being implemented to check dunal migration.	Yes		17.6	55.6	100
	No		82.4	44.4	0
Which agency is best carrying out the afforestation programme?	(a)	The Panchayat	14.7	77.8	11.5
	(b)	The State Forest Department	0	5.5	0
	(c)	Collective effort by the villagers	0	16.7	76.9
	(d)	Not sure	85.3	0	11.6
Number of families residing in the hamlet*			103	18	161
Number of families surveyed			34	18	52
Average size of the families (adults & children)			4.1+4.4	4.9+4.6	4.2+4.6
Number of families all of whose members are illiterate			5	8	8

**Source:* Panchayat office of the respective villages.

carrying out a particular management programme in a particular village. Thus the answers in this section may not be quite reflective of the actual situation. In Shibpur, however, the villagers did participate in a collectively launched afforestation programme which was very successful (in the village, all of the surveyed families are satisfied with it). Their response in this section is likely to be correct.

Conclusion

In overall review it seems that, there is a high dissatisfaction working among the villagers about the progress or maintenance of the management projects. Among the three villages, Beguakhali seems to be the most neglected one in terms of implementing any kind of project. On the other hand, afforestation and protective works to prevent dune migration and Coastal erosion have fallen short of expectation in Raspur-Gokulnagar and Shibpur respectively.

Acknowledgements

Since thanks are experessed to Mr. Kanailal Das, Calcutta University, for his help during the project. Thanks are also due to Prof. M.K. Dandyopadhyay, Calcutta University, for his helpful suggestions.

References

Bandyopadhyay, P. C. and Chakraborty, S. C. (1990) : Repeat beach mapping and survey profiling at Bakkhali, Sagar islands, 24-Parganas (South), West Bengal, *Records, Geological Survey of India.* 123 (3).

Bandhyopadhyay, S. (1990) : Evolution of the Sagar island coast-line, West Bengal, between 1923 and 1990, in Roy, B. K. (editor): *Cartography for Development of Outlying States and Islands of India,* Monograph No. 10 (1), National Atlas and Thematic Mapping Organisation: pp. 68-69

11

Social Forestry and Human Perception in Birbhum District

NANDINI CHATTERJEE

The recent spurt in the implementation of social forestry schemes has been a result of the increased awareness of deforestation and its adverse consequence. The country is now facing a major ecological and socio-economic crisis. Since 1980-81, several Social Forestry (SF) schemes have been initiated in selected fuelwood deficit districts. In order to circumvent this trend of land degradation, a National Wastland Development Board (NWDB), was set up in 1984. However, afforestation under social forestry is the basic programme, with emphasis on fuel and fodder.

Introduction

The western districts of West Bengal, laying in the plateau and plateau fringe tracts are confronted with this problem of environmental degradation. Relatively, this region may be considered as sub-humid with rainfall amounting to about 1,000-1,100 mm per year. This is less than the average for West Bengal. In the district of Birbhum, deforestation and land degradation began largely since the beginning of the 19th century. The enactment of the West Bengal Estates Acquisition Act in 1953 makes it obligatory for vesting of all forest lands with the government. Accordingly, all private forests vested in the government on April 1st, 1955 were at once handed over to the forest department for management.

Existing forests extend mostly over the acquired *danga* or high tracts, located along the western border of the district, as in Rajnager and Suri Police Station (PS) areas. The extensive lateritic and other *danga* lands subjected to deforestation, formed a second zone. Since

1964, these are being afforested gradually. Thus, the forest department has taken charge of the private forests, and brought about the subsequent rehabilitation of degraded forests as well as the reafforestation of barren waste-lands acquired by the department (Sarkar, 1964).

At the time when the lands were vested, these forests were in reality scrubby, degraded uplands, consisting mostly of *Sal (shorea robusta)* with a sparing mixture of piasal *(petrocarpus marsupium)*, mahua *(bassia latifolias)*, kend *(dispros melanoxolyn)*, pial *(buchanania latifolia)*, parasi *(cleistanthus collinus)* and so on. After the forest department took over the forests, fellings were continued with standard system under a rotaton period of ten years. Later on, the period of rotation was increased to fifteen to twenty years. Initially, a large number of species were tried such as arjun, *(terminalia arjuna)*, neem *(azardirachta indica)*, minjri *(cassia siamea)*, kaju *(anacardium occidentale)*, sirish *(albizzia lebeck)*, etc. Following this emphasis was laid on *sal, piasal* and teak. The recent trend is towards the planting of quick growing species like eucalyptus hybrid, *(E. tereticornis)* and *akashmoni/sonajuhri (acacia auriculiformis)* for the prime purpose or fuelwood and pulpwood supply. A close perusal of increase in forest area depicts a steady growth since 1960's but with a sharp increase since 1980's due to reasons already specified.

Social Forestry Schemes in Birbhum

The activities of social forestry schemes in Birbhum, as elsewhere, may be divided into the four components:

(a) *Strip plantations:* These are confined along the roads, railway tracks, embankments, canals and rivers. Such plantations constitute 29 per cent of all the components in Birbhum. Of this, each type accounts for 7.83, 2.37, 3.62, 13.79 and 0.68 per cent respectively. Emphasis is on canal bank afforestation in order to check bank slips and subsequent silting up of canal beds. Roadside plantations come next in importance as they provide shade besides possessing an aesthetic value. Riversides assume least importance as the immediate vicinity of rivers are preferred for cultivation.

(b) *Farm forestry:* This involves planting of trees on individual holdings for commercial purposes. Individual owners either meet the needs with their won resources or with minimum state support. Planting is usually carried along farm boundaries, farm roads, home-stead, lands. Share of this component in Birbhum is maximum, amounting to 54.15 per cent.

(c) *Village woodlot:* Plantations along government or community owned lands, by the forest department, fall under this category. Hence, such sites of forest land form a common unit to the villagers. School compounds, cultivable wastelands, playgrounds, etc. are chosen sites for such village woodlots. This component accounts for only 1.67 per cent.

(d) *Reforestation of Degraded Forests (RDF):* This constitutes a significant part of afforestation programmes and accounts for about 16 per cent of all the four components. RDF is mostly confined along the lateritic zones as there is a greater scope of coppicing. Reforestation by planting is resorted to in the other areas. Out of the total area under RDF, planting is carried out on over 67 per cent (Social Forestry Wing, 1987-88). Most of the private forests vested in the government fall under this category. RDF is an ongoing activity and is easier to adopt as it does not differ from traditional forest plantations. Moreover, the land is already occupied partly by a thin strand of forest cover.

Study Area

This study includes the two western police stations of Rajnagar and Suri. The area truly represents the drought prone, plateau fringe tract. Geologically, archaean granites and gneisses predominate. Quartzite veins are also observed. Depth to bedrock is considerably shallow. Latcritc cappings occur as detached exposures.

Terrain grades from highly undulating to imperceptibly sloping ground as one proceeds eastwards. Soil is predominantly coarse in texture. Climate is subhumid in nature. Thus, the environmental setting does not favour the development of considerable surface and groundwater resources. The streams are mostly seasonal, depth of groundwater levels are moderately deep. Groundwater potential is low. Hence, rainfed and irrigated agriculture do not appear to be successful ventures here. The present forest cover amounts to 11.5

per cent of total area in Rajnagar P. S. and 3.82 per cent in Suri P. S. The average for Birbhum district is 3.52 per cent. The study area forms an important zone of social forestry. The tribal inhabitants, which constitutes an important segment of total population, were traditionally accustomed to etching out a livelihood from forests. All these facts have led to the selection of the study area.

Methodology

Six villages were surveyed in order to bring into perspective the actual benefits accruing from social forestry and the response of the beneficiaries. The villages are Chhora, Salboni, Jahanabad, Lalboni, Karabad and Kharna. They are well distributed over the entire region. The first two are located in the east, near town of Suri. The second two are situated in the remote west, along the Bengal-Bihar border. The last two villages are located in the centre, within the Chandrapura-Tanipara *tasar* silk producing region (Table 11.1).

Table 11.1: Distribution of Landuse in per cent (1971 & 1981)

Village	Forest	Culturable waste	Unirrigated	Irrigated	Not avail. for cultiv.	Year
			Suri Police Station			
Chhora	11.86	1.07	44.89	14.85	27.31	1971
	12.22	–	78.69	1.78	7.31	1981
Salboni	24.91	11.49	59.77	–	3.83	1971
	24.85	6.88	56.22	–	12.05	1981
			Rajnagar Police Station			
Jahanabad	4.44	47.16	39.51	3.70	5.19	1971
	4.43	49.22	40.10	–	6.25	1981
Lalboni	26.55	28.62	34.90	6.12	4.01	1971
	21.80	28.60	20.93	–	28.67	1981
Karabad	–	18.60	76.25	–	5.15	1971
	–	15.79	59.44	15.48	9.29	1981
Kharna	9.48	29.33	12.00	9.60	39.59	1971
	9.66	29.32	11.87	9.59	39.56	1981

Source: Census Reports, 1971 & 1981.

Chhora

The village of Chhora is located a few kilometers west of Suri town, along a metalled road. The terrain is gently undulating with a veneer of stony loam. This soil type is locally termed *Kankure doash.* Regarding land use pattern, a slight controversy arises with respect to area under cultivable westeland. According to the villagers, about 25 per cent of the total area of 340 ha is under this category. Census reports, however, record a negligible value, even as early as 1971. Net sown area covers 80 per cent of the total area. Forest land registers a stable value of 12 per cent, whereas, area not available for cultivation depicts a decrease. If the estimate of the villagers is assumed to be right, net sown area amounts to 60 per cent.

From the point of view of agriculture, Chhora is not much different from the other villages of this tract. The main influencing factor has been the lack of a dependable source of irrigation. Dugwells do exist but they are mostly utilised for domestic purposes. Water table fluctuates from six metres during summer to 1m during rains. But the amount of water that can be assuredly withdrawn is not high. There are five mini deep tubewells at present at depths of 45m. There is a good aquifer at 6m depth that is being tapped by these mini deep tubewells. Their water is mostly used for drinking purposes. The chief source of irrigation are ponds, numbering thirty four. Water level is 1-12m in winter. Pumps are used for lifting irrigation water.

Rainfed aman rice constitutes the chief crop extending over 100 per cent of net sown area followed by wheat (2%), vegetables (2-5% NSA) and negligible amount of mustard and potato. Hence, cropping intensity is very low. Yield of aman is not considerable, only 2,000 kg./per ha. The present status of agriculture apprears to be insignificant. The presence of a ready market at Suri has been responsible of the cultivation.

Social Forestery

Chhora is a sample village amongst those that have been brought under the social forestry scheme. It lies adjacent to a protected forest area. According to census reports about 40 ha of land is under forest. It could be assumed that degradation of these forest lands is increasing.

This has contributed to an increase in the area under cultivable wastelands. Since 1981-82, about 15 ha of the village land have been converted into forest. Thus, in this village, afforestation of village woodlots has been undertaken by the forest department. Species selected have been Eucalyptus Hybrid and Sonajhuri. Hence, choice has been laid upon fuelwood species rather than timber or fruit crops. The quick growing and non-browsing nature of the planted species make them commendable species for afforestation in the waste lands. The deep penetrating roots are capable of extracting groundwater. Plantations under intensive management can be harvested as early as four-five years. Experiment was previously made with cashewnut trees, but this failed as the nuts were stolen by outsiders.

Benefits

Social forestry scheme at Chhora has not been utilised for commercial purposes. The chief benefit has been the provision of fuelwood. The demand of the villagers has been largely met. If this aspect is considered, then an indirect socio-economic benefit can be perceived. Expenditure on fuelwood has been reduced. Leaves shed by the trees form the major source of fuelwood. Illicit cutting of twigs is also carried on by the villagers as well as others.

Perception of Beneficiaries

The villagers do acknowledge the benefits of the forestry scheme as it has mitigated the problem of firewood supply. The basic drawback encountered is the lack of effective protection measures. The quantity of twigs available is reduced. Failure of cashewnut plantations was totally due to this. Preference of the villagers lies in the planting of both fuel and timber species such as *sal,* teak, etc. Even mahua and cane are preferred as alternatives.

Salboni

This village lies adjacent to Chhora and as such terrain and soil conditions are almost identical. However, the soils are sandy loams, with less content of *kankar.* Upto 1980, none possessed any land.

Since then, vested land of 1 bigha, (0.14 ha), each has been distributed among the landless labourers. The net cultivated land amounts to 60 per cent of the total area of 106 ha cultivable wasteland amounts to 7 per cent at present. This shows a decrease from 1971 when 11.5 per cent of land was under this category. Area not available for cultivation registers an increase due to building up of settlements. Forests cover 25 per cent of total area, similar to that of 1971.

The village is dependent on rainfed agriculture, due to lack of irrigation facilities. There are dugwells but they are used for domestic uses. The depth of water is 7m during summer and 0.6m during the rains. Monoculture is usually practised. Aman rice covers 100 per cent of net sown area and 80 per cent of gross cropped area. *Aus* rice has a minor coverage (2-5% NSA).

Social Forestery

Since 1970 about 5 ha of land have been brouhgt under the category of village woodlot by the forest department. In general, since 1971, forest area records a stable figure of 26 ha. Hence, it could be assumed that extension of forest has been along the cultivable wasteland. Species afforested are *sonajhuri, asan (terminalia alata)* and *sissoo (dalbergia sissoo).* Initially, experiment was carried out with *sal* trees by seedling method. But this failed as the seeds did not germinate in this dry terrain. Sonajhuri constitutes the main planted specie.

Benefits.

Impact of the forestry scheme is similar to that of Chhora. The shedded leaves are the major source of firewood. The twigs and branches are surreptitiously cut during the night. Nevertheless, the problem of fuelwood supply has been met to a great extent. A fact to be noted here is that farm forestry, which forms an important part of the social forestry scheme in Birbhum, has been a major failure in *salboni.* About two years back, some trees were planted along the *al* or earthen dykea separating one field from another. But this was discarded subsequently.

Perception of Beneficiaries

Althought the villagers agree that fuelwood problem has been solved their preference is towards *sal.* The failure of the seeding method has discouraged them. Moreover, there is considerable time lag between the planting of *sal* and reaping of benefits. Yet, if *sal* can be generated in a successful manner, the villagers are willing to wait prior to getting the returns. Undoubtedly their empahsis is on commercial platation of timber species. Next preference is for fruit trees, if necessary protection measures can be provided. Regarding afforestation of existing fallow lands, there is a confusion. Vegetable cultivation will take only 3-4 months, i.e. from May/June to September/October. If those *kharif* vegetables requiring the. low consumptive use of water are grown, then benefits will accure in a short time span. Tree planting usually requires 10-12 years prior to giving economic returns. The example of the palm tree may be cited here. It takes about 20 years before the tree is sold. At present the amount is only Rs. 600-800. The villagers have a varied response to farm forestry and village woodlot plantation. Planting of trees on individual fallowland constitutes one type of farm forestry. Where such individual fields are concerned, the villagers are reluctant to introduce social forestry. They prefer dryland farming. The time lag is a major constraint. In the case of village woodlots, their preference is towards, social forestry with longer growing species.

Jahanabad

This village is located within the Rajnagar Forest Range, near the Bengal/Bihar border. Terrain is undulating with detached rock exposures. There is a comparatively shallow soil veneer of sandy loams and loams. The presence of *kandos* or small seasonal streams is a characteristic feature. The present state of landuse can be described here. Net sown area is about 40 per cent of total area of 166 ha. Cultivable wastelands, according to census reports is around 49 per cent, i.e. 80 ha. However, villagers state that at present this is 135 ha. Forests cover 4.43 per cent of total area. An almost stable trend of land use is observed since 1971.

Jahanabad presents a better picture of farming. This may be attributed to the presence of the *kandos*. Precisely, the river lift irrigation is practised on a minor scale. *Dongas* or canoes are used to irrigate the fields adjoining the *kando* sides. Dugwell is present with water level fluctuating from 8-10m during winter to 2-3m during rains. There is one tubewell with a depth of 40m. But, both these sources are used for domestic purposes. River lift irrigation has slightly diversified the cropping pattern. *Aus* and *Aman* rice occupy 50 per cent of the net sown area, each. This is followed by mustard (12% NSA), wheat (4% NSA) and potato (4% NSA). Cropping intensity is 120 per cent. The *kando* sides, which form the lowest tracts, accord favourable sites for mustard and wheat cultivation due to available soil mosture. Further, wheat is a staple diet of these people, who are partly Biharis. Moreover, the village is located at considerable distance away from any town. Lack of alternative part time occupation leads to greater stress on cultivation.

Social Forestry

Jahanabad is a typical example of a social forestry village. Afforestation was undertaken over vested land about seven years back, (1982-83), over an area of 2 ha. Existing forests cover nearly 8 ha. Thus, at present forests account for 10 ha of land. Species planted are *eucalyptus hybrid* and *sonajhuri.*

Benefits

Commercial forestry is practised. The forest was taken over by a paper and pulpwood factory of Titagarh near Calcutta. In return, the beneficiaries would receive 25 per cent of the benefit. They were yet to receive the money at the time of survey. On an average they would receive Rs. 300/bigha (or Rs. 2222/ha).

Preception of Beneficiaries

Although monetary benefits are to be gained from handing over forest land to factory owners, this income is not a recurring one. The ultimate consequence is the total removal of the afforested land. In fact, at present, this land is included under the category of cultivable wastland. Preference of the villagers is toward the planting of *sal,* teak and grass. Inclination is more towards timber species of traditional forests. In case of grass, according to their estimate, the monetary returns would roughly amount to Rs. 3,000-4,000 in a seven year period. This may be compared to the commercial forests with the value of Rs. 300 per bigha during the same tenure. Moreover, the people can directly reap the benefits without having to depend upon outsiders to give them their due. Although bamboo cultivation is highly profitable, it is confined to the sides of ponds, where there is available soil moisture. Initially afforested area was over stony terrain. Hence, the villagers are not averse to the idea of social forestry. Their response is satisfactory but their preference of species differs from that of the forest department.

Lalboni

This village is located a few kilometers from Jahanabad, towards Rajnagar, the P.S. headquarters. The terrain is similarly undulating as in Jahanabad. The soils could be stated to be red sandy loams of a slightly coarser texture. There is little to be said about the agricultural development of this area. The village remains practically unirrigated as there is no reliable source of irrigation. Monoculture of aman rice is carried out. The level of dependence on agriculture is virtually nil.

Social Forestry

Lalboni well brings into perspective the impact of social forestry scheme as it is mostly dependent upon forestry. Afforestation was undertaken around 1982-83 on a common woodlot covering more than 20 ha. The species involved are again e*ucalyptus hybrid* and *akashmoni*. Forest area amounts to 16-20 ha.

Benefits

Benefits accruing from social forestry are twofold. In the first case, about 400 trees were planted in 1 bigha (0.14 ha), with free seedlings distributed by the forest department. Ultimately 200 trees remained. This was primarily utilised for fuelwood supply. Commercial forestry is practised as the product is sold at Rajnagar. Monetary returns from selling fuelwood amounts to Rs. 1,000 per bigha. The selling price amounts to Rs. 60 per quintal, whereas, the total produce amounts to 5 quintals per bigha. In case of *akashmoni,* the wood is sold for furniture purpose.

In the second case, about 16 ha of village woodlot have been handed over to the pulpwood factory of Titagarh. The responsibility of planting and rearing the *eucalyptus* trees lies with the factory. *Eucalyptus* has been selected as it gives two crops in one cutting. About 25 per cent of the value is given to the villagers. This roughly amounts to Rs. 4,000-5,000 per ha.

Perception of Beneficiaries

Preference of the villagers is again towards timber species like *sal*, teak, *piasal*, etc. In case of *piasal*, the benefit is Rs. 3,000 per tree is planted for twenty years. *Mahua* is not preferred as it takes a longer time. Although *Sisoo* is another choice of the villagers, its planting is hindered by the prevailing soil type. They are also interested in the *arjun* plantations for rearing of silkworms. This could be carried out profitably, even if 10 ha of land was made available. The afforested lands have already been cut, thus, increasing the area under cultivable wasteland by 4 ha. Nevertheless, the villagers are still interested in planting *eucalyptus* and *akashmoni* on common village woodlots if they receive free seedlings. Overall response to social forestry has been highly satisfactory in Lalboni. In fact, they are willing to accept forestry as their major source of livelihood even in the near future.

Chandrapura-Tantipara

The last two villages of Karabad and Kharna are located in the Chandrapura Forest Range, in Rajnagar P.S. They depict a diverse picture. To follow the impact of social forestry scheme, an insight into the detail of *tasar* silk culture is required. Chandrapura-Tantipara area situated between Suri and Rajnagar has been a centre for *tasar* silk production since previous times. At that time, raw silk was bought from Bihar and processed here. Hence the people of this area have a traditional skill in this field. The people thus prefer the plantations of arjun (*terminalia arjuna*) and asan *(terminalia alata)* as they provide fodder for the non-mulberry cocoons used in production of *tasar* silk. Precisely planting of arjun trees for this purpose began since 1981. The endeavour of a veteran freedom fighter of Chandrapura, Mihir Lal Chatterjee, needs to be mentioned. It was under his initiative and supervision that *arjun* trees were planted over 10 ha of forest land belonging to the Chandrapura Beat Office. The nearby Patelnagar Research Centre has been set up specifically for seed culture of silkworms. Since 1981, this aspect of commercial forestry is in progress, creating job opportunities for the adivasis or tribal population of this area. In 1984-85, about 30,000 cocoons had been reared leading to the production of 11.23 kg of silk thread. About 72 m of *tasar* cloth was derived (Forest Department 1988).

Karabad

The physiography of Karabad is characterised by gently undulating terrain and sandy loams. Upto 1981, there was a virtual absence of forest cover. Cultivable wastelend has a significant coverage of nearly 20 per cent of total area. Net sown area registered a stable value of 75 per cent, whereas, area not available for cultivation shows a slight increase since 1971, amounting to 10 per cent. The village remains practically non-irrigated. Dugwells are present with water level within 9m. There are few mini deep tubewell tapping aquifers at depths of 45m. Both types are used for domestic purposes. Aman monoculture is practised, signifying the stress on forestry for livelihood.

Social Forestry

The forest department has introduced Arjun plantations on 12 ha of vested land in 1981. Stress is on the rearing of these trees for *tasar* production. The Panchayat selected the beneficiaries who purchase eggs at the cost of Rs. 20 for the eggs of 100 Disease Free Layee (DFL) butterflies. These eggs are then kept at the Patelnagar Research Centre for 2-3 days until they germinate. The larvae are then transferred to the leaves. Usually 5-6 larvae are set to feed on one arjun tree. They generally consume all the leaves within 5 days and are subsequently transferred to another tree. The beneficiaries usually have to work for two-three months.

There are three rearings of silkworm in one year:

(a) May-June;
(b) August-September;
(c) November-January.

The commercial crop rearing of larvae for cocoon culture mainly applies to the third period. The rainy season rearing is a seed crop, i.e. the cocoon is allowed to reach the butterfly stage for egg production. The August-September rearing is partly a seed crop.

Benefits

The main benefit is derived from selling of the cocoons. This amounts to Rs. 15/16 per day during the commercial rearing period. An associated benefit is the selling of fuelwood from the pruned trees. Rearing is done after two years and then the trees are pruned. The discarded twigs are thus another source of livelihood.

Perception of Beneficiaries

The response of the villagers of Karabad to the social forestry scheme has been highly satisfactory. They are willing to continue commercial tree rearing for *tasar* culture. The programme introduced by the forest department has been a success. However, preference of the villagers lie in the planting and rearing of *asan* trees for *tasar* culture. *Asan* leaves are considerably larger than those of *arjun*. Hence, time of leaf consumption on one tree by the larvae is considerably slower. This would, thus, be a more economic venture. At present, the Arjun seeds are being bought from Bankura. Overall benefit has been satisfactory this is evident from the fact that the beneficiaries are not demanding any specific timber, fuelwood or fruit species.

Kharna

Kharna is located some 2-3 km away from Karabad. Terrain and soil conditions are similar to the latter. Land use, as recorded by the census, registers a stable pattern. There is considerable area under cultivable wastelands (29%) and a negligible area under forests (10%). The pattern of agriculture is somewhat identical to Chhora. *aman* occupies 75 per cent NSA and *aus* occupies 25 per cent NSA. A negligible amount of mustard and potato are cultivated around the tanks. The village resorts to 100 per cent tank irrigation. Prior to 1983, dryland farming was practised. Some amount of inferior pulses like *kulth* and *kalai* were grown.

Social Forestry

Since 1983, about 3 ha have been brought under social forestry by the forest department. Prior to this, these lands were under dryland farming. Species introducted are *eucalyptus* and *akashmoni.*

Benefits

Like Chhora, availability of fuelwood for local consumption has been the only benefit. Often raw twigs are also cut and then dried prior to selling. Such random destruction often results in artificial scarcity. The amount of fuelwood extracted is 14-15 kg per bigha (100-110 kg/ha). To supplement this source, grass is bought from adjacent forest lands.

Perception of beneficiaries

Response to social forestry is not remarkable as the villagers do not derive monetary benefits. This fact becomes more glaring when one compares the situation with Karabad. With similar soil and land conditions to that of Karabad, there is apparently no reason why Kharna cannot resort to *arjun* plantation for *tasar* culture. This would ensure greater cooperation on the side of the beneficiaries, viz. the tribal population. According to the research worker, the existing wastelands could be undertaken for this purpose on an experimental basis.

Conclusion

Giving due considerations to all the above stated facts, the following conclusions may be drawn. Firstly, social forestry appears to be the most important source of livelihood in the study area, if it is properly implemented. This could be supplemented by agriculture. In general, the villagers are highly responsive to the implementation of social forestry schemes, viz., those along village woodlots and degraded forests. They do realise the necessity of this venture for economic upliftment as well as environmental protection of wastelands. Till present, the forest department has mainly concentrated upon the quick-maturing varieties like *eucalyptus* and *sonajhuri* for the supply of fuelwood and pulpwood. Barring the prevailing controversy of groundwater and soil fertility depletion by eucalyptus plantations, other facts may be cited. The human response to these trees is slightly dissatisfactory. No doubt, problem of fuelwood is reduced and some monetary benefits are derived in case of pulpwood. Nevertheless, their preference is towards traditional species of sal, teak and fruit trees. If planted on village woodlots and degraded forests, they are willing to wait for returns as these trees are slow-maturing. Villagers around the Chandrapura-Tantipara region prefer *arjun* and still more Asan trees, for *tasar* silkworm culture. The introduction of *arjun* plantations in this region, by the forest department has been quite successful. The distribution of free seedlings accords an additional incentive in all the cases.

Acknowledgements

The research worker is grateful to her supervisor Professor Bireswar Banerjee, Department of Geography, Calcutta University. She is also grateful to the CSIR for financing the research. Finally, she acknowledges with thanks, the help rendered by DFO and other personnel of Divisional Forest Office,,Suri as well as the Forest Range Officer, Rajnager for in providing her with the transport and guidance.

References

Forest Department, Birbhum (1988?) *Shristhi Theke Aj*, 1954-87, Government of West Bengal, Birbhum, Suri.

Forest Department, Birbhum (1990?) *Bonsrijan*, Government of West Bengal, Suri.

Sarkar, N. (1964): Birbhum forests in *West Bengal Forests* : Commemoration Volume, Forest Directorate, Govt. of West Bengal.

Social Forestry Wing (1987-88): Social Forestry Project West Bengal: Highlights on Progress-1981-87. Forest Directorate, West Bengal.

12

Irrigational Water Use and Requirements:

A Case Study

V.K. RAI

As all life depends on water, it is an integral part of man's life and environment. Man's habits and activities, in all walks of his existence and forms of social organisations, are influenced more by his close association with water than with the land by which he earns his bread. Water has varied uses in human life. Our water needs are great. The total quantum of water available for use, however, bountiful at present, is more or less a fixed quantity. With an increase in population, more and more water is needed. Modern civilisation has placed heavy demand on water on account of urbanisation, modernisation of agriculture and industrial development. Unlike the other resources, water is rarely permanently depleted by use but gets replenished with time in the course of its movement through the hydrological cycle. There are various ways in which water is utilised but its largest use on global level is in agriculture, in the form of irrigation for the cultivation of different crops.

In India in 1974, irrigation used about 92 per cent of all the water consumed in the country. Domestic and industrial uses accounted for the remaining 8 per cent. Water scarcity in rural areas or the lack of access to it, forces people to consume less. Even, if all domestic and industrial needs are adequately met by 2025 A.D., irrigation withdrawls will be almost 73 per cent of total water communication. (Centre for Science and Environment, 1984-85). The present paper deals with the irrigational water use pattern in Ganga-Ghaghara *doab* east in spatio-temporal framework and further development of irrigation in future.

The Study Area

The Ganga-Ghaghara *doab* east extends approximately between 25°23′ to 26°16′15″N latitudes and 83°4′ to 83°39′E longitudes. It comprises the districts of Mau, Ballia and Ghazipur (excluding Zamania *tahsil)* of Uttar Pradesh. It embraces an area of about 7,409 km^2. Its maximum length, east to west approximately is 155 km whereas, its maximum width, north to south is 92.5 km. The *doab* area varies from year to year due to frequent shifts in the courses of Ghaghara and Ganga which form its northern and southern borders respectively.

The region is bounded by Gorakhpur and Deoria (U.P.) and Siwan and Saran districts (both in Bihar) in north east and Zamania *tahsil* and Varanasi (U.P.) and Buxar district of Bihar in the south. In the west, it joins the borders of Jaunpur and Azamgarh districts of Uttar Pradesh (Fig.12.1) For administrative purposes, the area under study is divided into ten *tahsils*, viz. Ballia, Bansdih, Rasra, Bairia (Ballia district), Ghosi, Mohammadabad, Mau (Mau district) and Saidpur, Ghazipur and Mohammadabad (Ghazipur district). These *tahsils* are sub-divided into 39 development blocks (Fig.12.1).

Climate of the area is sub-tropical. Summer is hot and dusty winds are frequent and winter is cool, dry and day and night temperatures vary considerably. The average annual precipitation for the area is 87.42 cm varied from 97.86 cm (Bansdih) to 65.50 cm (Ballia). The precipitation in the rainy season accounts for about 91 per cent of the mean annual rainfall.

Use and Requirements of Water in Agriculture

Irrigation is the artificial application of water to soil for crop production in dry as well as humid climates. Irrigation water is supplied to supplement water available from rainfall and contribution to soil moisture from ground water. Irrigation is either a supplement or complement to seasonal rainfall except in arid regions. Irrigation may be the only source of water available to plants for most of the period during the year in many arid regions of the world. The dependence of India's agriculture on the monsoon and its consequent vulnerability have been recognised from the earliest time. Irrigation has its special significance due to irregularity and uncertainty of monsoon rainfall

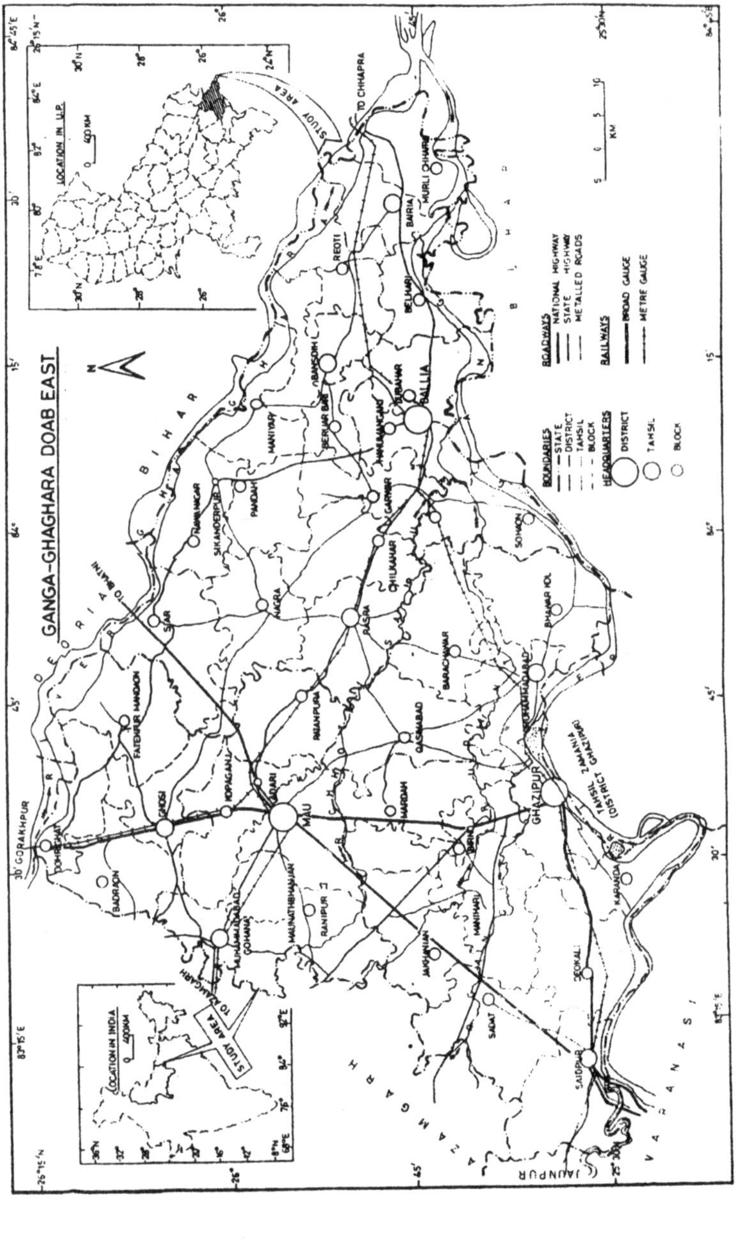

Fig. 12.1: Ganga-Ghaghara Doab East

which is almost wholly concentrated in the four summer months of June to September. Irrigation has not only played an important role in the development of Indian agriculture and in attainment of food self sufficiency, but has also helped the country to ensure that periodic fluctuations in rainfall do not disrupt output. Due to rapidly growing population of the country, it is necessary to produce more food. But land is a limited resource, because progressively more land is required for habitats, industries and other purposes. Therefore, the only solution appears to be multiple cropping, and it is possible only when there is assured irrigation. Water demand for irrigational use in the country has increased with an increase in agricultural intensity and changes in farming practices, like introduction of high yield varieties of seeds and increasing use of chemical fertilisers to meet the ever growing demands for food and fodder. The *kharif* crops are generally rainfed except rice whereas, the *rabi* and *zaid* crops, including vegetables, are invariably raised with the help of irrigation except in the *khadar* tracts of the Ghaghara river.

Area under Different Means of Irrigation

The irrigation in study area is done by different canals, state tubewells, private tubewells, pumping sets, persian wheels and masonary wells. Unfortunately separate data for the areas under different means of irrigation are not available. It is available under the categories like canal, tubewells, masonary wells, ponds and others.

Canals

Area under canal irrigation was 15.72 per cent of net area irrigated in 1980-81 and has registered an increase (3.6%) that is 19.32 per cent in 1990-91. Blockwise area under canal irrigation varies from 0.06 per cent (Dubhar) of the net area irrigated to 52.2 per cent (Pandah) in 1980-81 and from 0.1 per cent (Barachaur block) to 63.1 per cent (Kopaganj) in 1990-91 (Table 12.1). Decrease in canal irrigation in some blocks like Bansdih, Pandah and Ghosi is due to increase in well irrigation, mainly from private tubewells and pumping sets.

Tubewells

In this group area irrigated by state tubewells, private tubewells and pumping sets are included. Tubewell irrigation accounted for 79.91 per cent of net area irrigated in 1990-91 which was 79.44 per cent in 1980-81. Area under tubewells varied from 38.3 per cent (Siar) of NAI to 99.9 per cent (Pandah) in 1990-91 which was from 38.8 per cent (Ghosi) to 99.4 per cent (Muhammadabad) block in 1980-81 (Table 12.1). High percentage in Pandah and Muhammadabad was due to lack of canal irrigation in these blocks.

Masonary Wells

It is a traditional means of irrigation. It has registered a marked decrease, that is, from 2.96 per cent to 0.37 per cent during the period under consideration (Table 12.1). On spatial level it is nil in many blocks in 1990-91.

Ponds

They accounted for only 0.33 per cent of NAI in 1990-91 as against 1.11 per cent in 1980-81 registering a decrease of 66.48 per cent. It was highest in Kopaganj block (7.7%) and lowest in Muhammadabad, Karanda, Fatehpur Mandaon, Chilkahar and Garwar block (0.1%) in 1990-91 (Table 12.1).

Others

They accounted for only 0.07 per cent in 1990-91 which was 0.77 per cent of NAI in 1980-81.

Cropwise Irrigated Area

The Gross Area Irrigated (GAI) has increased from 44.12 per cent of Gross Area Sown (GAS) in 1980-81 to 57.11 per cent in 1990-91, showing an absolute increase of 36.23 per cent. This increase is mainly on account of a rapid increase in tube-well irrigation. The *rabi* crops account for about 74.81 per cent of GAI in 1990-91 while it was 82.54 per cent of GAI in 1980-81. This is due to the increase in kharif irrigation during the study period. The irrigation in *kharif* crops has increased from 17.46 per cent of GAI in 1980-81 to 25.19 per cent of

Table 12.1: Net area irrigated under different means of irrigation in Ganga-Ghaghara *Doab* East

(Percentage)

Block	Net Area	1980-81					Net Area	1990-91				
	Irrigated (ha)	Canal	Tubewells	MW	Ponds	Others	Irrigated (ha)	Canal	Tubewells	MW	Ponds	Others
Garwar	7,983	25.9	64.9	6.8	2.1	8.3	9,220	32.9	66.8	0.1	0.1	0.1
Hanumanganj	4,763	16.8	79.0	3.5	0.5	0.2	5,971	18.3	78.2	3.4	–	0.1
Dubhar	2,894	0.06	97.06	0.96	1.72	0.21	3,581	–	96.8	1.4	1.8	-
Belhari	1,018	–	94.3	5.6	–	0.1	1,984	–	98.1	0.8	–	1.1
Baria	2,871	–	94.3	5.6	–	0.1	3,334	–	97.8	–	–	2.2
Muralichhapra	1,823	–	99.7	0.2	–	0.1	3,410	–	100.00	–	–	–
Sohaon	4,636	7.8	91.3	0.8	–	0.1	5,383	9.3	90.7	–	–	–
Reoti	5,213	2.6	96.1	1.2	–	0.1	6,945	1.4	98.6	–	–	–
Banadih	6,693	1.5	73.1	4.8	0.4	0.2	0,417	12.6	87.4	–	–	–
Maniar	6,368	9.8	83.9	4.1	1.1	1.1	9,012	25.0	74.9	–	–	0.1
Pandah	8,291	52.2	43.9	2.5	2.7	0.2	10,023	42.6	53.5	3.8	0.1	0.009
Nawanagar	7,896	24.1	72.4	0.5	1.2	1.8	8,866	28.0	71.9	–	0.1	–
Bariabari	5,155	21.8	73.8	2.8	0.4	1.2	5,301	26.4	73.4	–	0.1	0.1
Rasra	10,387	–	87.6	9.2	1.8	1.4	12,595	–	99.9	–	–	0.1
Chilkahar	10,360	27.5	65.6	3.2	2.3	1.4	11,752	38.3	61.6	–	–	0.1
Nagara	17,708	38.8	58.5	1.9	0.7	0.1	19,392	41.3	58.6	–	0.1	–
Siar	12,491	41.5	55.3	2.9	–	0.3	15,840	61.7	38.3	–	–	–

(Contd.)

Table 12.1: *(Contd.)*

Block	1980-81						1990-91					
	Net Area Irrigated (ha)	Canal	Tubewells	MW	Ponds	Others	Net Area Irrigated (ha)	Canal	Tubewells	MW	Ponds	Others
Doharighat	8,374	42.2	43.4	9.2	2.4	2.8	8,596	41.1	58.3	0.1	0.5	–
Fatehpur	9,026	26.2	52.5	14.0	4.5	2.8	14,645	11.4	86.6	1.3	0.6	0.1
Handaon												
Ghosi	8,517	51.1	38.4	5.4	4.6	6.5	9,035	41.7	57.3	0.4	0.2	0.2
Badrson	9,315	10.4	76.3	6.9	4.8	1.6	9,557	8.7	88.7	1.4	0.6	0.6
Muhammdabad	10,219	–	96.8	0.8	0.2	0.2	10,433	–	99.7	–	0.3	–
Ranipur	13,068	2.9	96.3	0.3	0.1	0.4	15,007	1.1	98.9	–	–	–
Kopaganj	10,392	16.8	75.0	2.3	4.4	1.5	10,623	63.1	29.2	–	7.7	–
Pandaha	9,779	–	97.5	0.1	0.1	2.3	10,926	–	99.9	–	0.1	–
Ratanpur	11,259	18.5	79.7	0.4	0.9	0.5	11,501	9.7	90.3	–	–	–
Ghazipur	8,410	4.8	91.6	2.3	1.0	0.3	8,918	4.7	93.3	1.5	0.5	–
Karanda	5,186	8.7	89.4	1.6	0.1	0.2	6,500	9.0	90.7	0.2	–	0.1
Birno	8,710	–	98.7	0.2	0.9	0.2	9,744	16.9	82.6	–	0.3	0.2
Mardah	10,308	–	96.4	2.3	0.3	1.0	12,212	29.1	70.7	0.2	–	–
Saidpur	12,069	8.3	87.2	3.9	0.6	-	11,840	8.7	91.3	–	–	–
Deokali	10.947	21.1	75.4	3.3	0.1	0.1	10.973	8.9	90.3	0.8	–	–
Sadat	12,269	17.5	78.3	2.8	0.5	0.9	13,032	17.6	82.4	–	–	–

(Contd.)

Table 12.1: *(Contd.)*

Block	Net Area 1980-81						Net Area 1990-91					
	Irrigated (ha)	Canal	Tubewells	MW	Ponds	Others	Irrigated (ha)	Canal	Tubewells	MW	Ponds	Others
Jakhaniya	12,046	1.8	96.0	1.2	0.9	0.1	11,640	22.2	77.8	–	–	–
Manihari	11,999	8.5	89.8	1.4	0.2	0.1	11,790	19.7	1.2	–	–	–
Muhammdabad	9,008	–	99.4	–	0.3	0.3	10,194	–	99.7	0.2	–	0.1
Bhavarkole	5,363	13.8	85.8	–	0.3	0.1	6,016	15.8	84.2	–	–	–
Kasimabad	12,113	9.4	87.9	1.6	0.4	0.7	14,916	5.8	94.2	–	–	–
Barachaura	9,430	–	97.4	1.4	0.8	0.4	11,103	0.1	99.8	–	0.1	–
District average	3,34,357	15.72	79.44	2.96	1.11	0.77	3,78,327	19.32	79.91	0.37	0.33	0.07

Source: Computation, based on the data obtained from the *District Statistical Handbooks*, 1981 & 1992 of Ballia, Mau, Ghazipur and Azamgarh.

GAI in 1990-91 (Table 12.2). Amongst all the crops, wheat occupies the first place with 63.10 per cent of GAI in 1991 whereas in 1980-81 it was only 52.98 percent. Next is paddy with corresponding figures of 15.75 per cent and 2.91 per cent followed by sugarcane with 7.91 per cent and 9.98 per cent.

As regards the spatial distribution of irrigated area under different crops (in 1980-81 and 1990-91) wheat accounted for largest per centage of GAI and it varied from 46.79 per cent (Nawanagar block) to 90.96 per cent (Muralichhapra) and from 56.34 per cent (Doharighat) to 73.64 per cent (Deokali) in 1980-81.

Irrigation Intensity[1]

The irrigation intensity (that is the NAI as percentage of NAS) has shown a marked increase with time, for instance, from 61.76 per cent in 1980-81 to 68.32 per cent in 1990-91 for the whole region (Table 12.3, Fig. 12.2). Further, the intensity was highest in Nagara block (81.32%) and lowest in Belhari block (13.16%) in 1980-81. In 1990-91 it varies from 20.9 per cent (Belhari) to 90.0 per cent (Nagara). These variabilities and increase in irrigation intensity is due to the increase in number of minor irrigation means. For example, in Nagara block both canal and tubewell irrigation are available.

Utilisation of Water in Irrigation

Utilisation of water in irrigation has been calculated with the help of unpublished canal water data collected from the irrigation departments and groundwater utilisation with help of Appendix I and unit drafts of groundwater irrigation means. Unit draft for state tubewells was 16 ha m in 1980-81 and it varied from 4.00 ha m to 20.40 ha m in 1990-91 (Rai, V.K. 1996) (Table 12.4). The unit draft of PTW is 2.0 ha m in 1990-91, which was 2.20 ha m in 1980-81. The unit draft of PS is 1.4 ha m for both periods and for persian wheel the value is 0.9 ha m. The unit draft for masonry well was 0.55 ha m in 1980-81 and 0.5 ha m in 1990-91.

---

1. I = NAI/NAS x 100 where I = Irrigation Intensity
   NAI = Net Area Irrigated, NAS = Net Area Sown

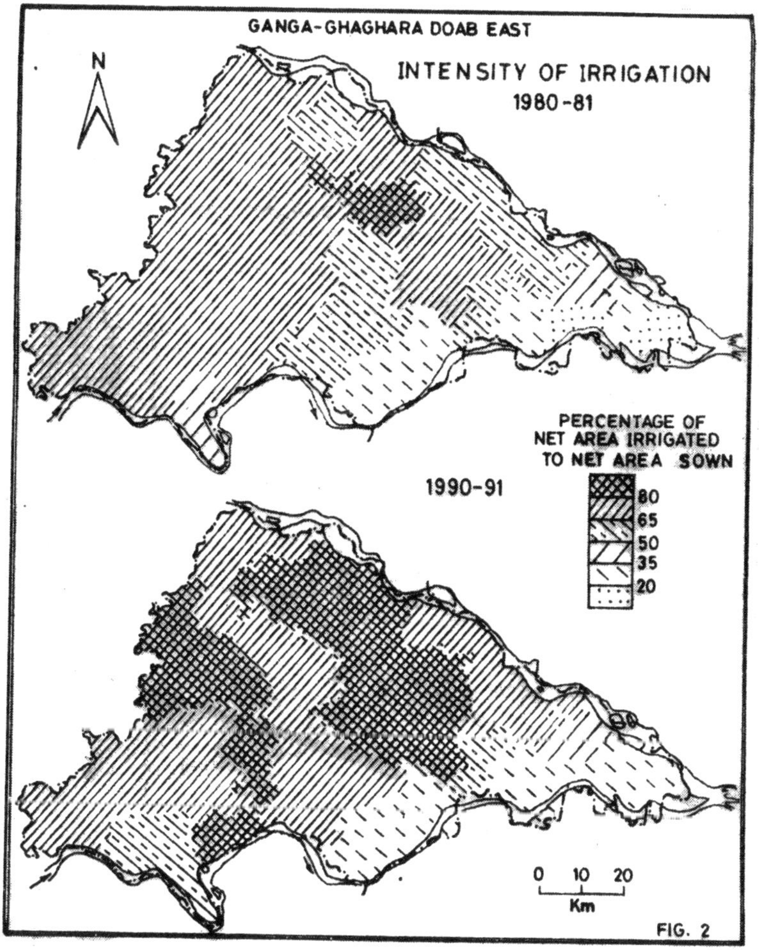

FIG. 2

Fig 12.2: Ganga-Ghaghra Doab East : Intensity of Irrigation 1980-81

Table 12.2 : Changes in cropwise irrigated area in Ganga-Ghaghara *Doab* East: 1980-81 to 1990-91 (Upper figure: 1980-81; Lower figure: 1990-91)

Crop	Area irrigated		Per cent of	Per cent
	Actual (ha)	Per cent of GAI	GAS	change
Kharif				
Paddy	10,517	2.91	1.28	+638.24
	77,641	15.75	9.00	
Maize	53	0.02	0.006	+792.45
	473	0.10	0.06	
Moong	5,660	1.56	0.69	-67.72
	1,827	0.37	0.21	
Sugarcane	36,118	9.98	4.40	47.96
	38,994	7.91	4..92	
Others	10,811	2.99	1.32	-51.71
	5,221	1.06	0.60	
Total	63,159	17.46	7.70	+96.58
	1,24,156	25.19	14.40	
Rabi				
Wheat	1,91,646	52.98	23.37	+62.27
	3,10981	63,10	36.04	
Barley	23,749	6.56	2.89	-52.76
	11,220	2.28	1.30	
Gram	26,014	7.19	3.17	-36.92
	16,409	3.33	1.90	
Pea	5,123	1.42	0.63	-6.73
	4,778	0.97	0.55	
Oilseeds	364	0.10	0.04	+284.69
	1,401	0.28	0.16	
Potato	10,466	2.89	1.28	+30.25
	13,632	2.77	1.58	
Others	41,245	11.40	5.03	-75.17
	10,243	2.08	1.19	
Total	2,98,607	82.54	36.42	+23.46
	3,68,664	74.81	42.72	
GAI	3,61,766	100.00	44.12	+36.23
	4,92,820	100.00	57.11	—
GAS	8,20,018	—	100.00	—
	8,62,846	—	100.00	—

Source: Computed from the data taken from *District Statistical Handbooks* for 1982 and 1992 of Ballia, Ghazipur, Mau and Azamgarh Districts.

Table 12.3: Irrigation intensity in Ganga-Ghaghara *Doab* East (1980-81 to 1990-91)

Block	Irrigation Intensity (per cent)	
	1980-81	1990-91
Garwar	73.40	82.30
Hanumanganj	54.41	52.60
Dubhar	27.27	34.10
Belhari	13.16	20.90
Bairia	23.93	31.20
Muralichhapra	18.66	30.40
Sohaon	30.13	22.80
Reoti	36.96	48.40
Bansdih	62.26	66.60
Maniar	53.50	67.20
Pandah	67.16	85.40
Nawanagar	61.68	70.80
Beruarbari	71.70	76.80
Rasra	62.12	82.40
Chilkahar	73.13	82.80
Nagara	81.32	90.00
Siar	71.91	85.10
Doharighat	6.79	67.30
Fatehpur Mandaon	51.47	83.90
Ghosi	72.79	81.00
Badraon	70.04	74.30
Muhammadabad	79.41	80.70
Ranipur	71.67	83.50
Kopaganj	72.45	72.60
Paradaha	68.69	81.00
Ratanpura	77.20	75.10
Ghazipur	73.10	81.90
Karanda	44.46	56.80
Birno	71.97	78.40
Mardah	70.83	81.80
Saidpur	71.01	68.80
Deokali	65.31	62.50
Sadat	73.82	72.00
Jakhaniya	76.93	71.60
Manihari	70.74	66.00
Muhammdabad	60.78	71.40
Bhavarkole	31.49	29.50
Kasimabad	68.81	7.80
Barachaur	59.12	69.50
Total	61.76	68.32

Tabel 12.4: Unit draft of State tubewells inGanga-Ghaghara *Doab* East (1990-91)

Block	Unit draft (ha m)
Garwar	20.00
Hanumanganj	20.40
Dubhar	15.56
Belhari	19.98
Bairia	8.90
Muralichhapra	5.79
Sohaon	20.35
Reoti	19.00
Bansdih	19.10
Maniar	14.58
Pandah	15.90
Nawanagar	13.26
Beruarbari	15.40
Rasra	9.37
Chilkahar	11.47
Nagara	15.12
Siar	15.00
Doharighat	7.41
Fatehpur Mandaon	5.97
Ghosi	10.79
Badraon	4.00
Muhammadabad	7.42
Ranipur	25.80
Kopaganj	9.60
Pardaha	17.55
Ratanpura	10.18
Ghazipur	16.64
Karanda	19.52
Birno	14.453
Mardah	17.61
Saidpur	17.19
Deokali	19.69
Sadat	14.92
Jakhaniya	15.43
Manihari	18.94
Muhammadabad	19.96
Bhawarkole	13.05
Kasimabad	16.29
Burachaur	13.96

Source: Rai, V.K., 1996, p.98.

There has been a well marked increase with time in the utilisation of water in irrigation, that is, from 141879.36 ha m in 1980-81 to 203294.20 ha m in 1990-91, registering an increase 43.29 per cent (Table 12.5). Out of the total water utilised for irrigation in 1980-81, the shares of different means of irrigation were: Canal=28 per cent STW = 12.96 per cent, PTW= 48.31 per cent, PS =2.59 per cent, PW = 0.62 per cent and MW= 7.52 per cent, whereas, the corresponding figures in 1990-91 were 21.45 per cent, 12.03 per cent, 39.46 per cent, 26.23 per cent, 0.06 per cent, 10.76 per cent (Table 12.5 & Fig.12.3). Thus we can say that share of ground water is 78.655 per cent of total utilised water in 1990-91 which was 72.0 per cent in 1980-81.

Table 12.5: Water Utilisation in irrigation in Ganga-Ghaghara *Doab East* (ha m) (Upper figure-1980-81; Lower figure-1990-91)

Source	Volume (ha m)	Per cent of total volume utilised	Per cent change
Canal	39,719.41	28.00	+9.80
	43,612.20	21.45	
STW	18,384.00	12.96	+32.97
	24,446.00	12.03	
PTW	68,536.60	48.31	+17.06
	80,228.00	39.46	
PS	3,675.00	2.59	+1351.21
	53,331.80	26.23	
PW	878.40	0.62	-85.35
MW	10,685.95	7.52	-85.52
	1,547.50	0.76	
Total	1,41,879.36	100.00	+43.29
	2,03,294.20	100.00	

Note: (a) Blockwise figures are given in Appendices.
STW =State Tube-well, PTW=Private Tube-well, PS=Pumping Set, PW=Persian Wheel, MW=Masonary Well

Source: Computed from the Appendix 1 and unpublished data collected from:
1. Office of the E.E., Irrigation Division, Ballia.
2. Office of the E.E., Irrigation Division Mau & Azamgarh.
3. Office of the E.E., Deokali Pump Canal, Div. II, Ghazipur.

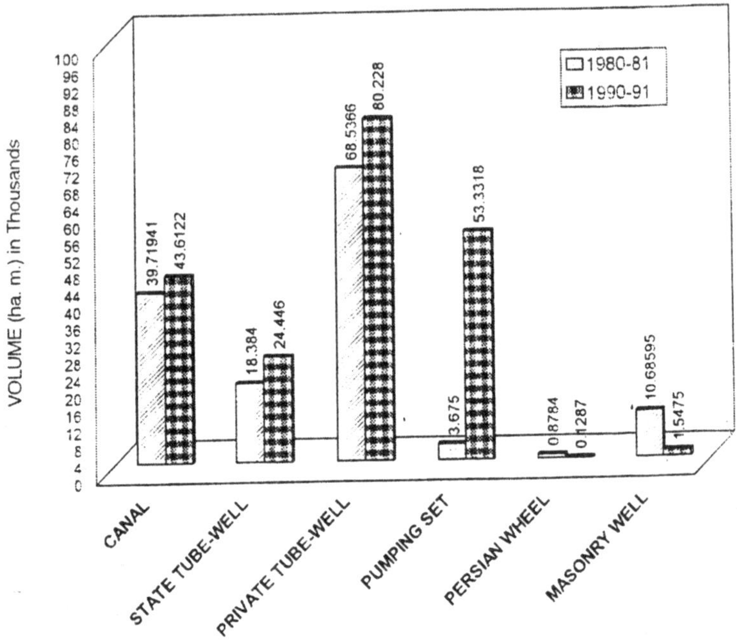

Fig 12.3: Ganga-Ghaghara *Doab* East : Utilisation of Irrigation Water Through Different Works

Water Requirements of Crops

Water is needed by plants to supply carbondioxide and oxygen in relation to the cells of photosynthesis and respiration, to transport raw materials, manufactured and waste products within the plant, and to maintain the rigidity (turgidity) of the plant structure (Rosemary J. More, 1967). Water is required to meet the transpiration losses from the plants and the evaporation losses from the soil, the total evapotranspiration losses being generally designated as consumptive use. Besides consumptive use, other losses are such as land preparation, transplanting and leaching etc. Much work has been done in the past by different scholars to find out the minimum water requirements of crops in order to ensure the optimum use of water to get the optimum yields (Sally 1968).

The water requirements of different crops in the region is 6,26,141.05 ha m in 1990-91 which was 5,66,317.05 ha m in 1980-81 registering an increase of 10.56 per cent during the study period (Table 12.6).

It is well known that *kharif* crops are mostly rainfed whereas the rabi crops are largely dependent on irrigation. Taking into account the rainfall of 71.99 cm during the *kharif* season (July-October) and the total *kharif* cropped area (7,16,418 ha. in 1990-91) the total amount of rain water available during the *kharif* season comes to 2,99,779.32 ha m, that is, 65.03 per cent of total water requirement of *kharif* crop (Table 12.5 & 12.6). On the other hand the rainfall amounts to 3.72 cm in *rabi* season (November to April) and taking into account the total *rabi* cropped areas (4,46,428 ha in 1990-91), the amount of rain water available for utilisation during the rabi season comes to 16,607.12 ha m , that is, about 10.05 per cent the total water requirements of the rabi crops (Table 12.6). Hence 34.97 per cent of *kharif* water requirements and 89.95 per cent of rabi requirements are to be met with from irrigation. At present about 65.63 per cent of total irrigational facilities are to be developed to meet the remaining 34.37 per cent of demand.

Conclusion

The study reveals that the present irrigation facility is not fulfilling the irrigational water requirements in the area and this is further applicable for the whole Ganga-Ghaghara *doab* region in Uttar Pradesh. Side by side it is clear from the above description that there is a major share of ground water irrigation that is about 80.34 per cent of total irrigated area and some development blocks have no surface water irrigation. In 39 development blocks, three blocks named as Beruarbari, Ratanpura and Muhammadabad are under 'Dark' categories, where stage of groundwater development is more than 85 per cent and other nine blocks named as Garwar, Dubhar, Reoti, Bansdih, Mardah, Saidpur, Deokali, Jakhaniya and Kasimabad are under 'Grey' categories with more than 65 per cent ground water development. Thus further groundwater irrigation development is not allowed by 'NABARD' (Rai, 1997). So there is urgent need for the development of balanced surface and ground water irrigation. Rapid increase in ground water irrigation has resulted in lowering of water table. In such a situation

Table 12.6: Crop water requirements in Ganga-Ghaghara *Doab* East (upper figure: 1980-81, Lower figure: 1990-91).

Crop	Cropped area (ha)	Water requirements	
		Depth(m)	Volume (ha m)
Kharif Season			
Paddy	2,69,302	1.14	3,07,004.28
	3,02,882		3,45,285.48
Maize	20,726	0.51	10,570.26
	15,565		7,938.15
Big Milet	18,087	0.30	5,426.10
	19,802		5,940.60
Arhar	27,524	0.30	8,257.20
	19,038		5,711.40
Urd	347	0.30	104.10
	439		131.70
Moong	5,855	0.30	1,756.50
	1,936		580.80
Sugarcane	38,258	2.24	85,697.92
	40,809		91,412.16
Others	24,718	0.16	3,954.88
	15,947		3,954.88
Rabi Season			
Wheat	2,60,386	0.41	1,06,758.26
	3,32,378		1,36,274.98
Barley	37,620	0.36	13,543.20
	20,164		7,259.04
Gram	55,701	0.14	7,798.14
	41,878		5,862.92
Pea	6,282	0.17	1,067.94
	6,275		1,066.75
Oilseeds	1,442	0.21	302.82
	2,366		496.86
Potato	10,325	0.69	7,124.25
	13,750		9,487.50
Others	43,445	0.16	6,951.20
	29,617		4,738.72

(Contd.)

Table 12.6 : *(Contd.)*

Crop	Cropped area (ha)	Water requirements	
		Depth(m)	Volume (ha m)
Total Kharif	4,04,817		4,22,771.24
	4,16,418		4,60,955.17
Total Rabi	4,15,201		1,43,545.81
	4,46,428		1,65,186.77
Total	8,20,018		5,66,317.05
	8,62,846		6,26,141.94

Source: Computed from data taken from:

1. *District Statistical Handbooks*, 1982 and 1992 of Ballia, Mau, Ghazipur and 1982 of Azamgarh.
2. Work carried out at the Institute of Agricultural Science, B.H.U., Varanasi (Sally, 1968; p.78).

shallow wells become dry and there is decrease in the draft of different wells. On the other hand excessive surface water irrigation has resulted in the development of waterlogging and soil salinity and alkalinity. In this situation, there is need for the development of surface water irrigation in 'Dark' and 'Grey' blocks or in other words there is urgent need for the development of conjunctive use of surface and ground water irrigation to meet the remaining 34.37 per cent of irrigational demand.

References

Bahadur, P. *et al.* (1979), Conjunctive use of surface and groundwater in Gandak Command Area, in *Proc. of the Workshop on Conjunctive Use of Surface and Ground Water,* Roorkee 1: c4-1.

Centre for Science and Environment, (1985), *The State of India's Environment 1984-85,* The Second Citizen Report, New Delhi, pp. 28-9.

Charlu, T.G.K. and Dutt, D.K. (1982), *Ground Water Development in India,* Rural Electrification Corporation, Technical Series No.1, New Delhi.

Dakshinamurti, C. *et. al.*, (1973), *Water Resources of India and Their Utilisation in Agriculture,* Water Technology Centre, Indian Agricultural Research Institute, Monograph No. 3, New Delhi.

Krishna, M.S., (1982), *Geology of India and Burma*, CBS Publishers & Distributors, Delhi.

More, R.J. (1967), Hydrological models and geography, in Chorley R.J. and Hagget,

P. (eds), *Models in Geography,* Methuen and Co., London, pp. 145-85

Navalawala, B.N., (1991), Waterlogging and its related issues in India, *Irrigation and Power,* Vol. 48, No.1, Jan. 1991.

Rai, V.K., (1993), *Water Resources Planning and Development,* Deep and Deep Publications, New Delhi.

Rai, V.K., (1996), Water Resources of Ganga-Ghaghara Doab East—A Study of problems and of prospects for development, U.G.C, Project Report, Department of Geography, Banaras Hindu University, Varanasi (unpublished).

Rai, V.K., (1997), Assessment of groundwater resources and its management: A case study of Ganga-Ghaghara Doab East, in A.K. Singh, V.K Rai and A.P. Mishra (eds), *Strategies in Development Planning,* Professor M.Shafi felicitation Volume, Deep & Deep Publication, New Delhi, pp. 228-64.

Sally, H.L., (1968), *Irrigation Planning for Intensive Cultivation,* Asia Pub. House, Bombay.

Sawant, S. D., (1991), Irrigation and water use, in M.L. Dantwala (ed) *Indian Agricultural Development since Independence*, M.L. Dantwala, Oxford & IBH Publishing Co. Pvt. Ltd., New Delhi.

13

Concept of Health of the People in Developing Countries

B.R.K.Sinha

Health of the people is no doubt a precious wealth of the nation as any sort of development of the concerned nation depends on the levels of health status of the people and their efficient and proper utilisation in productive sphere. From this point of view, health of the people is considered as an asset, wealth, capital, and factor of community development and country's progress. According to Madan, (1983) also, health of the people is fundamental to national progress in any sphere. It is the measure of energy and productive capacity in any country. Health is also a positive state of well-being of the people. Kuppuswamy (1975), mentioned, while dealing with the investment in human resources, health as an instrumental value which enables a person to work efficiently and ill health causes great unhappiness and constitutes a social loss. Sudha and Singh (1983) have observed that "the differences in health of labour force result in productivity differentials. Sound health gives greater stamina to stand the odds of occupation and to work overtime". Further, according to Kamble, (1984), level of living of the people in a nation is affected by the health conditions of the people in it.. Health according to him is an important asset of a community and healthy community is the foundation of strong nation. Health is an important determinant of economic and social development because disease creates vicious circle by deplenting human energy, leading to low productivity and earning capacity, deteriorating quality and quantity of consumption and standard of living. Therefore, a nation ought to give adequate attention to the health care of its people. In this sense, health is undoubtedly a welath that becomes a resource whenever and wherever it is utilised by an individual in any activity in the form of capacity, energy, strength, ability and other mental and physical capabilities.

Scenario of the Developing Countries

Health of the people in developing countries of the world is currently and increasingly becoming very important as the development of these nations depends to a very great extent on the health status of the people of those nations. From this standpoint our nation is committed to attain the goal of health for all by the end of this century. This has been accepted as one of the main instruments of human resources development and for accelerating economic and social development and also for attaining an improved quality of life of the people.

But the fundamental dilemma is that the developing countries of the world are presently passing through a variety of problems such as poor health and education, poor nutrition and caloric intake, nutritionally inadequate diet, hunger, undernourishment, malnutrition,undeveloped and underdeveloped human resources, inefficient utilisation of human, cultural and natural resources, low per capita income, high proportion of population employed in the primary sector, low agricultural and labour productivity, fragmentary industrialisation, poverty, limited purchasing power, over dependence on a small number of export cash crops, high incidence of unemployment, larger proportion of dependents, limited application of technology, dependence on foreign investment capital, dependence on imports of capital goods such as machinery, poor accessibility of health and educational institutions, poor quality of medical facilities, lower per capita and national expenditure on health and education, poor housing conditions, lack of quantity and quality of potable water, poor sanitation, presence of variety of diseases, increasing environmental pollution at local, regional and national levels, unstable political atmosphere, etc. In other words, it can be said that developing countries stand with weak economic, social, cultural and political structure. All these phenomena are directly or indirectly related with each other and have greater influence on the health status of the people. Consequently, under the above conditions of the developing world, the general health conditions of the people is much below optimum, rather poor which adversely affect the human resources (Sinha, 1987). Thus, concept of health includes the approaches of measuring such functional capabilities or health conditions and its other associated aspects. Conyne and Clack (1981) considered the concept of health as a measure of the population's capabilities to function biologically in a given environment.

Objective and Approach

In the light of the above importance and problems of human health the objective of the present chapter is to develop a conceptual framework of the basic ideas that are directly or indireclty related to the status of health of the people in developing countries. This study is exclusively conceptual in nature. Methodological aspect throws light mainly on the definitions, dimensions and development of human health and on fixing a yardstick to measure health status of the people in developing countries.

Approach of Definition of Human Health

Since health (the product of genetic and environmental factors) is a very complex phenomenon, it seeks proper approach of its definition. Meaning of health here is taken into account as an important aspect of the concept of health. Different scholars have defined human health differently in various ways. Accordingly to Ghosh (1978), "health suggests not only the absence of disease and weakness of the body, but the presence of the welfare of body, mind and society. Health implies a state of sound body and mind". This definition throws light on the disease free and perfect condition of body and mind of an individual. Valentei, Kvasha and Ivanov (1986) considered health as an ingredient of the quality of population. According to them population health is not only the absence of mass illness and population injuries but also man's concordant physical, psychic and moral development enabling him to carry on social and labour activities freely and without restrictions. This definition is more or less similar to the above definition. Madan (1983) defines that health does not mean merely the absence of disease but harmonious functioning of the body and mind of an individual in relation to his physical and social environment so as to enable him to enjoy life to the fullest possible extent and to reach his maximum level of productive capacity. This definition indicates the complete adjustment of body and mind of an individual to his external environment while functioning or carrying on certain activities. Rizvi (1986) understands health as right condition and the proper functioning of the body. In his concept of health he further mentioned, "those who are regarded as healthy, who can take heavy food, have a good muscular body and are able to work hard. For proper health it is necessary to be free from all kinds of mental tension too". This definition reveals the state of proper functioning capacity of the body which enables an individual to have heavy food and to work hard. Desai (1988) referred to World Health Organisation (WHO) which defined health as a state of complete physical and mental well-being and not merely the absence of disease. This definition emphasizes health as a right condition of body and mind and not only the absence of illness of the person. Dolfman (1974) views health of the people in the following manner. "An individual is healthy if he is functioning adequately in a stated environment and adapts the stresses of that environment within the range of normal functioning." Conversely, an individual is not healthy if he is not functioning adequately in a stated environment and does not adapt the stresses of that environment within the range of normal functioning. This definition reveals that health is a state or condition of normal functioning of an individual in a given environment.

After reviewing the above mentioned meanings of health the author tends to define health as a state or condition of functioning capacity of physical and mental faculties of an individual in a stated natural and human environment of a particular region and at particular reference of time. Such bodily and mental faculties should be free from disease or ailment too. Health of an individual may remain constant for a certain period of time in a given environment and thereafter due to certain changes in the components of biological composition of an individual's body and also due to change in the quality of natural, human and cultural environment and its influence on human body, health may change. In this sense, health can not be considered as stable state of body and mind of an individual as internal structure of body changes and weakens the resistant power due to change in age and also due to change in the quality of natural, human and cultural environment and its influence on human body. Thus, a person can remain healthy or of sound health till his/her physical and mental faculties remain free from disease or illness and also till his/her mental and physical faculties remain under the influence of congenial environment of different types. This situation enables a person to work adequately and properly without any stress in the existing environmental conditions and leads him/her to enjoy full length of life.

Dimensions and Development of Human Health

Human population of any region constitute two important dimensions—quantitative and qualitative. Quantitative dimension stands for size, magnitude, amount, sum etc. that can be measured directly in number and weight whereas, qualitative dimension, on the other hand, stands for nature, capacity, ability, knowledge, skills, energy, etc. that can not be measured directly in terms of number or weight but can be measured in terms of grade or levels. Here, number or size of people of a particular region is considered as quantitative components whereas nature, capacity, ability, knowledge, skills, talents, energy, aptitude, etc. either inherited in the people or achieved by the people are considered as qualitative components of the population. All these ingredients of quality are the product of health (physical and mental faculties) of the people. In this sense, quality of a person can not be separated at all from him/her or from the quantitative part of the people. Moonis Raza (1990), also considered health as an integrated quality of a person that can not be broken up into homogeneous parts and therefore, can not be measured directly. Trewartha, Robinson and Hammond, (1967) considered health (state of physical and mental well-being or health) as one of the most important qualities of a population. Valentei, Kvasha and Ivanov (1986) also considered health as a quality of population which is complex in its turn. Sudha and Singh (1983) also regarded physical and mental health as a quality of a person. It is, therefore, clear that health as defined above is a quality of the people of a particular region and can not be measured directly.

Finally, health has two aspects—physical or bodily and mental. Both are integrated and interlinked. Physical aspect refers to sound physical stamina, muscular capacity, race characteristics, etc. and mental aspect refers to right, fit and healthy creative abilities of mind. All these ingredients of these two aspects are also known as human resources of the people upon which all round development of a region or a nation greatly depends. Smith (1977) also made two categories of health—physical and mental while discussing the general criteria of social well-being in the U.S.A.

Human Health Development

Human health development is very important aspect of health which can be defined as the process of enhancing the individual's physical and mental functional capacity. In broader sense, it can be defined as life long process of increasing and maintaining sound stamina, strength, muscular capacity, knowledge, skills, talents aptitude, attitude and other physical or mental capacities actually or potentially available in human being.

How to develop human health, is another complex aspect of the concept of health. Different scholars have extended their views in a variety of ways. According to Mehta (1976) education, health and nutrition tend to enhance the individual's mental and physical capacities. In *Manpower Journal* (1973), it is highlighted that the development of human resources implies the development of both the physical and mental aspects of the productive capacity of workers. The physical aspect of the productive capacity can be developed by means of health services, nutrition, etc. and the mental aspect by education, practice and training. According to Jayachandran (1988), the basic requirements of a society are to have the potential for creative thinking by educating our mass of the people and to have a healthy mind and healthy body by providing nutrition food in sufficient quantity and providing health care facilities. Thus, it is evident that human health can be developed by means of food and nutrition, health care facilities of preventive, promotive and curative types, education and educational programmes of various types, optimum population, pure drinking water, healthy work conditions, well shelters, prohibiting the use of intoxicative materials, stopping environmental pollution, aleviating people from poverty, employing family and social welfare planning, etc.

Measurement of the Status of Health

How to measure health status of the people in a particular developing region is another significant aspect of the concept of health. In order to fix a yardstick to measure health status is not only a difficult task but also a very complex problem. Different scholars from different disciplines have employed various criteria in measuring health status of the people. According to Moonis Raza (1990), one has to select quantitative indicators of quality in order to make qualitative assessment, e.g. a doctor can use quantitative indicators of health for a tubercular patient: body weight, body temperature, blood pressure, counts of red blood and white blood corpuscles. Ghosh (1987) employed life expectancy at birth, infant mortality per thousand people, number of doctors per lakh people, average daily caloric supply, causes of death, homicide and serious assault per lakh people as indicators of health of the people. Smith (1972), stated infant mortality as an indicator of the levels of health of the people. Prema and Thomas (1992) made an assessment of the status of health of tribal women on the basis of clinical investigation and bio-chemical assessment of blood profile. They adopted height and weight according to age and haemoglobin level of the tribal women for measuring health status. Basu (1992) considered assessment of nutritional status as one of the most important health indicators of a population group. He further considered anthropometric and somatic measurement such as body weight, height, head circumference and various clinical signs of hair, eyes, lips, gums, teath, skin, gland and westing muscle as the indicators of health of the people. Trewartha, Robinson and Hammond (1967) regarded average length of life, increase in human stature, number of doctors and hospital beds per unit of population as indicators of measuring health status of the people. Pacione (1988) tried to measure health of the people employing infant mortallty rate, life expectancy, weight of new born baby, safe water supply, adult literacy, per capita income, public expenditure on health and other health care facilities and services.

Harbison (1973) considered life expectancy as a crude indicator of measuring health and suggested data on critical diseases, access to health facilities, infant mortality as better index of health measurement of the people. Madan (1983) considered birth and death rates, average expected life and diseases as indicators of health of the people. Jayachandran (1988) suggested medical institutions for health care, health education, nutrition food and caloric needs appropriate nutrition, education of elders in the family and proper identification of supplementary goods for the infants as indicators of measuring health. Mukherjee (1986) in his concept of health mentioned capacity to do manual work for atleast 4-5 hours a day, clear eye sight and memory, to be able to take active part in household activities and physical mobility as signs of good health.

It is obvious that most of the above mentioned scholars have considered, suggested and employed the quantitative indicators in measuring health of the people. Only last one tried to measure human health employing qualitative signs of health.

Since health itself is a quality upon which quality of population greatly depends, there should be some certain qualitative indicators also so that health status of the people can be measured with precission and accuracy. Therefore, author suggests to employ and involve both the criteria—quantitative and qualitative—in measuring health of the people. But for measuring health status of an individual particularly qualitative indicators should be employed.

Quantitative Approach

This method includes quantitative indicators such as life expectancy at birth, birth rate, infant mortality rate, mortality rate and its causes, number of diseases, height and weight of individual's body, number of medical institutions and hospital beds and also number of doctors including nursing persons per unit of population, access to medical and educational facilities, per capita expenditure on health and education and its related institutions, per day caloric intake and number of daily meals and breakfasts etc.

Qualitative Approach

This approach includes various types of qualitative indicators of health. These are nutrition, vitamines, types of food taken, quality of drinking water, quality of air, housing condition, types of diseases, high and low blood pressures, working atmosphere, clinical signs of blood, haemoglobin, structure of body, apetite and digestive capacity, problem in eye sights, mental tension, capacity to do mental and physical work regularly for standard working time of 6-8 hours in a given environment (i.e., in sun and rain), normal headload carrying capacity, state of sleeping, feeling of weakness and tiredness, laziness feeling, habit of smoking, habit of alcohol and drugs, achievenments of education and training, areal and occupational mobility of person, levels of skills, levels of overall productivity of an individual.

Some of the above mentioned quantitative and qualitative indicators are applicable for common people at regional, state and national levels and some are applicable for individual person at local level while measuring the state of human health. Therefore, after finding out the values or scores of certain indicators employed and applicable for common people or individual person, firstly a composite index can be prepared for making suitable grades of health of the people for a particular geographical unit during certain period of time and secondly, to prescribe the suitable set of perspective measures for the specific situation. Thus on the basis of the obtained values of the composite index, certain grades of health such as high, medium and low or sound health, average health and poor health, etc. can be prepared and used for representing the variation or differentiation in health status of the people across spatial units, occupations, different socio-economic strata age and sex wise. Thus, the above approaches appear to be more realistic and proper in order to make an assessment of the status of health of the people. The availability of data on the above quantitative and qualitative indicators is not so easy, and it requires a special field survey on the above mentioned quantitative and qualitative indicators of human health.

References

Basu, S. (1992), Nutritional status, physical growth trends and health problems among the tribal population of Bastor district (M.P.), in Tiwari, P.D. and Tripathi, R.S. (1992), *Dimensions of Scheduled Tribes Development in India*, Uppal Publishing House, New Delhi.

Conyne, R.K. and Clack, R.J. (1981), *Environmental Assessment and Design: A New Tool for the Applied Behavioural Scientist*, Praeger Publishers, New York.

Desai, V. (1988), *Rural Development: Issues and Problems*, Vol. I, Himalayan Publishing House, Delhi.

Dolfman, M.L. (1974), Towards operational definitions of health, *Journal of School Health*, Vol. 64, quoted by Akhtar, R. (1992), Spatial aspects of disease: A survey, in *ANNALS of NAGI*, Vol. XII, Nos. 1 & 2, June and December.

Ghosh, B.N. (1978), *Population Theories and Demographic Analysis*, Meenakshi Prakashan, New Delhi.

Ghosh, B.N. (1987), *Fundamentals of Population Geography*, Sterling Publishers Pvt. Ltd., New Delhi.

Harbison, F.H. (1973), *Human Resources as the Wealth of Nation*, Oxford University Press, London.

Jayachandran, S.; Vidyanath, V. and Rao, R.R.M. (1988), *Development of India's Resource Base, Pattern, Problems and Prospects;* Gain Publishing House, Delhi.

Kamble, N.D. (1984), *Rural Health*, Ashish Publishing House, New Delhi.

Kuppuswamy, B.(1976), *Population and Society in India*, Popular Prakashan Pvt. Ltd., Bombay.

Madan, G.R. (1983), *India's Social Problems: Social Disorganisation and Reconstruction*, Vols. Allied Publishers Pvt. Ltd., New Delhi.

Manpower Journal, (1973), A review of some methods of planning human resources development, Vol. VIII, No. 4, March.

Mehta, M.M. (1976), *Human Resources Development Planning*, the Macmilan Company of India Ltd., Delhi.

Moonis Raza, (1990), *Education, Development and Society*, Vikas Publishing House Pvt. Ltd., New Delhi.

Mukherjee, B.M. (1986), Aging members and their health in changing techno-economic condition, in Chaudhury, B. (1986), *Tribal Health: Socio-Cultural Dimensions*, Inter-India Publications, New Delhi.

Pacione, M. (1988), *The Geography of the Third World: Progress and Prospects*, Routledge, London & New York.

Prema, L. and Thomas, F. (1992), Nutrition and health problems faced by Kanikkar women, in Tiwari, P.D. and Tripathi, R.S. (1992), *Dimensions of Scheduled Tribes Development in India*, Uppal Publishing House, New Delhi.

Rizvi, S.N.H. (1986), Health practices of the Jaunsaris–A socio-cultural analysis, in Chaudhuri, B. (1986), *Tribal Health: Socio-Cultural Dimension*, Inter-India Publications, New Delhi.

Sinha, B.R.K. (1987), Human Resources Utilisation in Dhanarua Block, Patna Bihar (unpublished thesis).

Smith, D.M. (1977), *Human Geography: A Welfare Approach*, Edward Arnold Pub. Ltd.

Sudha, S.N. and Singh, A. (1983), *Fundamentals of Agricultural Economics*, Himalaya Publishing House, Bombay.

Trewartha, G.T.: Robinson, A.H. and Hammond, R.H. (1967), *Elements of Geography, Physical and Cultural*, MacGraw Hill Book Company, New York.

Valentei, D.; Kvasha, A. and Ivanov, S. (1986), *Population and Socio-Economic Development*, Progress Publishers, Moscow.

14

Appraisal of Family Planning Programme

M.B. SINGH, R.SINGH and S. PRIYADARSHI

In spite of remarkable economic progress made by India during the last 4 decades an annual population growth rate (2.47%) of 1971-81 decade has fallen down marginally to 2.35 per cent per annum in 1981-91 decade, showing a decline of only 0.12 per cent per annum. Such high population growth rate has very serious repercussions on socio-economic development of the country. For instance, high population growth is coming in the way of curbing the unemployment problem. The number of unemployed, which was 36.7 million in 1992, is estimated to cross 100 million by the turn of the century. Despite the new economic policy and an increase in economic growth rate, our per capita GDP in real terms has declined from 330 US dollars in 1991 to 310 US dollars in 1992 and 290 US dollars in 1993. Similarly, our foodgrain production has gone down from 575 gm per capita per day in 1988-89 to 551 gm per capita per day in 1993-94. Another grim situation arising out as a result of high population growth is high proportion of people living below poverty line (33.4% in 1987 88) and without shelter (36 million in 1991). This alarming population growth needs to be brought down by vigorous and effective promulgation of family planning programme.

Though India has been the first country in the world to adopt the National Family Planning Programme in 1951 (which later renamed as National Family Welfare Programme). Its achievements have not yet been very encouraging. The Government of India had proposed to bring down the birth rate to 30 per 1000 by the end of the Sixth Five Year Plan and 21 per 1000 by 2000 A.D. through adoption of family planning programme. But the country did not achieve this goal and it is still maintained around 29 per 1000 people.

Realising the consequences of the high population growth on socio-economic conditions of the country and little success achieved by the family planning programme in mind, the present endeavour aims to seek appraisal of the family planning programme and provide suggestive measures so as to make it more effective for efficient control of population growth.

Level of Family Planning Adoption

During 1965-66 the rate of sterilisation, IUD insertions and other contraceptive uses were 3.2, 1.7 and 1.2 per 1,000 respectively making a total of 6.1 per 1,000. These uses rose to 65.8, 37.0 and 10.6 amounting to 113.4 per 1,000 in 1988-89 respectively. Among different methods of family planning, sterilisation still plays a crucial role. Moreover, the sterilisation has become a women's programme only due to conservative thinking. Although the male sterilisation (vasectomy) is simple and easier than the female sterilisation (tubectomy) but due to false fear and prejudices it is less practiced. As a result, male sterilisation contributes only 3.5 per cent of total sterilisation and rest is compensated by females. Therefore, the breakthrough is needed in female sterilisation. The most remarkable fact in family planning adoption is that the northern developed states like Punjab, Haryana, Chandigarh and even Uttar Pradesh show higher proportion of IUD insertion among various methods of family planning while central, western and southern states are dominated by sterilisation method.

It has been observed that the states having good access to health facilities (see composite health index in Table 14.1) exhibit notable rate of family planning adoption in 1988-89. Such states are Punjab (250.7/1,000), Haryana (230/1,000), Pondicherry (203.7/1,000), Maharashtra (194.2/1,000), Gujarat (183.2/1,000), Chandigarh (177.6/1,000), Delhi (173.2/1,000), Tamilnadu (170.2/1,000), and Kerala (141.4/1,000). However, the central and eastern part of India experience slow level of adoption. For example, in Assam, Bihar, Jammu & Kashmir, Manipur, Meghalaya, Orissa, Rajasthan, Sikkim, West Bengal, Uttar Pradesh; Andaman & Nicobar, Dadra and Nagar Haveli, Lakshadweep, Mizoram and Arunachal Pradesh the level of family planning adoption is well below the developed states of India. These states are also characterised by high birth and death rates.

Table 14.1: Family Planning in India

States	Family planning adoption/1000 population (1980-81)	Composite health index 1990	Birth Rate 1988	Death Rate 1988
States				
Andhra Pradesh	137.2	0.90	27.4	10.2
Arunachal Pradesh	27.9	0.65	40.0	17.2
Assam	74.6	0.88	32.9	11.8
Bihar	81.8	0.96	37.3	12.6
Goa	85.6	3.62	17.7	7.9
Gujarat	183.2	1.21	20.5	11.0
Haryana	230.0	0.65	33.8	9.8
Himachal Pradesh	159.2	0.65	32.2	9.6
Jammu & Kashmir	74.4	0.78	33.1	8.4
Karnataka	133.5	1.13	28.7	8.8
Kerala	141.3	25.17	20.3	6.4
Madhya Pradesh	132.4	0.32	37.0	14.3
Maharashtra	194.2	1.75	29.4	8.9
Manipur	69.5	0.73	25.8	6.8
Meghalaya	21.0	0.60	36.4	9.1
Mizoram	68.2	0.82	–	–
Nagaland	8.8	0.63	22.3	5.0
Orissa	137.4	0.60	31.9	12.3
Punjab	250.7	1.24	28.5	8.4
Rajasthan	990.3	0.59	33.3	14.0
Sikkim	44.8	0.78	33.8	10.1
Tamilnadu	170.2	1.15	22.7	9.3
Tripura	43.4	0.79	26.6	8.1
Uttar Pradesh	116.7	0.79	37.2	13.2
West Bengal	977.8	1.52	28.4	8.4
Delhi	173.2	4.94	28.6	7.6
Union Territories				
A & N Islands	103.4	1.23	22.1	6.4
Chandigarh	177.6	3.66	22.4	4.7
D&N Haveli	117.5	0.95	–	–
Daman & Diu	39.6	5.88	28.1	8.9
Lakshadweep	39.7	8.97	25.5	6.6
Pondicherry	203.7	4.34	22.5	7.9
Total	113.4	1.00	29.9	9.6

Although there has been a substantial increase in adoption of family planning programmes in the country yet in no real sense it has been able to curb high population growth. This programme still lags much behind and has a lot to do considerably. Therefore, a sincere appraisal of family planning programme is required. Apart from this, attitude of people towards these programmes draws special attention for analysis so as to make the programmes more effective.

The Case Study of Varanasi District

In order to strengthen the family planning programme its association with socioeconomic conditions of various communities has to be analysed in depth. For fulfilling this need, a survey of five villages of Kashi Vidyapeeth Block, Varanasi district has been conducted in the year 1994. The villages under study witness tremendous population pressure (1,100 persons/km^2) and population growth of 2.3 per cent per annum.

In 1994 the villages under study comprise 11,742 population with only 36.48 per cent literacy (80.77% male, 19.23% female). On economic front, the position of villages looks very disheartening as less than 10 per cent families only own farming land above 2 acres; 20 per cent families are land less and 72 per cent families have their monthly income below Rs. 2000.

As compared to the average family planning adoption (113.3/ per 1,000) in the country, villages under study reveal poor rate of adoption e.g. 42/1,000. The Brahmin, socially most advanced caste, records higher rate of adoption (115/1,000) followed by Rajput (105/ 1,000) and Bhumihar (84/1,000) where as other castes are found much behind. It is noticable that more than 99 per cent adopters are females.

In the study villages the couples have adopted family planning after having 5 children that is 3 sons and 2 daughters. This shows the craze for son is still prevailing in the minds of people. Against this 5 per cent adopters adopted it only after 2 children that is after 2 sons or 1 son and 1 daughter. The hypothesis relating to adoption after less number of children in advanced castes and more children in socially and economically poor castes, as they take it granted for old age pension, does not confirm.

Age at marriage of female is one of the important demographic variables influencing the family planning adoption. Among study villages an average marriage age for males and females has been recorded as 19 and 16 years. The situation appears more painful when this is taken into account for so-called socially poor castes. About 70 per cent adopters were married before attaining the age of 18 years. Even after promulgation of Marriage Act, which envisages 21 years age for boys and 18 years for girls, sizeable number of marriages are still taking place well below the suggested age limit. Another remarkable fact in this respect is the lower birth interval (less than 18 months) among 55 per cent responds which stands dangerous for maternal and infant mortality.

Akin to marriage age, the age of couple at the time of adoption is also important because it has direct bearing on fertility, mortality, migration and use of family planning measures. The average age of adoption has been found around 31 years. Although Pal, Maurya and Bania castes adopted it below 30 years age yet it not comfortable sign as their marriages take place in early age. In fact, satisfactory situation arises in Bhumihar, Rajput and Kurmi castes where adoption materialised at the age of 30 years.

Of all adopters of 61.43 per cent adopters are illiterate. Among literate adopters (38.57 %) the percentage of educated adopters *e.g.* high school and above, is merely 5.24 per cent. Indeed this negative relationship between literacy and family planning adoption has occurred not because of the negative attitude of literates but due to low level of literacy. To obtain a real picture of such relationship the percentage of illiterate and literate adopters to total illiterates and literate people has been worked out (Table 14.2). The result confirms that the illiterates are less serious than literates. In the study villages the literacy has no direct and accelerative impact on the acceptance of family planning programme. For the positive role of education authors fully agree with B.K. Sahay's (1993) statement that literacy can effectively play a supportive role in promoting family planning programme under favourable socio-economic conditions such as low infant mortality and adequate old age-care, etc.

Another hypothesis regarding the service and business class and economically well off farmers who adopt family planning more numbers than traditional and poor farmers, is confirmed in these villages.

Table 14.2: Literacywise Adoption (1994)

Literacy Category	Adopted Percentage
Illiterate	61.43
Low Literate	33.33
Medium Literate	5.24
Percentage of Illiterate Adoptors to total illiterate	3.38
Percentage of Literate Adopters to total literate	15.19

Source: Personal Survey.

Attitude Towards the Family Planning Programme

Public opinion and attitude play deciding role in the success of any programme. Srivastava (1989) has rightly remarked that the success of any programme depends upon the people's participation.

An analysis regarding the level of satisfaction in terms of family size shows that about 42 per cent respondents are not satisfied with their present family size. Of this 40 per cent respondents treat 3-5 children as small family size. Only 17 per cent respondents realised their family size as big (Table 14.3). If former thought continues, there will be no breakthrough in the success of family planning programme. The number of children satisfying the majority of respondents is 4.

In the villages under study respondents having male preference with one daughter share 70.3 per cent of the respondents. It has also been found that the preference of male members goes upto 5 whereas for female it restricts to only one. This clearly portrays the male dominating nature of society. The highest number of respondents (24.4%) prefer 2 sons and 1 daughter followed by 17.8 per cent respondents who like 2 sons and 2 daughters whereas 9.3 per cent respondents prefer 3 sons and 1 daughter. But the ray of hope is still alive with the fact that there are 23.2 per cent respondents who like only 2 children either only 2 sons and 1 son and 1 daughter (Table 14.4A). Such respondents belong to economically and socially well off castes like Kurmi, Rajput and Brahmins.

Table 14.3: Opinion about the Family Size

Family Size	No. of Children (Respondents%)					Percentage of Respondents
	1	2	3	4	5	
Big	-	-	9	26	65	17
Balance	-	20	25	27	28	41
Small	25	35	21	12	7	42

Table 14.4A : Sex-Wise Preference of Children (Respondents %)

Male	Female			Total
	0	1	2	
1	3.7	9.7	0.6	14
2	9.8	24.4	17.8	52
3	3.8	9.3	5.9	19
4	4.2	2.8	5.0	12
5	0.8	1.8	0.4	3
Total	22.3	48.0	29.7	100

In response to know the reasons of disliking the female child, 51 per cent respondents put dowry as number one while 27 per cent respondents mention daughter as an economic burden upon the family as they go to their husband's home and do not become the old age pension. Hence, dowry needs to be curbed by stern action.

To make family planning programme a grand success it is essential to know the reasons of low adoption. It is clear from the Table 14.4B that the highest number of respondents (37%) have not adopted family planning programme on account of ill health caused by its adoption, whereas 22 per cent and 14 per cent respondents blame for inadequate supply of family planning materials and no follow up actions after sterilisation respectively. Thus, about 82 per cent respondents make government responsible for low adoption due to its failure in supply of materials, follow up action and adequate motivation.

It has been observed that the rate of family planning adoption is quite high in the states where availability and accessibility of health facilities are better. Moreover, the family planning services required before and after adoption are rendered by the health care networks. Thus, it would be be fitting to say that in order to enhance the family planning adoption, the comprehensive analysis of health care facilities and utilisation is utmost task and forms an important aspect of family planning programmes appraisal.

Table 14.4B: Reasons of low adoption of family planning programme

Reasons	Percentage of Respondents
Adverse effect on Health	37
Inadequate supply of family planning materials	22
No follow up	14
Less popularity of programme	99
Need of working hand	13
Social reasons (impotency, social taboo, etc.)	5

Utilisation of Health Care Facilities

The services, supply of materials and diffusion of knowledge of family planning programmes are performed through the network of health care system. Moreover, there is direct impact of health care utilisation on family planning adoption because better the health utilisation greater the adoption. Hence, the evaluation of health care system is must in order to attain the high level of effectiveness of this system.

Looking into the inequality in the availability of health services the member countries of the WHO adopted a resolution in the 1978 to achieve the goal of "Health for All by 2000 A.D." It was also reiterated that the primary health care is the key to achieve health for all. In view of the above, it has been proposed to expand the health facilities in India further in a phased manner so as to have for every primary health centre with a 30,000 population (20,000 in hilly and tribal areas) and every sub-centre with a 5,000 population each (3,000 in hilly and tribal areas) by 2000 A.D. Till 1951, there were no primary health centres and sub-centres in rural areas. During the past four decades in India concerted efforts were made in providing health care facilities to its huge population (Khan, 1992). At present these services are rendered through a network of 22,229 primary health centres (PHCs),

1,31,379 sub-centres (SCs) and 1923 community health centres (CHCs) in 1991. As such the country has got a primary health centre for about 38,000 population and a sub centre for 6,5000 population, showing considerable variation from the targeted figure in achieving the health for all (Table 14.5).

The increase in the number of doctors (16,480 in 1951 to 3,65,000 in 1990), hospitals (2,694 in 1951 to 10,157 in 1989), hospital-beds ratio (1.13 lakhs in 1951 to 6.19 lakhs in 1991), dispensaries (6,515 in 1951 to 28,565 in 1989), nurses (16,550 in 1951 to 2.65 lakhs in 1989), midwives (19,281 in 1951 to 1.81 lakh in 1987) and health visitors (578 in 1951 to 14,273 in 1987) indicates the improvement in the health services of the country. The present bed-population ratio (1 bed/1,000 population in 1991 which was 1 bed for 4,166 population in 1951) and doctor-population (1 doctor/2,318 population in 1991 which was 1 doctor for 17,000 population in 1951) ratio also show noticable improvement. The resultant improvement becomes obvious after having a look on the death rate, birth rate, infant mortality and life expectancy.

The life expectancy at birth has increased from 32 years in 1941-51 to 58.2 years in 1991-92. The infant mortality rate has come down from 146/1,000 in 1951 to 79 in 1992. Similarly, the maternal mortality has restricted to only 47/,1000 live birth. An increase in absolute number of these indicators clearly reveals improvement achieved in this sector. A significant declining trend in birth and death rates also confirms progress obtained in the field of health care. Birth rate has declined from 49.2 in 1901-11 to 29.9 in 1991-92. Similar reduction has also been witnessed in death rate from 42.6 in 1901-11 to 9.6 in 1991-92. But the rural areas revealed little improvement as compared to their urban counterpart. For example, in urban areas infant mortality is around 58/1,000 live birth while in rural areas it is 98/1,000. A considerable gap also continues to persist in birth and death rates in urban areas and country sides. Besides, there are inter-state variations in life expectancy. In Kerala, life expectancy of 69.5 years at birth has reached high levels compared to Bihar, Madhya Pradesh, Orissa, Rajasthan and Uttar Pradesh where concerted efforts are needed for improving the quality of life. Kerala's infant mortality rate of 14/1,000 is less than a fifth of the all India estimate of 74 and less than one-eighth of the rates prevailing in Madhya Pradesh and Orissa.

Table 14.5: Existing and Required No. of PHCs and Sub-Centres in 1990

States	Existing		Required		Difference			
	P.H.C.	Sub-centre	P.H.C.	Sub-centre	P.H.C.		Sub-centre	
States								
Andhra Pradesh	1,283	7,894	2,217	13,300	934	–	5,406	–
Arunachal Pradesh	28	155	29	173	1	–	18	–
Assam	440	5,110	747	4,483	307	–	627	+
Bihar	2,001	14,799	2,879	17,275	878	–	2,476	–
Goa	20	145	39	234	19	–	89	–
Gujarat	706	6,433	1,377	8,262	671	–	1,829	–
Haryana	366	2,299	549	3,293	183	–	994	–
Himachal Pradesh	201	1,502	172	1,034	29	+	468	+
Jammu & Kashmir	266	1,460	257	1,544	9	+	84	+
Karnataka	1,133	7,793	1,499	8,995	366	–	1,202	–
Kerala	886	5,094	970	5,820	84	–	726	–
Madhya Pradesh	1,181	11,910	2,206	13,236	1,025	–	1,326	–
Maharashtra	1,646	9,248	2,631	15,787	985	–	6,539	–
Manipur	68	420	61	367	7	+	53	+
Meghalaya	71	315	59	355	12	+	40	–
Mizoram	35	220	23	138	12	++	82	+
Nagaland	33	201	40	242	7	–	41	–
Orissa	924	5,426	1,055	6,332	131	–	906	–
Punjab	2,036	2,853	676	4,056	1,360	+	1,203	+
Rajasthan	1,040	8,000	1,467	8,801	4,27	–	801	–
Sikkim	20	132	13	81	7	+	51	+
Tamilnadu	1,386	8,681	1,862	11,172	476	–	2,491	–
Tripura	49	472	92	551	43	–	79	–
Uttar Pradesh	3,103	21,653	4,637	27,822	1,534	–	6,169	–
West Bengal	1,544	7,873	2,269	13,625	725	–	5,742	–
Union Territories								
A & N Island	16	97	9	56	7	+	41	+
Chandigarh	0	12	21	128	21	–	116	–
D&N Haveli	5	34	4	28	1	+	6	+
Daman & Diu	4	30	3	20	1	+	10	+
Delhi	8	42	314	1,884	306	–	1,842	–
Lakhsdweep	7	14	2	10	5	+	4	+
Pondicherry	22	73	27	161	5	–	88	–
India	20,536	1,30,390	28,210	1,69,260	7,674	–	38,870	–

Although when a comparison is made with more developed countries, where annual population growth, birth rate, death rate, infant mortality and life expectancy are 0.61 per cent, 15/1,000, 9/1,000, 15/1,000 and 74 years respectively in 1985-90 (UNO, 1992), the progress made in the country does not appear comfortable. This means the achievement in terms of mere absolute number in health services is of no significance unless and until it curbs the menace of rapid population growth (2.3% per annum during 1981-92). Therefore, there is a need to seek ways of making appropriate use of the existing network of health facilities for the benefit of majority of the women and children, if the goal of health for all by the 2000 A.D. has to be achieved.

A study conducted by the Indian Council of Medical Research (ICMR) is 8 primary health centres of Uttar Pradesh, Madhya Pradesh, Rajasthan, Haryana, Maharashtra and Gujarat has shown that infrastructural facilities in terms of space, equipment and general drugs available in these centres were very poor (*The Hindustan Times*, 1994). The recording system and quality of maternal and child care services were also not satisfactory. The study highlighted that the incidence of birth interval of less than 18 months as a major risk factor of infant and maternal mortality. Similar higher incidence of risk factor was identified in women who were either in age group of less than 18 years or over 34 years or who had a birth interval of less than 18 months. The ICMR has emphasised the need of training of health functionaries and community education and argued that it is possible to improve the quality and coverage of the health care facilities within the existing norms of the primary health care delivery system using the comprehensive maternal and child health care package along with high risk approach strategy. The study also stressed the need to undertake an operation research to evolve the appropriate intervention to decentralise the responsibilities, transfer of technology, mobilisation of the local human resources and to develop a system specially for the referral of high risk mothers and new borns so that the health services provided by the primary care level could be sufficiently strengthened and reorganised.

Authors also agree with the opinion of Dr. Satnam Singh (1994) that compared to the United States (126 medical schools and 23 independent schools of public health) the situation in India is very discouraging. Today, we have only one institute i.e. the All India

Institute of Hygiene and Public Health at Calcutta. A chain of such post graduate institutions in public health should be established in the country which were badly realised at the outbreak of plague in Surat City (Gujarat state) in September 1994 and Malaria in Rajasthan state in October 1994. Besides, the typhoid fever, which is fifth common communicable disease, causes very high morbidity and mortality in the country. About 7 million mandays are lost every year in India because of the infection caused to man by typhoid fever.

Another point, which needs to be discussed in depth in appraisal of family welfare and utilisation of health care facilities, is attitude of people of different socioeconomic groups. Several researches have indicated that most of the facilities like education, health, housing or public distribution systems are mostly monopolised by higher socio-economic sections of population, who, therefore, are the largest beneficiaries of such provisions than those who belonged to the lower consumption fractiles or lower socioeconomic sections (Kundu, 1993; Prasad, 1979; Sivnandan, 1979; Pandit, 1979). The study conducted by Sinha and Rajeswari (1993) on "Social Structure and Utilisation of Health Care Facilities in Rural Haryana" is worth quoting in direction of attitude of recipient of health care facilities. The study reflects that in Bhiwani district, upper caste households are the least beneficiaries of health care facilities due to dominancy of Jats and the Brahmins whose traditional values have made them more resistant to change i.e. due to lack of proper health education. A higher level of health care among the service castes in the villages of Bhiwani and Kurukshetra districts indicate that they have been more responsive to change. Contrary to these in Kurukshetra district the households of upper caste are good recipient of health care facilities.

The Case Study of Varanasi District

A survey regarding an utilisation pattern of health care facilities in rural areas (Mangolpur, Raunakhurd, Tilampur and Katari villages) of Varanasi district by Singh and Kumra (1993) reveals that although a sizeable proportion of respondents are aware of the services like medical care (83.07%), maternal and child health care-MCH (71.28%), immunisation (78.97%), family welfare (71.79%), control of communicable diseases and laboratory facility (38.48%) rendered by PHC but they do not use it to satisfactory extent. For instance, out of the total respondents of 195, only 50.25 per cent respondents have utilised medical care facilities followed by immunisation (45.64%) of vaccines such as DPT, Polio, BCG, Measles, etc. MCH care (31.78%), family welfare (25.64%), control of communicable diseases (8.20%) and laboratory facility (5.64%) (Table 14.6). It has been observed during field investigation that the health workers and Auxiliary Nurses and Mid-wives (ANM) have neither effectively tried to diffuse information (only 18.98%) about health care facilities available for villagers nor helped in persuading villagers to avail of these facilities. In order to know reasons for utilising PHC services, 26.7 per cent respondents visited PHC owing to confidence in trained and specialised doctors, 25.30 per cent because of nearness, 14.36 per cent due to nearness as well as cheap treatment, while 14 per cent respondents have utilised PHC services on account of the availability of free medicine.

In order to strengthen health and hygiene through PHC facilities it is essential to know that satisfaction level of respondents. About 38 per cent respondents were happy with the treatment, 39.33 per cent were partially happy while 16.66 per cent were unhappy. The remaining 6 per cent did not clearly express their opinion (Table 14.7). In health care delivery system inter personal communication plays a significant role (Jain *et al.* 1992). It is obvious from this study that only 20.51 per cent respondents have confirmed regular visit of health workers (once in a month). However, 40 per cent respondents did not receive their services regularly. Besides, 39.49 per cent said that health workers never visited their families (Table 14.8).

Table 14.6: Utilisation Pattern of Primary Health Care Services

Health Care services	Knowledge of services to the respondents		Utilisation of services by the respondents	
	Number	Percentage	Number	Percentage
Medical Care	162	83.07	98	50.25
MCH Care	139	71.28	62	31.79
Immunisation	154	78.97	89	6.64
Family Welfare	140	71.79	50	25.64
Control of Communicable diseases	94	48.20	16	8.28
Laboratory Facility	77	39.48	11	5.64

Source: Survey conducted by Dr. J. Singh and Dr. V.K. Kumra, 1991.

Table 14.7: Reasons of Dissatisfaction with the Functioning of PHC

Reasons	Percentage of respondents
Non-availability of medicine	21.57
Poor attention	57.84
Demand for money	18.63
Ill maintenance	1.96

Source: Survey conducted by Dr. J. Singh and Dr. V.K. Kumra, 1991.

Table 14.8: Visits of Health Workers and Level of Happiness

Visits	Percentage of respondents	Happiness level	percentage of respondents
Regular	20.51	Happy	38.00
Irregular	40.00	Partially Happy	39.33
Never Visited	39.49	Unhappy	16.67
—	—	Indifferent	6.00

Source: Survey conducted by Dr. J. Singh and Dr. V.K. Kumra, 1991

Summary

It is clear from the foregoing discussions that due to the lack of sincerity on part of the government and irresponsibility of the officials concerned, the family planning measures have not been adopted considerably in the countryside. Not only this the poor and inadequate network of health care facilities also hampers the national interest. Trained medical officers and proper health education are pre-requisite for achieving the goal of family welfare as well as health care. In this regard, study conducted by ICMR (1988) is worth quotable. This indicates that only 15 per cent of the PHCs had the requisite number of health personal for the population they were catering to (*The Hindustan Times,* Feb 14, 1996). Another aspects which need to be emphasised are education in general and female education in particular, old age security, better socio-economic conditions, attractive incentives to couples adopting family planning measures, insurance scheme in case of deaths due to sterilisation, necessary disincentives for having more than two children, stringent action against quality persons performing the marriage of their daughter before the prescribed age limit and medical officers who are not visiting their centres regularly. The population control is most urgent problem of the country and it is at the root of almost all problems. This should be taken as political issue and be included in the election manifestoes of all political parties.

Acknowledgement

Authors are very grateful to Dr. K.K. Dube for his critical remarks and suggestions given in. Authors are also thankful to Dr. V.K. Kumra and Dr. J. Singh for providing some relevant data of their own survey conducted in 4 villages of Varanasi district.

References

Jain, R. *et al.* (1992) : Interpersonal channels—A media of communication between PHC and its beneficiaries, *Indian Journal of Preventive and Social Medicine,* Vol. 23 (2), p. 23.

Khan, M.E., (1992) : Population dynamics and the family welfare programmes in India: An overview, *The Journal of Family Welfare,* Vol. 38(3), p. 64.

Kundu, A., (1993) : *In the name of urban poor*, Sage, Delhi.

Prasad, P.H., (1979) : Caste and class in Bihar, *Economic and Political Weekly,* Vol. XIV (7&8), pp. 481-4.

Pandit, N., (1979) : Caste and class in Maharashtra, *Economic and Political Weekly,* Vol. XIV (7&8), pp. 425-36.

Rele, J.R., (1974) : Trends in Fertility and Family Planning, in Bose, A. *et al.* (eds.): *Population in India's Development,* 1947-2000, Delhi, pp.343-51.

Sahay, B.K. (1993) : Family Planning: Only Hard Option Now, *The Hindustan Times,* 1-7-1993.

Singh, S., (1994) : More doctors, less health, *The Hindustan Times,* 23-10-1994.

Sinha, S., and Rajeshwari, 1993: Social structure and utilisation of health care facilities in rural Haryana, *Annals,* NAGI, Vol. XXX, No.1, pp. 37-47.

Sivanandan, P., (1979) : Caste, class and economic opportunity in Kerala: An empirical analysis, *Economic and Political Weekly,* Vol. XIV (7&8), pp. 425-86.

Srivastava, S.K. (1989) : *Family Planning and Population Problem in India,* Common Health Publishers, New Delhi, p. 121.

Statistical Abstract of India, (1990) : Central Statistical Organisation, Department of Statistics, Ministry of Planning, Government of India. *The Hindustan Times,* 10-9-1980, 25-9-1980 and 26-7-94.

15

Profiles of Population at Micro-Levels of Murshidabad District

N. D. BHATTACHARYA

Murshidabad—an arena of mighty political events during the 18th century, is one of the most industrially backward districts in West Bengal. The district is situated on the northernmost apex of the *Ganga delta* which comes under the grip of several natural hazards and other calamities almost every year (Bhattacharya, 1997). The Bhagirathi river divides the district into almost two equal micro-regions i.e., the western portion is known as the *rarh* and eastern as the *bagri*. The above two facets have further been delineated into seven mini-micro regions (Bhattacharya, 1961). This multi-problemed district, under the second category of the levels of development, embraces an area of 5,324 sq km with a population of 4.7 millions in 1,918 villages and 18 urban centres that are spread over the 26 community development blocks. The decennial growth-rate is 28.20 per cent (1981-91). Of the total population of the district, 13.40 per cent are scheduled castes whereas, 1.30 per cent are tribal, community.

Murshidabad had once a pride of place in the international markets in respect of her excellent silk products, ivory works, bell-metals and *atr* (perfume manufacturing). But the decline of *Nawabi-rule* in Bengal, synchronised with the deteriorating of such flourishing industries, mainly due to lack of royal patronage, are responsible for the decay. Unfortunately, no proper steps have been taken even after the 50 years of our independence to revive these dying industries which can easily earn huge foreign exchange and provide employment opportunities to the local youths.

Murshidabad is a major agricultural district in West Bengal where 62.73 per cent of total workers are actively engaged in agrarian activities. It may also be noted that out of 1.4 million of total main workers 69.76 per cent of male workers alone are closely associated with cultivation and allied activities whereas, the female participation rate is less significant (14.26%).

For correct understanding of social morphology (Kayastha, 1956), an attempt has been made to study a few facets of population of the district and we shall discuss the decennial growth, spatial distribution, density, working-force, influx of displaced persons etc. in brief.

Early Estimates of Population

The demographic information are scanty for the ancient and medieval periods of the district. But the ancient sites, accounts of the foreign travellers, old plans etc. are indicative that Murshidabad had attracted a large number of people since the days of eastward march of the Aryans. We have testimony that ancient capital Karnasauvarna, Panchastupa, Mahipalnagar-Gaysabad, Kiriteswari, Kusumkhola, Saligram, Kankagram, etc. were highly populous. At the dawn of the 18th century, Murshid Quli Khan, selected his capital on the banks of the Bhagirathi, on account of central position, natural barriers and healthy climate and Murshidabad welcomed an enormous influx of various castes, creeds and professional groups. Even after the battle of Plassey (June, 1757), Colonel Clive, the mighty man of East India Company, amagingly observed that "the City of Murshidabad was as extensive, populous and rich city as the City of London". For the first time, W. Adam counted 62,037 persons in the *Daulatabad Thana* in connection with his report on vernacular education in 1837. The Statistical and Geographical Report (Gastrell, 1860), recorded 1,100,080 persons, of which 704,717 were Hindus. The evidence of systematic enumeration of population, to some extent, was taken in 1872 which revealed that the district had 1,214,104 persons. Prior to this, some efforts were made to count population that were "entirely unreliable, judging by modern standards" (Bagchi, 1944).

Decennial Change Since 1872

The growth of population of Murshidabad district has not been steady due mainly to sporadic factors like natural hazards, famines, influenza epidemi, etc. Within a span of 120 years, the population significantly increased, between 1872 and 1991, from 1.2 millions to 4,2 millions (Table 15.1).

Table 15.1: Growth of Population.

Year	Area in Sq.km.	Population	Density in sq km	Percentage variation
1872	6,677	1,214,104	182	–
1881	5,553	1,226,790	221	+ 1.0
1891	5,553	1,250,946	225	+ 2.0
1901	5,550	1,333,184	238	+ 6.6
1911	5,550	1,372,274	242	+ 2.9
1921	5,493	1,224,181	223	– 9.0
1931	5,416	1,370,677	253	+ 12.0
1941	5,343	1,640,530	307	+ 19.7
1951	5,423	1,715,759	316	+ 4.6
1961	5,325	2,290,010	430	+ 33.5
1971	5,341	2,940,204	553	+ 28.3
1981	5,424	3,697,552	695	+ 28.2
1991	5,324	4,740,149	890	+ 28.2

The growth was remarkably slow between 1872 and 1911, due to spread of a verulent type of malaria, notoriously known as *Burdwan Fever* (1881) and other epidemics of small pox, cholera etc. that took heavy toll of lives (Mukherji, 1938). The ravages flood of 1904 submerged a vast area which spread an epidemic of cholera in 1905 and caused over 8,000 deaths (O'Malley, 1914). According to the Bengal Drainage Committee (1906-07), the growth was far from uniform in the malarious lowlying areas of *bagri* than least malarious tracts of the *rarh*. The decade (1901-11) witnessed the spart of immigrants in the Sagardighi and Nabagram areas from the adjoining districts of Birbhum and Santal Parganas. A demand for considerable labourforce for the construction of three railway lines was significant percentage variation (6.6) during the decade (1901-11). The population sharply declined by (-9.0%) in 1921 due to floods (1918), influenza and other viral diseases, bad harvests and first world war (1914-18).

Abrupt rise (19.7%) was noticed between the decades 1931 and 1941. The movement of large number of panic-striken, people from Calcutta to Baharampur during the Second World War (1939-45) took place. The experts opine that "War-condition hampered the proper enumeration and the people of undivided Bengal intentionally furnished wrong figures" because of communal passion (Spate, 1954). The accompanying map shows (Fig. 15.1) influx of "displaced persons" from the erstwhile East Pakistan during the period 1947-51. The growth of teeming population was noticed since 1961 onwards. The percentage of decadal variation of total population (1981-91) was +28.20 whereas, the growth rate of urban population was much higher (+42.87) than the rural C.D. pockets (+26.68). Such spectacular growth is very indicative of infilteration of the people (*The Telegraph*, 1999) from the adjoining Islamic state of Bangladesh, apart from the natural growth and migration from other states of India.

Spatial Distribution of Population

The human wealth, as controlled by geographical phenomena, is not uniformly distributed all over the district of Murshidabad. The distributional pattern within the framework of the two micro-regions, The *rarh* and the *bagri*, shows a striking contrast to each other (Fig. 15.2). There are certain areas that are swarming with population whereas, some areas restrict human habitats. The physiographic variations have their profound influence on the present pattern of distribution. There are two distinct types of clusterings—consolidated and linear. The former is predominant in the *rarh* and in the *Kalantar* areas of the *bagri* whereas the latter is conspicuously seen along the natural levee and flood plains of the Padma, Bhagirathi, Bhairab, Sealmari and Jalangi rivers. The main concentration of rural population are to be found "along the swing of the streams" (Bhattacharya, 1978), both old and new banks, particularly in the *bagri* and around the irrigational tanks of the *rarh*. The shifting of river courses and sandy soils are the determining factors in local variation. Apart from the physiographic considerations, another cause in the variation in population distribution is also "related to the process of settlements". In this district, the human beings have found themselves with "ample opportunity in the course of centuries to migrate from less favourable

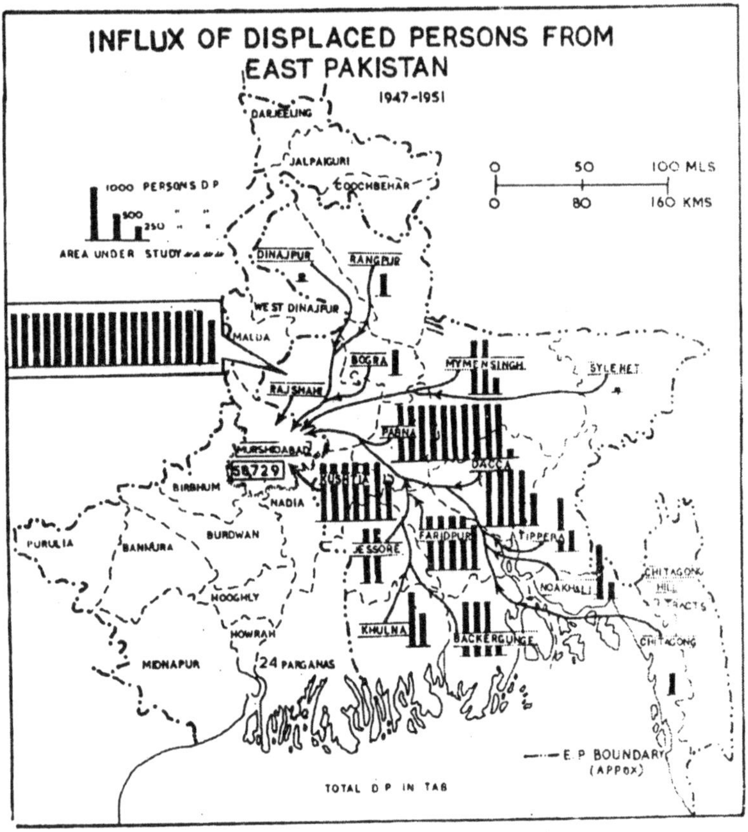

Fig. 15.1: Influx of Displaced Persons from East Pakistan

to more favourable spots and multiply" (Chatterjee, 1962). A thick mantle of alluvium along the Padma, stretching from Farakka to Jalangi, is a favourable factor for the largest concentration of population in the area. The dendretic pattern of settlements due to drainage and transport lines are noticed. The cottage industries, river-borne trade, lac cultivation etc., are some of the important factors for such concentration. The Bhagirathi basin, starting from the Suti entrance to the Nadia district border, "attracted the people since ancient times, due to availability of fresh drinking water, easy communication for trade and commerce and irrigation facilities. The sacred water of the Bhagirathi river was convincing religious factor for attraction of

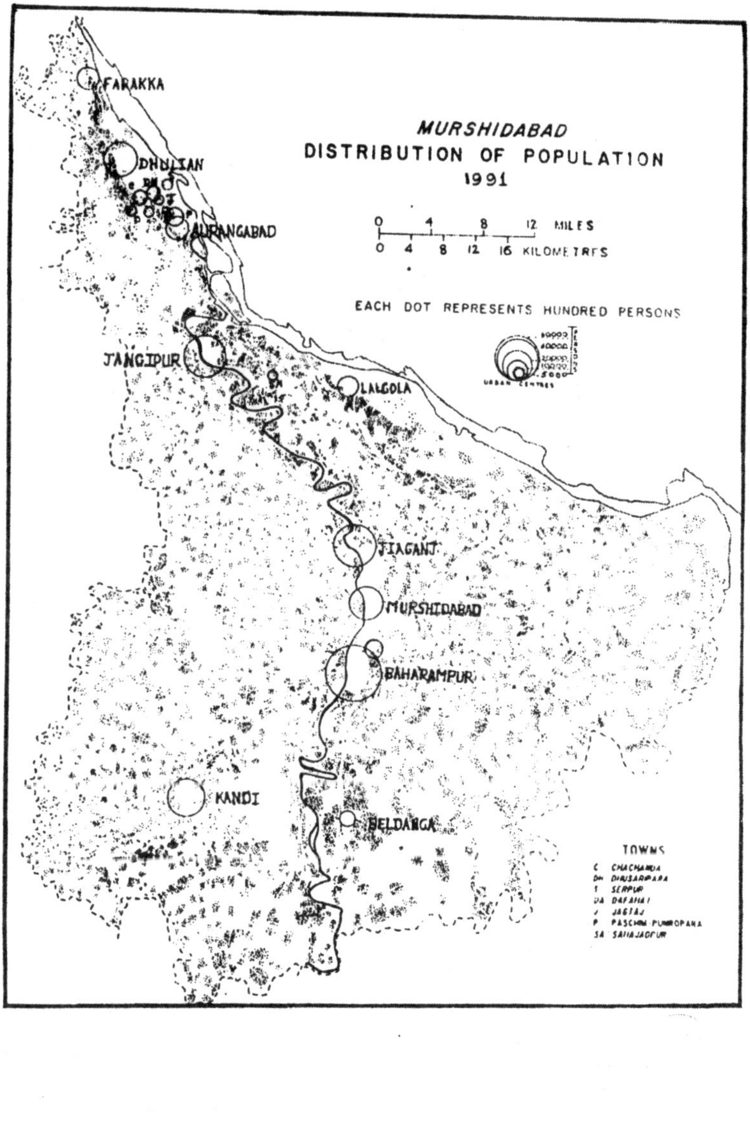

Fig. 15.2: Murshidabad : Distribution of Population 1991

the people to settle on the banks of the river" (Bhattacharya, 1972). It has been observed that the concentration is the highest where the land is above 15m along the both banks of the river. The current census (1991) reveals that 4.2 million of rural population are unevenly distributed over the district. The high concentration is noticed in some parts of Khargram, Kandi and Bharatpur C.D. blocks and both flanks of the *Badshahi Sarak* (old royal highway), connecting Burdwan-Morgram road. Such concentration is mainly due to tank irrigation and fertile agricultural lands. The serpentine river banks of the Bhairab, Choto Bhairab, Gobranala, Sealmari and the Jalangi—all in the *bagri* where considerable human concentrations are noticed on the concave banks. The fertility of new alluvia, close proximity to the rivers are helpful for agricultural activities. These areas are home of *aus* paddy, jute, sugar cane, and mustard cultivation.

The areas of moderate concentration comprise of Burwan, Raghunathganj, Suti, Nawda, Hariharpara, Nabagram and Sagardighi C.D. blocks. Areas of undulating nature show sparsed distribution of population. The high grounds are mostly occupied by human habitats and the lower grounds are devoted to mono-crop i.e. *aman* paddy cultivation. Some water-logged areas, strewn all over the district, are almost uninhabited. The *Hizal, Patan, Belunbils* and marshy-areas of the *Kalantar* restrict human agglomerations. Due to shrinkage in size of these water bodies, a large number of people have encroached the lands. Certain areas that are usually flooded during the month of July to November, are practically devoid of human habitats. It may be noted that recent encroachment by the Padma river from Farakka to Jalangi C.D. areas, already created havoc in destroying populous settlements (Bhattacharya, 1978).

Urban Corridors

According to 1991 Census, there are eighteen urban centres, of which seven originated with municipal status prior to independence, representing three distinct urban corridors where 4.9 lakhs people reside in the district of Murshidabad. All these urban centres, including two urban agglomerations—Baharampur and Aurangabad, are mainly concentrated either on the levees of the rivers or along the transport lines. The first corridor with eight urban centres, stretching from the planned Farakka Barrage township (21,485) with super thermal power station to Aurangabad (25,861) — the largest *bidi* manufacturing centre in West Bengal, situated to a very close proximity of the Padma river. This cluster includes Dhulian (33,191)—an old entrepot and a 'receding town' due to heavy encroachment of river since 1901; and other group of new urban centres such as Chachanda (7,358), Dhusaripara (9,703), Serpur (5,576), Dafahat (7,828), Jaitaj (7,194) and Paschim-Punropara (23,359) are located within this zone which support 1.3 lakhs of people. The second corridor consists of three urban areas with a population of 94 thousands, such as, Jangipur-Raghunathganj (55,981)—a twin-town and sub-divisional headquarters; Sahajadpur (13,610) an important mart; and Lalgola (24,469)—terminus of Eastern Railways and one of the important cattle-markets in the district. Most important urbancorridor with a group of five urban settlements of historic significance are Murshidabad (30,327)—old metropolis of Bengal and tourist destination; Jiaganj-Azimganj (42,104) another 'twin-town' and trade centre; Baharampur (1,15,144) — the district headquarters and a large collecting and distributing centre; Kashimbazar (8,753)—mediaeval commercial centre; and Beldanga (20,249)—the largest cattle market in West Bengal—all along the right bank of the Bhagirathi. Truly, Murshidabad is economically sandwitched between two rapidly growing urban centres like Baharampur and Jiaganj. Conspicuously, this corridor alone support 2.2 lakhs, almost half of the total urban population. Amidst the vast agricultural surroundings, Kandi-on-Mayurakshi, acts alone as a focus of socio-econonic and cultural activities in the *rarh* region. In this connection it may be noted that most of the small towns are "urban" in nature and the author carried out intensive research on Lalgola, Beldanga and Aurangabad towns (Bhattacharya, 1972) for implementation in the Census 1971 programme. Table 15.2 shows the year of origin, progress of population with civic status, growth-rate, scheduled caste and tribe population, sex-ratio and literates, in each urban centre in the district.

Inter-censal changes in Densities

The indepth study on the inter-censal changes of densities at micro-level (C.D. Block) was made by the author earlier for the pre-independence (1901-1941) and post independence periods since 1951 for the correct understanding of demographic situation of the area concerned (Bhattacharya, 1972). A perceptible variation in densities from 246 sq km to 890 sq km recorded in 1901 and 1991 respectively. As already indicated earlier that the local relief, soils, climate, natural calamities, trade and commerce, transport net works, commissioning of barrage and super thermal plant at Farakka and other socio-political factors account for the variation in density pattern in the rural and urban areas. The total density of population, as indicated above, recorded more than the state average (767 sq km) and rural density being 813 sq km whereas the urban density was 462 sq km. An appreciable changes in density pattern is noted along the revarian tracts, stretching from Farakka to Jalangi, where Muslim concentration is the highest and an enormous infiltrations from Bangladesh international border, are very common, unabated and affecting the economy of the region. Davis (1951) rightly observed that— "Muslims do multiply faster than the Hindus". A dozen of C.D. blocks, out of 26, have cent per cent of rural population. The highest density areas are Shamsherganj (1,525 persons/sq.km), Beldanga (1,452), Lalgola (1,452), Raghunathganj (1,385), Berhampur (1,452), Farakka (1,074) and Suti (985) which recorded much higher density than the district average (813). The lowest density areas are Nabagram, Khargram, Sagardighi and Kandi — all are in the remote *rarh* region. According to Mitra (1951) the density with "500 persons is the limit in the rural areas that can support from agriculture."

The urban density of the district is recorded 462/sq km during the last census. Significantly, all new small urban settlements that originated between 1961 and 1991 as non-municipal, have a very high density. All municipalities, such as Jangipur (7,205), Baharampur (7,057), Dhulian (4,272), Lalgola (4,133), Jiaganj Azimganj (3,611) and Beldanga (3,149) recorded considerable decadal changes in density patterns but the decaying old capital city of Murshidabad shows the lowest growth of density during 1991.

Working Force

Prior to discuss the working force in the rural-urban sectors of the district, it is desirable to note the concept of workers, in accordance with the recommendation of I.L.O. that has been dividing into 'main workers' and 'marginal workers' since 1981 for the propose of comparability of data (Nanda, 1991). Main workers are those who work more than six months or 183 days in a year and the marginal workers are those who work less than the said period. The total working forces are however, further divided into nine broad industrial classifications since 1951 (Mitra, 1951).

Physical environment that has a sovereign effect on the occupation of human beings and their material conditions (Kayastha, 1964). It has profoundly influenced the economic activities, social institutions and cultural patterns of any area. The material conditions of the people of *rarh* and *bagri* are wholly depended on cultivation of *Aman, Aus* paddy, jute and sugarcane which are closely connected with the physical as well as the climatic conditions of above two micro-regions. It has rightly been observed that "under agricultural operations, the rate of growth is slower than the spur of industrial and commercial activities" (Singh, 1956). In Murshidabad, agriculture is by far most important occupation that employ about 62.73 per cent (4.5 million) of total rural workers. This percentage includes all *munish* or marginal workers who serve the well-to-do *bhagchasi* (cultivators) on daily or weekly wages basis. Although the lands are chief source of occupation, nearly one-fifth of the people derive their livelihood from household industries and other services. It has been observed that the young generations, belonging to agricultural community with marginal modern education, are gradually giving up their ancestral professions and becoming service minded. They dislike to plough the fields under the scorching sun and torrential rains, but prefer small services either in the offices or in the commercial establishments. Some younger people even prefer to work as domestic servants, rickshaw peddlers, tonga-drivers, and shopkeepers. Insignificant percentage of main working people are employed in the household industries. This clearly points out the lopsided nature of economy of the area concerned. Cultivation now admittedly the remunerative occupation but it is also true that "agriculture alone cannot pull out of the mire of the poverty". (Dewett, 1952). The modern economic progress has been marked by a diminution in the numbers engaged in agriculture. Of all the C.D. blocks, Shamsherganj, Farakka and Suti (II) record the high percentage of main workers to total population (40%). These areas have certain trade centres, educational institutions, government offices, medical

Table 15.2: Urban Population Characteristics in the District of Murshidabad

Urban centres (with status/ward)	Year of origin	Area in sq km	Population with size 1961	Population with size 1991	Density 1991 sq km	Growth rate 1951-1961	Growth rate 1961-1971	Growth rate 1981-1991	Sex ratio 1997	S.C. Pop. 1991	S.T. Pop. 1991
1	2	3	4	5	6	7	8	9	10	11	12
1. Aurangabad (NM)	1961	2.77	12,783 (IV)	25,861 (III)	9,336	+36.60	+8.73	+31.27	980	1,907	—
2. Baharampur (M)-30	1876	16.67	63,317 (II)	1,15,144 (I)	7,057	+12.01	+17.75	+23.97	958	13,850	674
3. Beldanga (M)-17	1961	3.98	7.841 (V)	20,249 (III)	3,149	+24.70	+5.70	28.34	952	1,438	—
4. Chachanda (NM)	1991	1.15	—	7,358 (V)	6,398	—	—	—	992	452	—
5. Dafahat (NM)	1991	2.72	—	7,828 (V)	2,878	—	—	—	965	799	—
6. Dhulian (M)-14	1909	7.77	17,220 (IV)	33,191 (III)	4,272	+8.10	NA	+38.33	992	4,162	4
7. Dhusaripara (NM)	1991	3.31	—	9,703 (V)	2,811	—	—	—	1,011	1,002	—

(Contd.)

Table 15.2: *(Contd.)*

Urban centres (with status/ward)	of origin	Year in sq km	A-ea with size 1961	Population 1991	Density 1951-1991 sq km	Growth rate 1961-61	1981-71	Sex ratio 1997	S.C. Pop. 1991	S.T. Pop. 1991	Remarks
1	2	3	4	5	6	7	8	9	10	11	12
8. Farakka Barrage Township (NM)	1971	3.70	—	21,485 (III)	5,807	—	—	+46.66	863	7,057	230
9. Jagtaj (NM)	1991	0.70	—	7,194 (V)	10,132	—		—	989	3,956	—
10. Jangirpur (M)-15	1869	7.77	24,201 (III)	55,981	7,205	+32.57	+27.44	+27.83	954	7,587	13
11. Jiaganj-Azimganj (M)-14	1896	11.66	23,673 (III)	42,104 (III)	3,611	+23.64	+14.93	+28.66	953	9,852	184
12. Kandi (M)-13	1869	12.95	19,780 (IV)	39,652 (III)	3,062	+29.96	+32.80	+21.70	932	8,117	84
13. Kashimbazar (NM)	1971	2.78	—	8,758 (V)	3,149			+18.04	965	2,363	206
14 Lalgola (M)	1961	5.85	10,657 (IV)	24,469 (III)	4,183	11.00		38.03	955	5,036	—

(Contd.)

Table 15.2: *(Contd.)*

Urban centres (with status/ward)	of origin	Year in sq km	Area with size	Population	Growth rate			Sex	S.C. ratio	S.T. Pop.	Pop.
		1961	1991	Density	1951-1991 sq km	1961-61	1981-71	1991 91	91	91	
1	2	3	4	5	6	7	8	9	10	11	12
15. Murshidabad (M) -12	1869	12.95	16,990 (IV)	30,327 (III)	2,342		(-) 19.00	+ 48.1	9,895	314	—
16. Paschim Punropara (NM)	1981	7.72	—	23,359 (III)	3,026	—	—	+ 32.90	974	1,466	—
17. Sahajadpur (NM)	1991	1.57	—	13,610 (IV)	8,669	—	—	—	994	1,896	—
18. Serpur (NM)	1991	0.63	—	5,576 (V)	8,851	—	—	—	1,010	1,061	—
Dist. (Urban)	-	106.66	1,964.64	4,94,347	462.00	—	—	48.87	957	82.755	1,710

Note: (a) Beldanga Originated in 1921-1941 Declassified in 1951 and revived in 1961 as NM and became Municipality in 1981.

(b) Srimantapur declassified in 1991

M = Municipality: NM = Non-Municipal Area; Roman Fig. Size of Pop.

Source: 1. (DCH-1991, Murshidabad, Part XII - A, 200 p.

2. DCH-1991, Murshidabad, Part XII B, Table IB.

3. All-India Congress Committee, Journal p. 32.

centres, thermal plant which offer employment opportunity to the people. Female working populations in the rural areas are less significant (Table 15.3). A glance at the Tables 15.3 & 15.4 reveal the decadal changes (1981-91) of working force by sex in the rural-urban sectors of the district.

Table 15.3: Rural/Urban and Male/Female distribution of Population

Sectors	Total population (1991)	Percentage of total population					
		Total workers		Main workers		Marginal workers	
		1981	1991	1981	1991	1981	1991
Rural:							
Total	42,45,802	28.87	31.50	27.11	29.79	1.76	1.71
Males	21,86,950	50.64	51.78	49.30	51.14	1.34	0.69
Females	20,58,852	6.14	9.98	3.95	7.14	2.19	2.34
Urban:							
Total	4,94,347	27.65	32.48	26.95	31.17	0.70	1.47
Males	2,52,392	44.77	47.63	44.08	47.16	0.69	0.47
Females	2,41,955	9.98	16.67	9.25	14.49	0.73	2.18
District:							
Total	47,40,149	28.75	31.60	27.09	29.98	1.66	1.67
Males	24,39,342	50.09	51.35	48.81	50.73	1.28	0.62
Females	23,00,807	6.50	10.69	4.45	7.92	2.05	2.77

Sources: Census, Paper 3 of 1991, p. 423 & District Census Handbook, Part XIIB, P. 2

Out of 4.7 million of people, only 31.60 per cent were counted as workers, of which 29.98 per cent were under the category of main working force and the rest were marginal workers. In respect of percentage variation to total population a very slight change is noticed in the rural-urban pockets during the decade (1981-91). The work participation rate is much higher in the rural than urban areas. This is true of both males and females, the gap being considerably more in case of female working force.

Table 15.4 Rural/Urban and Male/Female Distribution of Workers

Sectors	Main workers	Percentage of main workers							
		Cultivators		Ag. labourers		Household industries		Other workers	
		1981	1991	1981	1991	1981	1991	1981	1991
Rural									
Total	12,70,502	38.56	35.02	36.26	32.04	5.51	12.65	19. 67	20.29
Males	11,24,527	41.16	39.09	37.71	34.61	3.52	5.18	17.61	21.12
Females	1,45,975	4.76	4.15	17.37	12.54	31.49	69.30	46.38	14.01
Urban :									
Total	1,53,987	4.23	3.78	6.72	5.96	8.14	28.39	80.91	61.87
Males	1,19,180	4.96	4.76	7.79	7.26	6.23	16.23	81.02	71.75
Females	34,807	0.62	0.45	1.45	1.52	17.03	69.70	80.40	28.23
District:									
Total	14,24,489	35.37	31.64	33.51	29.21	5.76	14.35	25.36	24.80
Males	12,43,707	28.11	35.79	35.19	31.98	3.75	6.25	22.95	25.98
Females	180,782	3.95	3.44	14.26	10.42	28.76	69.38	53.03	16.76

Source :District Census Handbook, Murshidabad, Part XIIB, p. 3 and Census, Paper 3 of 1991, p. 533.

Table 15.4 presents the work participation of males and females of the district with rural-urban break up as at 1981-91 and also indicates four broad categories of main working force, such as, cultivators, agricultural labourers, household industrial and other workers. The decadal trends in the different sectors of works seem not to be very encouraging. The perceptible change is observed in the household industry (69%) and other services, particularly for the women falks (69%) of the rural urban tracts. This is possibly due to welfare programmes for the females taken up by the government. A very perceptible decline in the work participation, specially in the service sector, both rural-urban units, during 1981-91, are interesting characteristics of working force of the district of Murshidabad.

Influx of Deplaced Person

In the demographic history of Murshidabad, the influx of deplaced persons from Bangladesh, erstwhile East Pakistan, is a striking phenomenon. A large number of permanent settlers, mostly agriculturists and artisians, being afraid of *Noakhali massacre* left their *bastu-vitas* (ancestral homes), and rushed to India for security and shelters. Between the years 1947 and 1955, Murshidabad district become the venue of displaced persons who flocked from different districts of Bangladesh. The maximum numbers of families, mostly agriculturists, came from Rajshahi (more than 18,000 persons), Pabna (10,000), Kushtia (6,000) and other neighbouring districts. (Fig. 15.1) It may be noted that these displaced persons came out of their country, batch by batch, since 1947. Seven transit camps were opened by the Government of West Bengal in the different parts of the districts *viz.*, Kurmitola (Murshidabad town), Begdohor (Jiaganj town), Nupur (Suti P.S.), Nimtita (Shamsherganj P.S.), Khidirpore (Hariharpara P.S.), Sadikhanr-Diar (Jalangi P.S.) and Goraimari (Domkal P.S.) to provide temporary shelters to the displaced persons. The Government of West Bengal had also encouraged these displaced persons (DP) to resettle at the Dandakaranya region. Within one year from 1st March, 1960 to Feb. 1961 about 20,000 DP sought refuge in the district (Bhattacharya, 1979). It may be noted that hundreds of DPs have already settled within the various parts of the district.

Another wave of human migration came in just after the 25th March, 1971 when the military junta of West Pakistan deliberately conducted an untold genocide in East Pakistan. Thousands of terrified and innocent people fled from their motherland in order to save lives and honour of women and sought security and shelters from the neighbouring people of India. As a result, the impact on Indian economy was tremendous and the socio-political tensions were at the top. Within the district, a number of temporary camps were opened and the displaced persons were provided with all facilities that were readily available. Out of the mire of the military whimers, Bangladesh—a new state, came into existence. After the recognition of Bangladesh on 6th December, 1971 by the Government of India, some of the displaced persons safely returned to their respective *bastu-vitas*. The impact of partition is not so evident now, but the unauthorised infiltration is unabated. Such growing trans-border migration is the serious concern for the state authorities (*The Telegraph*: 1999)*. This new demographic phenomena is telling upon the economy of the area under study (Mazumdar, 1998).

---

*A displaced person as defined by the Registrar General, India (1955) is as follows "A displaced person means any person who has entered India having left or being compelled to leave his or her home in Western Pakistan on or after the 1st March, 1947 or his or her home in East Pakistan on or after the 15th October, 1946 on account of civil disturbances or the fear of such disturbances or on account of the two dominions of India and Pakistan."

References

Bagchi, K.G., (1944), *The Gange Delta*, Culcutta University.

Bhattacharjee, B.K., (1979), *Calcutta*. West Bengal Dist. Gaz. p. 93.

Bhattacharya, N.D. (1959), Murshidabad—A study in urban geography, *Nat. Geog. Journal of Ind.* Vol. V, Part-I, pp. 38-51.

— (1961), A study in settlement geography of murshidabad district, *Dist. Census Handbook*, pp. 137-75

— (1966), Rural settlements of Murshidabad, *Nat Geog. Journal of India*, Vol. XII, Part IV, pp. 239-51

— (1972). A few facets of population of Murshidabad, *Women on the March, AICC Journal*, July, p. 32.

— (1972) Glimpses of urban settlements, *Women on the March, AICC Journal* Feb. pp. 38-41.

— (1978), Changing Courses of the Padma and human settlements, *Nat. Geog, Journal of India*, Vol. XXI, Parts 1 & 2, March-June, pp. 67-76.

— (1997), "Environmental hazards—A micro-regional study" in P. Nag *et al.* (eds) *Geography and Environment*, Local Issues, Vol. III, pp. 269-84.

Chakraborty, B. (1999), Borderline Case.Study— A Review article by *The Telegraph* dt. 19th Feb. *see* also. "Master Plan for Displaced Persons in West Bengal", in West Bengal, Vol. XVIII, No. 10. Dec. 1973.

Chatterjee, S.P. (1943). *Geography and Population Indian Culture*, Vol. X., No. 4., p. 59.

— (1962), Regional pattern of density and distribution of population in India, *Geog. Rev. of India*, Vol. XXIV, No. 2, June, pp. 1-38.

Davis, Kingsley (1951), *The Population of India and Pakistan*, London, pp. 43-69.

Dewatt, K.K. & Singh, G. (1952), *Indian Economics*, Delhi, p. 28.

Gastrell, J.E. (1860), *The Statistical and Geographical Reports of the Murshidabad*, Calcutta, p. 5.

Gupta, A. (1979), *'Floods-1978'*, District Magistrates' Report on Murshidabad, 9th Nov.-1997.

Kayastha, S.L. (1956). Demographic features of Himalayan Beas basin, *Nat. Geog. Journal of India*, Vol. II, Part I, p. 14.

— (1964), *Himalayan Beas Basin—A Study in Habitat, Economy and Society*, BHU, Varanasi. p. 21.

Mazumdar, A. (1998), Article on infilteration in W. Bengal (Bengali), *The Bartaman*, Dt. 22nd Dec. 1998. p. 4.

Mitra, Asok, (1953), *District Census Handbook*, 1951, Murshidabad.

Mukherji, B.B. (1938), Final Report on the Survey of Settlements Operations in the District of Murshidabad (1923-24), Calcutta.

Nanda, A. (1991). *Census of India*, Paper 3 of 1991—Workers and Their Distribution, New Delhi.

O'Malley, L.S.S. (1914), *Bengal District Gazetteers*, Murshidabad, Reprint in May 1997, Ed by K.K. Biswas, Calcutta, pp. 80-95.

Singh R.L. (1955), *Banaras—A study in Urban Geography*, Nand Kishor & Sons, Varanasi, p. 65.

—(1956), Mirzapur—A study in urban geog. *Geographical Outlook*, Vol. I, No. 1. Jan. pp. 16-27.

Spate, O.H.K., *India and Pakistan*, Methuen & Co. London, p. 97.

16

Dynamics of Urban Phenomena in Indian Cities

KAMALA KANT DUBE

Indian cities during the last four decades, have exhibited vertigeneous urban population growth and resultant expansion in their size. The country had only 74 class one cities of over one lakh population in 1931, which rose to 223 in 1991. Till 1951, there was no city of 50 lakh population, whereas four cities of over 50 lakh population emerged in 1981 census. There was not a single urban centre in India accommodating one crore population till 1981, but by 1991 three cities attained this population. In 1901,z the urban population of our country constituted 11 per cent of its total inhabitants, which rose to 17 per cent in 1951 and 25.7 per cent in 1991, According to the National Institute of Urban Affairs, in 1981, India ranked fourth in the world with respect to urban population. Then, the percentage as well as the total number of urban population in China and U.S.A. was greater than that of India. But in 1991, India grew to second rank in this regard. It is estimated that by 2001 it's urban population will exceed 44 crores and it will attain the first rank (Bose, 1978). Such a sporadic increase, as stated by demographers and economists, is a sign of unhealthy development which may retard the economic progress of the nation.

The gain in population of Indian cities, to a greater extent, is due to immigration of unemployed youth from rural areas in search of better livelihood. Larger the city, greater is the population of outborn persons in it. On the contrary, among small towns migration did not contribute much to their population growth. Another striking feature is that there has been predominantly male irunigration in our cities. As such, the sex-ratio in urban areas has become quite adverse, ranging between 553 to 893 females per thousand males (RGI, 1991). This phenomenon, however, has unconsciously emerged without realising seriousness of the problem and exercising any substantial effort to solve it. There appears to be a keen competition among large urban centres to attain the cadre of metro-city and among metro-cities to become a mega-city. Such activity has not only given way to spontaneous, haphazard and unwanted sprawl of cities beyond their corporate limits in the countryside but also has partially paralysed the urban way of life causing considerable deterioration in the living conditions especially in their central parts. Lack of serious regulation and control has posed variegated problems like dearth of residential accommodation, growth of slums and blighted areas, traffic congestion, overcrowdedness, intermingling and inappropriate urban land uses, inadequate public facilities, pollution etc. Also consequent upon persistent and ever-increasing population pressure in our metro-cities there are multiplying ill-effects and evils like serious urban diseases, anti-social activities and accidents.

This research work apart from focussing upon some evolutionary aspects and morphological characteristics of Indian cities, aims at highlighting the changing urban phenomena which have brought about manifold urban and civic problems resulting in deterioration in the age-old pleasing scenareo of our urban centres. Having referred to the attempts of urban development by different planning agencies, here, rational planning proposals have been put forward, taking into cognizance the regional and environmental perspectives in the frame of applied urban geography.

Evolution and Distribution of Cities

Most of the Indian cities, especially in the northern plains, are riverine. Prior to incoming of railways, the agro-products of the northern plains were transported through the navigable Ganga, Yamuna and their important tributaries. Their navigable and sacred character helped emergence of the sites of settlements along the banks. Thus, many Indian towns and villages clung to river banks since hoary past. Among them Varanasi, Allahabad and Mathura are religious centres, whereas, Patna, the present capital of Bihar, has been the headquarters of Magadh empire. Delhi, Agra, Lucknow and Faizabad represent the relics of the muslim rule in India. Calcutta, Mumbai and New Delhi still reveal the strong British impact. Kanpur, Ghaziabad, Noida and several other towns of Hoogly industrial belt like Dhanbad, Bokaro, Sindri, Titagarh as well as Bhubaneswar, Bhillai, Tatanagar etc. are of relatively recent origin.

As regards the distributional pattern of urban centres, the majority of towns is concentrated along the Ganga and its tributaries. The metropolises of Calcutta, Delhi, Kanpur, Lucknow and Patna form the nucleii of such concentrations. These cities do contain more than 40 per cent population of the northern plains. The 'pull factor' exercised by them appears to be greater than that of other centres in the region.

All these cities witnessed tremendous growth in their population as well as in size during the post independence period. The Delhi metropolitan region in the west and the Calcutta conurbation in the east have shown the polynuclear character to which a number of towns have intermingled. The large cities, however, record rapid population growth whereas, the small and medium sized towns exhibit slow and moderate growth.

Morphological Characteristics

Extensive expansion of built-up areas in Indian cities have taken place since the first war of independence (1857), when the Britishers established 'cantonments' and 'civil lines' among the large towns to maintain their strongholds. Viewing at the morphological set up in most of the Indian cities, some distinct features may be identified. To illustrate, the old and haphazardly grown cities have expanded around a physico-cultural nucleus and business centre characterised by narrow and zigzag streets and lanes as well as multistoreyed residential structures of which a few are renovated while many are dilapidated and obsolete (Brush, 1961).

On the outer rim of the inner zone of cities extend the monotonously planned but relatively more open civil lines with mostly single storeyed tiled roofed bungalows enclosed by illmaintained big gardens and lawns, lying along wide and straight roads usually accommodating government officials, advocates and other urban elites. The cantonments, contiguous to the civil lines occupy quite spacious campus and vast open land studded with military officer's bungalows, barracks, offices, canteens, rifle shooting range, parade grounds and other isolated structures lying beyond the approach of civilians. Long and narrow independent walled colonies developed near railway stations among all Indian cities appear like 'a city within city'. The emergence of cultural, religious, educational and medical institutions, e.g. clubs, churches, temples, libraries, colleges, universities, hospitals etc. has led to increase in the built-up portion of cities. Of late, with the development of industrial estates and residential colonies on the outskirts along the major roads diverging from city centre the star-shaped urban sprawl has taken place in most of Indian cities (Taneja, 1971).

Consequent upon the 'filling in process' guided by 'azonal factors' as well as with a view to accommodating the growing population, a number of housing schemes and various other structures have sprung up in different parts of cities, exhibiting an unsystematic and haphazard growth in the new built-up area. In this phase of urban sprawl, the process of 'chewing up' of the adjoining rural area is quite common (U.N.P.D., 1990). Also, following the development of new industrial establishments in the peripheral section of cities and extension of their municipal limits, there has emerged a rural-urban fringe along the major thoroughfares on the outskirts of the urban area.

The Fringe Phenomenon

The fringe zone lying around cities is an area wherein extensive tracts are often taken from rural to urban land uses. Beyond municipal limits of a city, along the main roads the land, formerly under cultivation and studded with rural dwellings, is slowly but continuously being converted into new urban structures comprising of both residential as well as non-residential buildings. In fact, with the growth of a few cultural, educational and medical centres, increased trade and commerce as well as the expansion of civic functions, the process of suburbanisation is activised and strongly expressed through ribbon shaped sprawl of settlements along diverging roads among Indian cities (Jayarajan, 1965).

Needless to mention that such an outgrowth is the result of appalling congestion and overcrowding in the city premises and decentralisation of business as well as industrial activities from the core to outer rim. Automobiles have played vital role in nullifying the distance factor and in causing industrial developments in the fringes. However, the intensity of economic impact of Indian cities is feebler than that of western cities. Ellefsen (1962) has rightly remarked that "Much of the area close to an Indian city, which have been an effective zone for a western city of the same size remains isolated and tied to agriculture." In this portion lie the old villages and hamlets which although remain independent and isolated but are considerably influenced by the city.

Thus, the growth of rural-urban fringe in Indian cities, being a dynamic phenomenon, neither perfectly rural nor urban, is the basic result of suburbanisation. And with a view to overcoming regional imbalances and initiating planned urban development, this process has to be regulated and organised.

Urban India is not only witnessing very spontaneous growth of fringe zone but also its constant encroachment by cities without any concerted effort for planned use of its land. Bigger the city, larger is the size of its rural-urban fringe. Here, it is rather sad to point out that our cities are regularly swallowing the first class fertile farmlands of their surroundings, irrespective of the increasing demand of food and other agro-products in the country. Simultaneously the loss of agricultural land is yielding to rural unemployment, rural poverty and rural-urban migration. It is, therefore, enevitable to explore the ways and means to stop and restrict such encroachments. In general, more vertical rather than horizontal expansion has to be encouraged so as to prohibit and control the continuing urban invasion of the countryside.

In the sequel, it will not be out of place to mention that even the remote villages of developed countries like U.S.A. and Japan are well-equipped with all modern amenities and look like 'rural-urban' whereas many Indian villages lying within municipal limits of a large city are deprived of basic urban facilities and public utility services. More and more of our fringe villages are likely to fall victim of urban invasion in the years to come. The rural-urban fringe areas of Indian cities, as a result of urban encroachments, suffer from multifarious problems pertaining to land use, slums, environmental degradation etc. Due to lack of proper administration and taxation, homogeneous development of the fringe zone is out of question. As a matter of fact, this zone paradoxically lies at a time under the influence of three administrative agencies—the municipal board, the district board and the block, among which there is hardly any coordination and integration.

Gottman's statement that "the problem of land use changes caused by urban explosion become chronic" is quite valid in case of rural-urban fringes of Indian cities, where part of the cultivable land is sold for houses and factories and much of the farmland lies derelict and abused as a consequence of urban interference into the farming system, enhanced land values through competition between colonisers and industrialists (Gottman, 1963; Dickinson, 1964). Sen Gupta rightly supports Wehrewien's view that the land in the fringe zone of Indian cities is getting frozen and use of farm is being converted into non-farm urban or quasi-urban uses (Sengupta, 1987; Wehrewien, 1942). Most of the farmlands are abandoned and remain vacant until their new use is introduced.

Emerging Urban Problems

The built-up area of Indian cities have swelled even more than ten times during a span of hundred years. For instance, the four largest cities of the country have registered growth in their built-up area measuring even more than 100 km^2 showing a huge concentration of industrial and commercial population. The polynuclear development of Calcutta metropolitan region incorporated about 28 towns and eight cities, exhibiting a density of above 2,500 persons per 50 km^2 area. Kanpur, the largest metro-city of Uttar Pradesh has enlarged its built-up area from 5 km^2 (1965) to 49 km^2 (1991).

Most of the riverine cities of India frequently suffer from floods. Vast tracts of riverside urban land are inundated during heavy rains, which causes loss of property and hinders the usual functioning of the city life for several days. The problems experienced during the peak floods of 1916, 1948, 1956 and 1978, when the waters of the Ganga, the Yamuna, the Gomati and their tributaries intruded into central portions of Varanasi, Allahabad, Lucknow and many other cities are examples in point.

The traffic congestion among cities is rather a more alarming problem. The traffic survey of Varanasi (1989) reveals that about two hundred auto-vehicles, five thousand bicycles and cycle rickshaws pass through Godowlia road crossing every hour. Roads of the down-town due to unauthorised encroachments, have turned into bottlenecks at places causing traffic Jams for hours together. Overcrowdedness in city buses and overloading in trucks are common phenomena among Indian cities.

The overgrowing size of urban centres has affected both habitation and life in Indian cities. The major problem is that of residential accommodation. The prices of land in urban areas have been increasing vigorously, even more than that of the gold. House rents are also shooting up very fast. According to a government survey, in 1990 one-third of the total urban population of India used to live in huts having only one dark room with a curtain in place of door, without any facilities of lavatory and drinking water (Mitra, 1983). The survey further reveals that among large cities about 45-50 per cent of the families reside in single room accommodation. In Calcutta and Mumbai the situation is worse with this figure exceeding over 60 per cent. About 74 per cent population of Calcutta lives in slums. In Kanpur alone the slum dwellers constitute about 20 per cent of its population. Here 30 per cent houses are still *kachha* and form horrible slums (Amrita, 1985). As indicated in a Planning Commission's report, in Lucknow city 15 per cent of its inhabitants have no shelter. (Planning Commission, 1990).

In metro-cities of Calcutta, Mumbai, Delhi, Chennai, Bangalore and Ahmedabad in 1991, the people living in *jhuggi-jhonpadis* accounted 118 lakh, 111 lakh, 58 lakh, 57 lakh 56 lakh and 36 lakh respectively. If this trend continues, almost all metro-cities of India will have more than half of their population living in hutments. These cities have witnessed illimpact of poor and substandard habitation in two ways: (a) the slums and blighted areas are increasing very fast and (b) the suffering from the scarcity of light and water disposal is becoming more and more difficult day by day. The pollution of water, noise and air has become a common phenomena in Indian cities. Simultaneously, the urban administration has become loose, inefficient and corrupt. Consequently, the antisocial elements as well as unsocial activities are increasing day by day, making the city life unsafe and unhealthy.

The existing living conditions in large urban centres of India are quite pitiable. In Calcutta 37.1 per cent, in Mumbai 22.2 per cent, in Delhi 24.4 per cent, in Chennai 34.6 per cent, in Bangalore 27.7 per cent and in Ahmedabad 24.2 per cent houses are deprived of electricity. Similarly, 66.9 per cent houses in Calcutta, 41.7 per cent in Mumbai, 48.4 per cent in Delhi, 65 per cent in Chennai, 56 per cent in Bangalore and 33.9 per cent in Ahmedabad do not have their own drinking water facility. In fact, there the water requirement of hardly 50 per cent urban community is met with. Even worse is the state of lavatory facility. Latrines are non-existent among 50 per cent houses of Calcutta, 26.6 per cent of Mumbai, 31.8 per cent of Chennai, 43.7 per cent of Bangalore and 26.2 per cent of Ahmedabad. Open spaces, public places, playgrounds, platforms, footpaths, streets and roadside vacant land are used for the purpose of evacuation. It gives very ugly and dirty look to the city premises.

Due to uncontrolled urban population growth, the law and order situation in cities has got worsened, resulting into high propensity of various crimes like thefts, murders, kidnapping, robbery, terrorism etc. All such anti-social activities have become a part and parcel of the urban way of life, causing mental tension and a sense of insecurity among the citizens. Acid and bomb blasts are recent additions to such ghastly acts. On the other hand, the frequency of road accidents has also gone very high as a result of breach of traffic rules, rash and drunk driving etc.

In this context a reference may be made to the fast growth of capitalism in metro cities. The handful of urban elites are becoming wealthier and wealthier mostly through their unfair occupations and evasion of income taxes. At the same time, the urban services are extravagantly utilised by the rich and well to do people, who often avoid paying taxes and use their resources to manipulate it. Conversely, a large portion of urban population is found to be suffering from non-availability of even basic amenities of life like food, cloth, shelter, education and health. Due to low oxygen level, air and noise pollution, contaminated water supply, poor sanitation, malnutrition as well as lack of health facilities and awareness, several respiratory and water borne diseases like asthma, tuberculosis, jaundice and anemia etc. are prevalent in the inner zone of cities, mostly among labourers, *thela* and rickshaw pullers, sweepers, beggers etc. These have'nots also fall victim to insecurity, accidents, drug addiction and oppressions from antisocial elements. Moreover, this weaker section of urban society is adversely affected by the burden of municipal taxes.

Small Towns: Hope of Future

It has been mentioned earlier that the small and medium sized towns of India have been registering stagnation in terms of growth of population as well as development of economic activities, due mainly to the fact that they lack in infrastructure, industrial development and generative force for employment opportunities. Although occupying strategic and nodal position in the socio-economic system of the respective regions small towns more often than not fail to attract the rural migrants. But they act as service centres to the surrounding countryside and transmit growth impulses helping in utilisation and development of resource potential of their hinterland.

Recently, the need to promote a network of small and medium sized towns is being, strongly felt in India. This is mainly with a view to restricting the evils of overurbanisation, resulted from the growing concentration of population in large cities on the one hand and establishing a link between the rural hinterland and the highly urbanised area on the other. All the district headquarters of the country, in this context, may prove to be ideal nodes for attainment of these objectives because they contain a manageable population size as well as semi-urban characteristics. They can, thus, be appropriately utilised for controlling the problems arising out of excessive growth of large cities. Moreover, they may efficiently provide a suitable link between the large city and the surrounding rural area.

If the small and medium sized towns succeed in diverting even a fraction of migrants from villages entering into large cities, there will enevitably emerge a good prospect for proper development of urban centres in India. Among all such towns, especially of the Ganga plain, which are endowed with homogeneous terrain and good agricultural hinterland, latest technology based agro-industries may be suitably developed. Simultaneously, their organised and planned development is also essential. It is needless to emphasise that the time has come to strengthen, renovate and revitalise small and medium towns by way of improving their transport and communication network, power supply system and the education and health facilities. Such developments will not only ensure balanced urban growth in the region but also will generate fresh employment opportunities and attract large rural population therein. In due course, these towns will serve as efficient liaison or 'bridge' between rural service centres and metro cities of India. In this context the statement of Ruth Glass that "the perennial fear of metropolises has produced a perennial hope for small towns" is quite relevant (Dube, 1988).

Urban Planning Strategy

It is an established fact that the towns reflect the socioeconomic organisation of a region; they not only shape and mould the life style, behavioural pattern and culture of the people but also play a vital role in future development of the nation at large. In developing countries, the urban settlements although forming a minor portion of the entire settlement coverage, indicate an advanced economic society. Major characteristics of a good city are the same throughout the world, *viz.* enough residential accommodation, adequate space for shopping, offices, educational and health services, open spaces for recreational centres, harmonious road pattern and good employment avenues. So, ideal cities should be attractive centres possessing all essential amenities and public services.

In the opinion of Kingsley Davis (1955) the crux of the problem of Indian cities is 64 neither urbanisation nor rural-urban migration but human multiplication and city growth. Hence merely a few show schemes and paper plans would not solve the problem at all. Any organised feasible programme for the planned development of Indian cities need be comprehensive. The haphazard character of growth and unhygienic conditions would necessitate improvements in and around it. The need of the hour is to regulate our cities by providing efficient transport network, better housing facilities, improved infrastructure and adequate utility services so as to make the urban life comfortable and pollution free.

For many Indian cities, with a view to rationalising the development of their built-up area, efforts of planning were exercised but so far fruitful outcome could not be achieved. Official master plans, prepared by town and country planning organisations of the government are partially effective. Most of them are not even properly implemented. In fact, the works of all the three development agencies, viz. improvement trusts of the municipality, city development authority and town planning organisations of the government, have so far been limited to the widening a few roads and launching some housing schemes for high, middle and low income group people. Some slum clearance schemes, although well designed, were never attended seriously. True, the urban renewal of the central part of cities is a tough task involving heavy finance, legislation and public coordination etc. Moreover, it is not a very successful measure for urban development planning. The development of new residential communities in the peripheral zones of Indian cities is more feasible and practical, as it will not only ensure the economically gainful and appropriate use of space but also create the scope for rational development of public utility services in the city. Continuous encroachment in the surrounding countryside, i.e. urban invasion of the rural tract by cities has become a conspicuous phenomenon. This kind of urban growth at the cost of the valuable cultivable land is quite precarious particularly in the country of predominating agrarian economy like ours.

It is in fitness of things to emphasise the vertical rather than horizontal expansion of even medium sized Indian towns, so as to control the process of suburbanisation. Economic apartments of Japanese design may be recommended in the outskirts of Indian cities. Establishments of a number of new towns at 'growth points' near most of the metropolises should form a part of scheme of regional development. This process will also involve the setting of satellite towns linked with the 'primate city' by roads and railways. Industrial zones need to be earmarked at the intersection points of these roads. For instance Ramnagar, Mughalsarai, Chunar and Bhadohi near Varanasi; Naini, Handia, Phulpur, Shaidabad, Sarai Akil and Shankargarh near Allahabad; Jajmau and Bithoor near Kanpur may be selected as such growth points. These centres may be developed as planned townships which could share part of migration to large and crowded cities from the adjoining rural area. Through intensive urban pollution surveys, risk zones of cities need to be identified and proper action plans be prepared so as to restrict environmental degradation in cities and their surroundings.

Concluding Remarks

Indian cities require four remedial measures: (a) the renovation of the city core, (b) regulated development of suburban zone and fringe area, (c) control of building activities in the city centre and check on environmental degradation through clearance of slums and eradication of pollution, and (d) revitalisation of small and medium towns. The process of urban renewal should be activated in order to create open spaces and decongest the oldest parts of city by demolishing obsolete and dilapidated buildings and replacing them by modern ones. No doubt, such crucial work needs a meticulous endeavour including a lot of specific budget, legal action, motivation and compensation to the affected people. Many urban renewal programmes have not achieved much success even in the developed countries, so what to talk of the feasibility of this programme in the cities of poverty striken developing countries like ours, where there is a serious lack of coherence and coordination between local and national development agencies.

It has to be borne in mind that all Indian cities, for at least the century ahead, will continuously grow in size as well as in population until the population of nation ceases to grow through concerted efforts. Their spontaneous and haphazard sprawl may, however, be controlled to some extent by decentralising urban activities and initiating an organised development in the newly incorporated areas of cities. But these strategies may be effective only for the time being. What we anticipate and need is a permanent solution to the problem of overurbanisation, disorganised and misdirected developments and increasing rural-urban migration in our urban areas.

References

Amrita, Abraham (1985), Housing for the poor, *Economic and Political Weekly*, vol. 20, No. 6, February 9, New Delhi.

Bose, Ashish (1978), *India's Urbanisation: 1901-2001*, (second ed.), New Delhi p. 85.

Brush, John E. (1961), Morphology of Indian cities', in Roy Turner (ed.), *India's Urban Future*, University of California, Berkeley, p. 129.

Davis, Kingslay (1955), The origin and growth of urbanisation in the world, *American Journal of Sociology*, Vol. LX., No. 5, p. 429.

Dickinson, R.E. (1964), *City and Region: A Geographical Interpretation*,

Routledge & Kegan Paul, London, pp. 171-6.

Dube, K.K. (1988), Role of small towns in Urban development in India, in K.K. Dube and A. Singh (eds.), *Urban Environment in India,* Inter-India Pub., New Delhi, p. 66.

Ellefsen, R.A. (1962), City hinterland relationship in India, in Roy Turner (ed), *India's Urban Future,* University of California Press, Berkeley.

Gottman, Jean (1963), Megalopolis or the urbanisation of the north-eastem Sea-boards, *Econ. Geog* Vol... p. 196.

Jayarajan, C. K. (1965): 'What is happening to the fringe areas of our cities?' *The Civic Affairs,* Oct. Vol. 13

Mitra, Asok (1983), Key-Note Address, National Seminar on Metropolitan Development, held at Calcutta.

Planning Commission (1990), *Eighth Five Year Plan,* Govt. of India, New Delhi.

Registrar General of India (1991): *Census 1991,* Series 1.

Sen Gupta, Smita (1987), *Residential Pattern of Suburbs,* Concept Publishing Company, New Delhi, p-205.

Taneja, K.L. (1971), *Morphology of Indian Cities,* NGSI Pub. No. 8, Varanasi p. 129.

U.N. Population Division (1990), *Pattern of Urban and Rural Growth,* Department of International Economic and Social Affairs, New York.

Wehrewien, G.S. (1942), The Rural-lirban Fringe, Mayer and Kohn (eds.) *Readings in Urban Geography,* Chicago, p.542.

17

Demographic Characteristics of the Metropolitan Areas of India

SISIR KUMAR BISWAS

Demographic Status

Metropolitan areas (M.A.'s) having population more than one million each exert tremendous influence in the urban scenario of India. They are "the cream of urbanisation process as a whole in the countries where they occur" (Biswas, 1977). Out of a total of 3,768 urban centres (1991 Census), the metropolitan areas number 23 only but they constitute as much as 32.54 per cent of India's urban population. The remaining 67.46 per cent of total urban population of the country is covered by 3,745 urban, or agglomerations centres. A striking feature to note is that the average population of an urban centre in the country is 57,637 while the same for a metropolitan area becomes 30,86,814. Hence, it may be said that the position of an M.A. is sky high compared to the other urban centres of the country.

The Metropolitan Areas of India experienced a phenomenal growth since independence. Starting with a meagre one in 1901 represented by Calcutta U.A., this number had increased to two in 1911 and remained constant during the census periods 1921, 1931 and 1941. After independence however, the M.A.'s had undergone a steady onward rise with 5 in 1951 to 7 in 1961, 9 in 1971, 12 in 1981 and finally 23 in 1991. Similarly, the percentage share of M.A. population to India's total urban population bad increased from below 12 per cent in pre-independence era to 3254 per cent in 1991. The growth rate of M.A.'s was only 8.86 per cent in 1931 but suddenly became a gigantic 121.32 per cent in 1951 due to heavy influx of refugee population in metropolitan areas. Since 1951 and upto 1981, there was a downward growth rate but took a leap in 1991 when it was 67.76 per cent compared to earlier 51.35 per cent of 1981 (Table 17.1).

Table 17.1: Trend of Urbanisation of Metropolitan Areas in India

Census	No. of M.A.	Total population	Percentage share of M.A. population to India's total urban population	Growth Rate (per cent)
1901	1	15,10,008	5.84	–
1911	2	27,63,586	10.65	83.02
1921	2	31,29,518	11.14	13.24
1931	2	34,06,869	10.18	8.86
1941	2	53,07,540	12.02	55.79
1951	5	1,17,46,616	18.81	121.32
1961	7	1,81,01,748	22.93	54.10
1971	9	2,78,31,065	25.51	53.75
1981	12	4,21,21,700	26.41	51.35
1991	23	7,06,61,259	32.54	67.76

Source: Census of India, 1991.

In the next census year of 2001 some more M.A.'s are most likely to be included in the list. Eleven potential centres which had more than 7 lakh population each in 1991 are earmarked as prospective centres to become metropolitan areas. These are Vijayawada, Jamshedpur, Dhanbad, Thiruvananthapuram, Kozhikode, Jabalpur, Agra, Allahabad, Meerut, Asansol and Nashik.

In the global sphere, India has secured a place among the top ten Metropolitan Areas (M.A.'s) of the world. Greater Mumbai with 12.6 million population in 1991 ranked sixth while Calcutta stood tenth with 10.9 million population. Again, if the first twenty M.A.'s are considered Delhi becomes included in the list (Table 17.2).

Sex Ratio

Sex composition is a very important element of population (Siddiqui, 1984). Among the 23 metropolitan areas, sex ratio (number of females per 1,000 males) is found to be highest in Kochi (996) followed by Madurai (968), Vishakhapatnam (938), Chennai (927) and Hyderabad

(924). On the other extreme, the lowest sex ratio occurs in Ludhiana (798) preceded by Coimbatore (808), Calcutta (827), Greater Mumbai and Patna (829 each), Delhi (831), Kanpur (837) and Surat (839). Lower sex ratios are generally encountered in cities with preponderance of industrial workers or in very large metropolises where people often live without a family. This is no exception in the present case too. Taking the all India urban sex ratio as 893 (1991 Census) it is noticed that more than 50 per cent of M.A.'s of the country—12 out of 23 fall short of national average—a fact justified in respect of bigger cities of a country.

Table 17.2: Top Ten Metropolitan Areas of the World (Year 1990-91)

Name of M.A.	Population (in Million)	Rank
Mexico City	20.2	1
Tokyo	18.1	2
Sao Paulo	17.4	3
New York	16.2	4
Shanghai	13.4	5
Bombay	12.6	6
Los Angelis	11.9	7
Buenos Aires	11.5	8
Seoul	11.0	9
Calcutta	10.9	10

Source: Census of India, 1991.

Literacy

It is a decisive factor for evaluating urban potentiality (Nag, 1995). This is more true for an underdeveloped country like India. In respect of literacy rate of M.A.'s, Kochi has bagged the top position by scoring 83.43 per cent literacy, while the second and third ranks go to Coimbatore (74.04%) and Chennai (72.17%). Mumbai with 72 per cent literacy had followed closely the third rank holder i.e. Chennai. The other three M.A.'s in the 70+ group are Vadodara, Nagpur and

Pune respectively. Ahmedabad, Calcutta, Bangalore and Indore lie in the next group of 65 to 70 per cent. There are however, as many as seven M.A.'s who could not attain national average of urban literacy—a very pathetic affair for the country's highest urban centres.

In the case of female literacy, it is Kochi again (81.18%) to top the list of M.A.'s. She is followed by Coimbatore (71.32%), Chennai (66.83%), Greater Mumbai (66.39%) and Vadodara (66.32%). The other M.A.'s in the 60+ group are Nagpur, Calcutta, Pune, Bangalore and Ahmedabad. Only five M.A.'s are below the national average of 54.01 per cent. Female literacy is a pertinent issue of today and in this regard the M.A.'s have fared better than in the case of overall literacy (Table 17.3).

Working Force

For study of metropolitan area economic base, working force assessment provides a vital clue (Mattila and Thompson, 1955). In this case, a striking.fact is that 13 out of 23 M.A.'s are lagging behind the national average of 30.45 per cent in respect of urban working force (1991 Census). This indicates a huge dearth of employment as prevailing in the metropolitan areas. This is indeed a gloomy picture for the topmost group of urban places of the country. The ten M.A.'s who could cross the national average are Madurai, Ludhiana, Coimbatore, Greater Mumbai, Surat, Bangalore, Delhi, Pune, Chennai and Calcutta respectively. Of these, Madurai has the unique distinction of scoring 47.9 per cent followed by Ludhiana with 39.31 per cent. Working force is a good indicator of economic prosperity of cities, and in this respects the M.A.'s of India barring a few show signs of economic slump.

Population Growth Rate

It is a highly significant indicator of potential status of a M.A., since it reflects the cumulative effect of all individual parameters (Biswas, 1982). It is also a strong determinant of degree of urbanisation (Gibbs, 1961). In the present case Vishakhapatnam has attained a growth rate of 74.27 per cent, and this has been due to large scale development of industrial sector in and around the M.A. Followed closely are

Table 17.3: Demographic Traits of Metropolitan Areas of India

Name	Population (1991)	Sex ratio	Population growth rate (1981-91)	Literacy per cent			Working force (per cent)
				Total	Male	Female	
Greater Mumbai U.A.	12596,243	829	33.43	72.00	77.61	66.39	34.26
Calcutta U.A.	1,10,21,918	827	18.73	69.28	74.03	64.53	30.63
Delhi U.A.	84,19,084	831	46.18	64.68	70.46	58.91	31.76
Chennai U.A.	54,21,985	927	24.29	72.17	77.51	66.83	30.67
Hyderabad U.A.	43,44,437	924	67.04	57.59	64.05	51.13	27.31
Bangalore U.A.	41,30,288	903	39.87	68.95	74.45	63.45	32.66
Ahmedabad U.A.	33,12,216	890	28.94	69.41	75.40	63.42	28.91
Pune U.A.	24,93,987	905	47.38	70.03	76.26	63.80	31.32
Kanpur U.A.	20,29,889	837	28.81	62.02	67.05	56.99	26.67
Lucknow U.A.	16,69,204	867	62.97	61.60	68.31	54.89	26.69
Nagpur U.A	16,64,006	916	36.24	71.00	76.70	65.30	28.04
Surat U.A.	15,18,950	839	64.21	52.93	57.68	48.18	33.93
Jaipur U.A.	15,18,235	868	49.18	61.07	67.22	54.92	28.24
Kochi U.A.	11,40,605	996	38.14	83.43	85.68	81.18	29.39
Vadodara U.A.	11,26,824	898	42.54	71.31	76.30	66.32	29.26
Indore U.A.	11,09,056	895	33.13	65.17	71.93	58.41	29.90
Coimbatore U.A.	11,00,746	808	23.39	74.04	76.76	71.32	35.67
Patna U.A.	10,99,647	829	19.55	63.22	70.51	55.93	26.67
Madurai U.A.	10,85,914	968	20.49	56.72	64.84	48.60	47.97
Bhopal M.C.	10,62,771	894	58.51	60.25	66.78	53.72	28.73
Vishakha patnam U.A.	10,57,118	938	74.27	63.22	71.66	54.78	28.44
Ludhiana M.C.	10,42,740	798	66.72	62.58	65.76	59.40	39.31
Varanasi U.A.	10,30,863	860	28.77	53.04	61.23	44.85	27.01

Source: Census of India, 1991
U.A.= Urban Agglomeration ; M.C. = Municipal Corporation.

Hyderabad (67.04%) and Ludhiana (66.72%)—the two industrial giants amongst the M.A.'s of India. The two other M.A. in the 60+ group are Surat (64.21%) and Lucknow (62.97%). Taking 36.19 per cent as national average of urban growth rate (1981-1991), it is seen that altogether 13 M.A.'s including the five mentioned here are above the national average while 10 M.A.'s are much below the national average. Development inertia of the metropolitan areas of India in general seem to be very bleak (Table 17.3).

Geographical Distribution

The metropolitan areas are not well distributed geographically. Before independence, Calcutta in the east and Mumbai in the west were the only M.A.'s throughout the span of the country. In the post independence era, Delhi, Chennai and Hyderabad came to represent north and south central India was a void till 1981 when Nagpur emerged as a M.A. Based on the last census count of 1991, northeastern India is completely blank, while eastern India has only two i.e. Calcutta and Patna. Western India has a good concentration of six M.A.'s led by Greater Mumbai and joined by Pune, Surat, Vadodara, Ahmedabad and Indore. Northern India has again six M.A.'s though distributed in two zones, (a) Varanasi, Kanpur and Lucknow; and (b) Delhi, Ludhiana and Jaipur. The southern peninsular block with seven M.A.'s has the highest zonal concentration with Hyderabad and Vishakhapatnam lying in the northern end followed southward by Bangalore, Chennai, Coimbatore, Kochi and Madurai. The large geographical space starting from central India and stretching upto the northeast has a poor concentration of only two M.A.'s and controlled by the majestic Calcutta mega city.

Size Hierarchy

Urban size hierarchy reveals deviations if any, from the standard hierarchy expected for the urban centres of a region (Browning and Gibbs, 1961).

The standard "rank-size rule" (Zipf, 1949) helps to understand the existing and expected size-hierarchy that guides the planners to plan for the urban settlements of a country. In the present case, it is observed that the average that the M.A.'s would have to increase or decrease (in percentage) to fit the rank-size comes to 13.21. In an earlier study (Biswas, 1977) based on 1971 data this figure was worked out to be 15.40 which means that the overall deviation from norm has become less in a span of twenty years. As regards individual cases, it may be noticed that actual population is short of expected population for nine M.A.'s headed by Greater Mumbai in whose case the difference is 50.94 per cent of actual size. Similarly, for the remaining fourteen M.A.'s, actual population is more than expected population.

Table 17.4: Application of Rank-Size Rule to the Metropolitan Areas of India

Name of M.A.	Rank	Reciprocal of rank	Actual population	Expected population	Difference between actual and expected population	Difference as per cent of actual population
Greater Mumbai	1	1.00000	1,25,96,243	1,90,12,362	64,16,119	50.94
Calcutta	2	0.50000	1,10,21,918	95,06,181	15,15,737	13.75
Delhi	3	0.3333	84,19,084	63,37,454	20,81,630	24.72
Chennai	4	0.25000	54,21,985	47,53,090	6,68,895	12.34
Hyderabad	5	0.20000	43,44,437	38,02,472	5,41,965	12.47
Bangalore	6	0.16667	41,30,288	31,68,727	9,61,561	23.28
Ahmedabad	7	0.14286	33,12,216	27,16,052	5,96,164	18.00
Pune	8	0.12500	24,93,987	23,76,545	1,17,442	4.71
Kanpur	9	0.11111	20,29,889	21,12,485	82,596	4.07
Lucknow	10	0.10000	16,69,204	19,01,236	2,32,032	13.90
Nagpur	11	0.09091	16,64,006	17,28,397	64,391	3.87
Surat	12	0.08333	15,18,950	15,84,364	65,414	4.31
Jaipur	13	0.07692	15,18,235	14,62,489	55,746	3.87
Kochi	14	0.07142	11,40,605	13,58,026	2,17,421	19.06
Vadodara	15	0.06666	11,26,824	12,67,491	1,40,667	12.48
Indore	16	0.06250	11,09,056	11,88,273	79,217	7.14
Coimbatore	17	0.05882	11,00,746	11,18,374	17,628	1.60
Patna	18	0.05555	10,99,647	10,56,242	43,405	3.95
Madurai	19	0.05263	10,85,914	10,00,651	85,263	7.85
Bhopal	20	0.05000	10,62,771	9,50,618	1,12,153	10.55
Vishakhapatnam	21	0.04761	10,57,118	9,05,351	1,51,767	14.36
Ludhiana	22	0.04545	10,42,740	8,64,198	1,78,542	17.12
Varanasi	23	0.04347	10,30,863	8,26,624	2,04,239	19.81
ΣX	—	3.73424	7,09,96,726	7,09,97,702	1,35,29,994	303.95
ΣX/N	—	—	3,086,814	30,86,856	5,88,261	13.21

After Browning and Gibbs, 1961.
Source: Census of India, 1991.

In either cases the difference of actual and expected sizes would have to be adjusted by increasing or decreasing the sizes to conform to rank-size rule.

Conclusion

Demographic characteristics of urban settlements are always changing with passing of time, and this is no exception for the metropolitan areas of the country. For examples, thère were one M.A. in 1901, five M.A.'s in 1951, 12 in 1981 and 23 in 1991. Again, Calcutta had enjoyed the topmost rank amongst the M.A.'s till 1981 but relegated to second position in 1991. Similarly, the individual demographic parameters like growth rate, literacy, working force, sex ratio etc., have also been changing quantitatively year to year. Since the M.A.'s of India form a system, it is to be noted that recurring studies are essential to see how far the system works and what are its deviations. In India, the metropolitan areas and other higher order cities control the entire process of urbanisation (Alam, Rao and Gopi, 1974). Experts are of opinion that the M.A.'s are not functioning properly as a system (Gopi, 1982). Qualitative improvement of urban environment and simultaneous development of economic base of the metropolitan areas which dominate the urban scenario of India are absolutely necessary for overall development of the country. The changing pattern of metropolitan system *vis-a-vis* urbanisation process in general is a matter of utmost importance to demographers and urban planners, and hence systematic evaluation is needed through time to time empirical studies and other academic exercises.

References

Alam, Rao and Gopi (1974), *Trends and Patterns of Metropolitan Development in India,* Occasional Paper No. 3, Centre for Urban Research, Dept., of Geography, O.U., Hyderabad.

Biswas, S.K. (1977), Spatial and demographic behaviour among the million cities of India and other leading countries of the world, *Indian Journal of Regional Science,* Vol. IX, No. 2.

Biswas, S.K. (1982), Quantitative evaluation of the development potentialities of the towns of Bijnor district in S. Manzoor Alam and K. N. Gopi (eds), *Settlement System of India,* Oxford & IBH Publishing Co., New Delhi.

Brownir 7, H.L. and Gibbs, J.P. (1961), Some measures of demographic and spatial relationships among cities, in Jack P. Gibbs (ed.), *Urban Research Methods,* D. Van Nostrand Co., Inc., Princeton, U.S.A.

Gibbs, J.P. (1961), Some demographic characteristics of urbanisation, in J. P. Gibbs (ed.), *Urban Research Methods,* D. Van Nostrand Co., Inc. Princeton, U.S.A.

Gopi, K.N. (1982), Metropolitan systems of India, in S. Manzoor Alam and K.N. Gopi (eds.), *Settlement System of India,* Oxford & IBH Publishing Co., New Delhi.

Mattila, J.M. and Thompson, W.B. (1961), The measurement of the economic base of the metropolitan areas in J. P. Gibbs (ed.). *Urban Research Methods,* D. Van Nostrand Co., Inc., Princeton, U.S.A.

Nag., P. (ed.) (1995), *Hospet,* Urban Studies Series 5, National Atlas and Thematic Mapping Organisation, Calcutta.

Siddiqui, F.A. (1984), Regional pattern of sex composition of the population of U.P., *Geographical Review of India,* Vol. 46, No. 2.

Zipf, G.K. (1949), *Human Behaviour and the Principle of Least Effort: An Introduction to Human Ecology.* Addison-Wesley Press, Inc. Cambridge, M.A., U.S.A.

18

Impact of A Deteriorating Lake System on The Physical and Cultural Environment:

Management Strategies for Bangalore City

DAKSHA C. BARAI and P. PRABHU

Bangalore the fifth largest and the fastest growing cities of South Asia has grown over an undulating terrain in the heart of the Deccan plateau. The present population of 4.5 million is likely to reach 7 million by 2011. The city has grown over a ridge and valley topography with the high ground at 923m above sea level. The important streams of Arkavathi and South Pinakini flow along the strike of the main ridge. The streams flow down the high grounds to form a radial pattern of drainage. On the eastern side flow the Koramangala, Ulsoor and Naravannahalli. The physiography had tended to give raise to form tanks such as Sunnekal (Shanthi Nagar), Sampangi and Shule. Over 262 existed providing a network of interconnected tanks and streams, which finally join the Arakavathi and South Pinakini. These tanks were the main sources of water supply to the residents of Bangalore.

The water system was sustainable till the late 19th century. In the 1890's the city's, civil and military stations tried to examine new sources and studied the potentialities of Malligal valley, Hebbal, Rachenahalla and Hessaraghatta tanks. A project drawn for Rs. 5.5 lakhs on Sankey tank failed. In 1899 water supply management was badly lacking and the Hessaraghatta tank, supplemented by Kakol and Byatha tanks was set up to supply water by setting up the Hessaraghatta reservoir to tap water of the Arkavathi. This catered to the needs of 2.5 lakhs people at 55 lt. per person per day, when planned in 1896 (Jayashree, 1996). With a further growth of population, supply fell short of requirements. In 1933 the Tippagondanahalli reservoir was commissioned. It stored the water of the Arkavathi and Kumudhavathi. In 1962 the demand for Cauvery water to Bangalore was initiated and the project took of in 1969 and by 1974 the Cauvery water was brought to Bangalore. Between 1941-1994 the population grew seven fold, i.e. from 4.11 lakhs to 29.22 lakhs, and area expanded from 70 sq km to 360 sq km. In 1991 the city had an area of 466 sq km with a population of over 40 lakhs. With such a phenomenal growth of numbers, demand for water outstripped its availability. By 2001 it likely to be eight million. Even the fifth stage of the Cauvery will not meet its requirements of 540 million lt daily. As population grew, lakes declined in size and numbers. With further growth of urbanization and industrialization many of the lakes dried up. Today the tanks have been subject to encroachments, landfills, sewage disposal and land grabs for construction, thereby posing a serious threat to the extinction and deterioration of the environment. Bangalore city had over 262 tanks in the early part of this century, but today has only eighty. Hence a cause of concern in terms water availability and environmental degradation (Fig. 18.1) has come up.

The existence of these lakes in the city was significant owing to its impact on the city's hydrological, socio-economic, cultural, ecological and recreational needs of the population and as they declined their importance also declined as indicated below.

(a) *Hydrological Significance*

The lakes were primarily used for domestic and agricultural needs of the people. In the 19th century unfiltered water was supplemented with the Karanjes systems of tanks like Dharambudhi, Sampangi, Ulsoor, Sankey etc, supplemented by wells and step ponds *(The Times of India,* 1994). These tanks also offered drinking water to the houses. The lakes also play an important role in recharging ground water. Lying on a plateau, the water tends to flow outwards. Hence if the outgoing water is not trapped and wells recharged the situation becomes precarious. The rains could trapped the ground water which could be harvested to one-third of the demand (Subramaniyan, 1985). With tanks being breached, the system is failing, and both surface and groundwater have been steadily declining.

(b) *Socio-economic Significance*

Lakes primarily met the domestic and agricultural needs of the people.

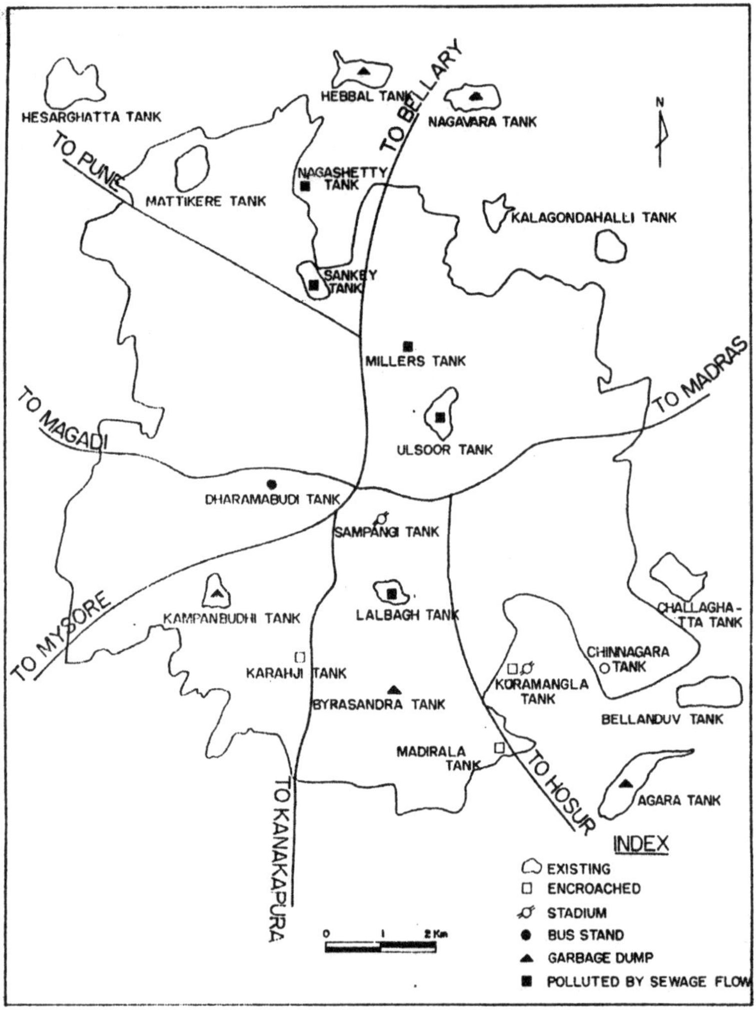

Fig. 18.1: Bangalore City : Distribution of Natural Lakes—Existing, Encroached and Extinct

The advent of obtaining water from long distances, converted the lakes to settling ponds. Industrialization during middle of the 20th century virtually engulfed the lakes.

(c) *Cultural Significance*

Tanks have played an important role in the social and cultural life of the city. The Sampagai tank was linked with the Kharga festival. Regular fairs are held on the banks of Kempanabhudhi tank. The Karanjkere was attached to Karanj, Anjana Swamy temple etc. The catchment areas of tanks houses huge slums that defecate the lake and also let out sullage. To prevent pollution it is necessary to provide basic sanitation to slum dwellers.

(d) *Ecological Significance*

On an area of 4.8 per cent of the water surface, the city has 40 per cent of the bird species in the wet lands. The bunded tank with varying depth being a maximum near the bund, promotes the growth of variety of emergent, floating, anchored floating, each of which show a preference to a particular range of depth. Plants can thus be stratified because of growth habits and structure. The fresh water communities accommodate a large number of small organisms, like periphyton insects, mollusca etc., which form a staple diet of many species of invertebrates and vertebrates in the lakes foods chain (Jayaram, 1996). The sediments of the lake offer an environment rich in organic matter, from decaying plants and animals. Thus a long food chain develops with fish, bird and man at the head of the food web.

(e) *Recreational Significance*

The lakes have been a source of scenic beauty. Some lakes like the Ulsoor and Yediyur offer boating services. Bird watching for pleasure, instructional and educational purpose is another significant form of recreation.

(f) *Environmental Significance*

The lakes in the past were surrounded by vegetation, thus making the adjoining places cool and moist. The vegetation and water make the place healthy thereby attracting people to the pensioners paradise.

Status of Lakes in the Bangalore Region

The rapid rate of urbanization in the last three to four decades and sprawling growth of the city has resulted in the extinction of a number of tanks. The degree of degradation and extinction, and the loss of lakes' capacity to sustain themselves have reflected the extent of human interference. Interference has been mainly in the form of construction, encroachment, landfills, sewage disposal, land grab etc.

Construction

Over the years, residential localities have sprung up on tank beds. Basavanagudi and Hanumanthana Någar are located on the Karanji lake bed and Sampangiramanagar on extension of the Sampangi tank. Some of the construction on tank beds have been for public utilities, such as the bus station, stadium, market etc. The Subhash Nagar bus stand is built on the reclaimed Dharamabudi tank, while the State Pollution Control Board is constructing its main building and laboratory on a tank in Basaweshwara Nagar (Jayashree, 1996). The Sooloy tank houses the football stadium and the Kanterava Stadium is on Sampangi tank bed. At present the Koramangala tank bed has a major housing project and the national games complex (Jayashree, 1996). The Pudupacherry tank bed now forms parts of Cookes Town and St. Thomas Town and the City Market. The Ashrayra housing project stood in dispute with activists over the construction on the Madivala tank.

Encroachment

Most tanks have been used for the construction of temples, burial grounds and roads, e.g. Agraharam and Hebbal tanks. These are instance of some tanks having fallen in the hands of land grabbers. Further land filling and garbage dumping have ruined several tanks. Solid waste are dumped into 25 large and small tanks including some important ones such as Kempanbudhi, Byrasandra, Tovarakere and

Agara. Domestic wastes flow in to 56 tanks including Sankey, Lalbagh and Hebbal. Industrial effluents are seen in 14 tanks—the worst case scenario is Chikkasandra in Pennya area. The Gulakomale on the out skirts of the city is on the verge of decay from effluents of a distillery that over flow even treatment plants (*Indian Express,* 1997). When partly on fully filled with waste tanks can have disastrous effects of siltation, pollution and eutrophication. Dried tanks, such as the Koramangala, filled with wastes was converted to an indoor stadium. Garbage dumping, both degradable and non-degradable, can affect the ground water quality and the tank is unable to recharge the ground water.

Brick and earth quarrying has taken place in 23 tanks where mud is lifted to manufacture bricks. The material of the tank bed has been used with cement for building purposes. Mud lifting is practiced in vast proportions and brick manufacture has affected the land and the lake ecosystems.

When fresh water was trapped from other sources to meet the city's requirements, the tanks have been breached. Curiously some tanks had been breached as a part of the malaria eradication programme. Lakes and tanks where the water could not flow, were cut off leaving behind stagnant water, which formed breeding grounds for mosquitoes that caused malaria. Subashnagar, Miller Akkithemanahalli. Nagashetty Halli and Sinivagilu were some of the tanks that were breached for breeding mosquitoes *(Times of India,* 1994).

Lakes get their inflow during monsoons. Over its formative years, Bangalore had 276 lakes playing an important role in recharging ground water apart from trapping and storing water. A grid system of storm water drains which had been prevailing in the past, is regarded as an engineering wonder of significance (Jayashree, 1996). Today pollution and breaching storm water is no longer effectively drained, but drains get blocked tending to flooding the roads during rainy season.

Status of Fauna and Flora

Several species of birds, plankton, aquatic macrophytes and fish are being found in the lakes. Harvesting the aquatic flora and fauna had let to problems. There have been over fishing, introduction to exotic fish which is detrimental to nature diversification of species and eutrophication as a result of sewage let in to the lakes have led to the extinction of several lakes. Poaching of protected species of birds and animals, besides loss of species has let to the declining lake system. The aerial growth of the city has affected the habitation of birds such as babblers, resident shrikes, the red headed mirlin and red start (Ramesh, 1996). When Bangalore was doted with 400 tanks there were 12 species of birds in an area between Shivaganga and Domlur which was then covered with jungle. The jungle bush, quail, woodshrike, white blowned bulbul, crow pheasant and grey partridge were regular denizers (Ramesh, 1996). Indiscriminate quarrying, boating, fishing condimentation with sewage and boating in lakes like Hebbal, Yediur and Ulsoor have hit the water birds.

Thus due to several factors, the physical structure and ecology of the city has been affected. Today efforts are being made by ISRO, Center for Science and Technology, the forest department, The Indian Institute of Horticultural Research along with BDA, BCC, KSTDC and BWSSB to conserve the tanks through NGO's.

Measures to Conserve the Tanks

There are several measures to conserve tanks, which are applied considering the dominant problems. These are:

(a) *Desilting*

The lakes are highly silted especially those along the margins of the city where agricultural practices of the past and the continued sedimentation of organic matter due to the inflow of sewage have reduce the holding capacity of the lakes. Desilting not only increased the holding capacity of water but can open up block capillaries to activate the aquifers. The desilted earth can be effectively used for the formation of jogging track, islands and the formation of the wet lands (Jayaram, 1996).

(b) *Deeweeding*

Lakes seem to be choked with the profuse growth of water hyacinths and weeds due to eutrophication. Lakes that are free of contaminates are easy to deeweed. The forest department has attempted to deeweed the Madivalla Lake as a pilot project where water on 60 ha has been submerged with weeds (Mehu, 1995). A boat like weeding machine from Finland the R-S. 2,000 gobbled up weeds in nature, but the cost 38 lakhs was prohibitive (Mehu, 1995). The forest department has also difficulty in deeweeding, since sewage removal requires multi departmental and multi agency involvement. The forest department is faced with funds crunch . Lakes free of contamination are easy to deeweed. Deeweeding the Ulsoor would cost Rs. 15 crores (Chandrakanth, 1995) .

(c) *Fencing*

The lake beds require physical protection in the form of barbed wire fencing as found in Ulsoor Lake. Apart from problems of proper survey and demarcation of boundaries has become inevitable due to obvious litigation. It is neither productive nor aesthetic. It has been observed that a fence along the perimeter of Ulsoor lake has not been able to prevent sewage from fouling the water. The Dredging Corporation of India has indicated that Rs. 4.4 crores will be required for dredging operation and another Rs. 5 crores for preservation of the lake by diversion of sullage water by BWSSB and BCC (Chandrakanth, 1995).

(d) *Water Treatment*

The continued inflow of effluents into tanks from industrial and domestic wastes has generated environmental problems. The physical and chemical methods are not only expensive, but also cost higher in terms of regular maintenance. Besides the chemical methods are also not ecofriendly. The chemicals themselves add to the pollution. A combination of both methods will be ecofriendly and cost effective. Raw sewage in the lakes contaminates ground water as well. An effective treatment measure should take into account, both monetary consideration and environmental change.

(e) *Afforestation and Gardening*

The lakes around the city require both afforestration and foreshore and formal landscaping in the catchment area, i.e. planting trees along roads, public premises civic amenity and open spaces. The forest department has taken developmental works around 75 lakes. About 25,000 saplings have been planted since 1989 in foreshores and peripheries of water bodies *(Deccan Herald,* 1993). The survival of the saplings range from 20 to 80 per cent and are dependent on several factors, such as unhindered heavy cattle grazing, toxic elements in the industrial effluents flowing into the tank, vandalism etc. *(Deccan Herald,* 1993). Forest nurseries have been established in Hebbal, Sankey and Madivala tank beds. A deer park has been formed by the department at the Kempanabudhi tank.

Management of Lakes

Apart from the conservation measures suggested, remedial measures to manage lakes would involve a process of continuous supervision to avoid a reversion to the present or worsening situation. It is essential to involve people. To form lakes protection committee, peoples participation is essential. Some good examples at peoples participation are at Tindlu, Madivala, Donekere and Vasanthapuram lakes. Lake protection committees have been constituted at Madivala, Hebbal Dorekere and Tavare kere lakes. The Thoreau Foundation with government agencies is involved in conserving Hebbal lake *(The Hindu,* 1997). A 1.25 crore Master Plan has been finalized to divert sewage water, desilt the water and beauty the ecosystem. The forest department and the BWSSB are involved in the joint effort *(The Hindu,* 1997). A plan is also drawn to approach the centre to review 17 lakes under the (NLCP) National Lake Conservation Plan *(The Hindu,* 1997).

Further, environmental education can prove to be a positive step towards environmental protection and conservation. Boating can be provided by planning according to the size of the lake. Jogging tracks, gardens, woodlots established around the lake would add to the recreational as well as conservation measures.

Lakes and wet lands which once provided man's needs in the city and formed the ecological blood streams are now exposed to human interference and extinction in several cases and a degradation of land resources. Attempts have been made at revival and conservation for over 17 lakes, especially the Hebbal, Dorekere, Sankey etc., under the National Lake Conservation Plan (NLCP). As a multi-departmental approach is needed, it is essential that a holistic approach be made, where agencies and people initiatives recognize an ecofriendly lifestyle at the individual, community and a city level.

References

Chakarapani, B.K. & Krishna, M.R, (1994) "Plankton and bird species richness of Sunkey tank". *Perspectives on the Ecology and Development of Sankey Tank,* Centre for Science and Technology, Bangalore.

Chandrakanth, R., (1995) "Cleaning up Ulsoor lake: A herculean task" *The Hindu,* Monday 3rd April, p. 3.

Deccan Herald (1993), "Taking the Plunge", Action Plan: Center for Science and Technology, pp. I & II.

Dev, Nisha (1993), Thirst things first - Spectrum (Supplement), *Deccan Herald,* 7th August, pp. I & II.

Jayaram, C. (1996) Management of lake system and surface water bodies in Bangalore, *Seminar on Environment Related Issue of Bangalore, Proceedings— Environment Association of Bangalore,* pp. 94-102.

Jayashree, Alladi (1994) "Save your water pockets", *The Times of India,* 15th April, p. 6.

Jayashree, Alladi (1996), "Vanishing Lakes, who is to Blame", *The Hindu,* 15th July, p. 1.

Mehu, Sowmya Ali (1995), "Forest departments pet machine to deeweed city tanks", *The Times of India,* Monday, 5th June, p. 3.

Ramesh, B.S. (1996), "Degradation of lakes drives rare birds", *The Hindu,* Thursday, 11th July, p.3

Subramaniam, D.K. (1985), "Bangalore city's water supply — A study and analysis, *Essays on Bangalore,* Vol. 4, pp. 51-147.

The Hindu (1997), 'Conservation plan for Hebbal lake', 24th March, pp. 3.

19

A Study of Slum Environment in Varanasi

V.K. Kumra and B.L. Singh

In India, the urban population has made rapid strides in recent decades. But our policy-makers have not given much thought to match housing, civic and sanitary facilities transportation etc. with the ever increasing population. As a result many of the incoming population find their way either in the slums or they themselves develop another slum or squatter settlements on government or privately owned open spaces of their own choice. In the city area at many points one can easily visualise slums of different types. Some of them are new or emerging slums and many of them have become slum in due course of time. Although slum population provides many services to non-slum areas but they remain neglected on many counts by our policy makers. Ultimately our negligence deteriorates the urban environment owing to day to day increasing slum population. Keeping these facts in view, here an attempt has been made to analyse the housing and housing condition and availability of civic and sanitary facilities in slums of Varanasi metropolitan city. In the last policy implications and key issues in making the slum environment cleaner and healthier, have been dealt in reference to urban slums of Varanasi.

Study Area

Varanasi is situated on the crescent shaped left bank of the River Ganga. This ancient city having over a million population in 1991 (10,26,467 persons in 1991 Census) attracts an estimated 70,000 persons everyday, almost all of them are pilgrims and tourists. Modern Varanasi maintains its traditional glory for trade and commerce. The core of its commercial activity conforms more or less with the heart of the city, the *Chowk* and its contiguous areas. Besides maintaining superiority for silk fabrics, brass wares, wooden toys and tobacco products, the city has recently made great headway in large industries employing about 10,000 workers. But the cottage industries are still predominant which employ over 85,000 people. With the increased functional activity, the population of Varanasi has increased out of proportion which in turn has not only over burdened the existing housing stock but also resulted in the growth of slums and squatter settlements in the city. Invariably, it has deteriorated the city environment. In addition, mushrooming of population has generated tremendous pressure on civic and sanitary facilities available in the city.

Genesis and Growth of Slums

Rapid urbanisation has brought an unmangeable, inevitable and persistant problems of slums in almost every city of the country. The periphery of the city becomes overcrowded as the slums develop at any time during the evolution of the city (Aruna, 1985). This is due to habitation of low income group people including hawkers, labourers, rickshaw pullers and other daily wage earners and more so by recently arrived poor migrants as they cannot pay higher rents. Owing to less rental income, owners find difficulty help in maintaining the building condition which further causes unhealthy environmental condition in the neighbourhood. In due course, it takes the shape of slums and blighted areas.

The major growth of slums in the city has taken place within a span of 44 years. In 1941, only 10.09 per cent of the total population of the city was living in slums and blighted areas which became about 20 per cent in 1991. After the partition of the country in 1947, thousands of refugees from Pakistan settled in the city mostly occupying the commercial area. Consequently, the city experienced maximum (85.05%) growth in its slum population during 1941-51 whereas urban population grew by 32.19 per cent. Further, after the 1971 Indo-Pak War, thousands of refugees came to the city and settled along the railway lines and roads, near the railway stations and outskirts of the city (Singh, 1990). Statistics revealed that the decadal growth of slum population has been always higher than the city's population growth (Table 19.1). It may be attributed to the fact that housing development has always mismatched with the growth of population. As a result, people get it easy to settle in existing slums or wherever they find open spaces. According to a report of Varanasi Development Authority (1987), about 21.5 per cent of the total population lives in 72 slums in and around the city. A study conducted by Singh (1990) revealed that there are 115 slum clusters in Varanasi with a total population of 1,54,738 persons. Many of the slums and squatter settlements have developed at the periphery of the city as also inside the old city in dilapidated houses (Singh, Kumra and Singh, 1988). The slums of Varanasi were classified into four categories on the basis of housing density, living density, population density, nature of dwellings and availability of civic and sanitary facilities. The first category of slums is characterised by extreme overcrowding and have a high population and accommodation density. It largely encompasses the old commercial areas of the city characterised by dilapidated houses with common toilet facilities and narrow lanes (4-6 feet in width). Slums of second category consist of areas which have evolved in a haphazard manner disregard of building regulations and devoid of sanitary facilities. It is spread over nearly one-third of the area under slums and inhabited by 25 per cent of the total slum population of the city. The third category of slums are basically the rural pockets called as *abadies*, living within the municipal limit of the city. These localities do not possess underground drainage and even open drains are inadequate. Waterlogging is a regular phenomena which results in unhealthy environment for human living. The fourth category of slums, having a population of 20,438 persons is primarily consists of temporary hutments situated mostly along the railway lines, roads and government owned open spaces. Although such localities are characterised by low density of houses but have high accommodation density. There are about 145 temporary hutments in the Cantonment area only. More than 50 hutments have been constructed on the sides of G.T. road near Andhrapul. There are complete absence of sanitary facilities which make the habitants more prone to diseases. Looking various categories of slums, it was thought better select one locality from each category of slums for the present study (Fig. 19.1).

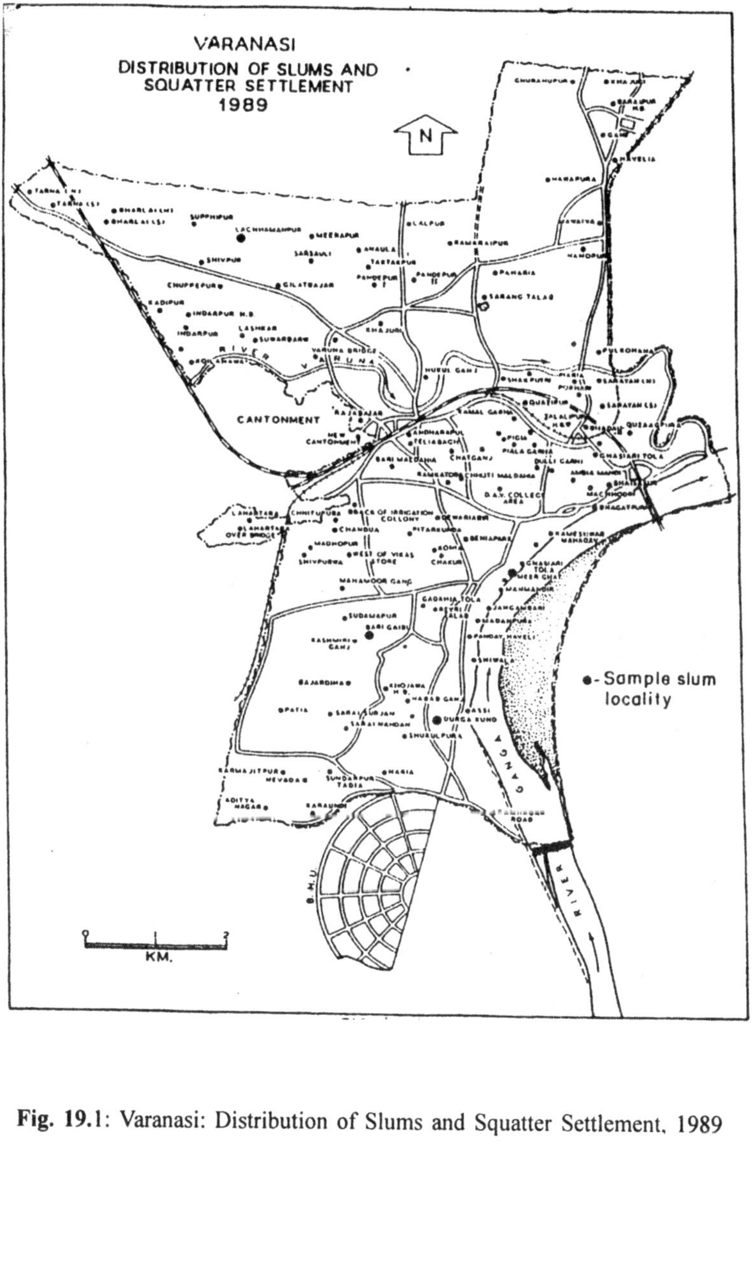

Fig. 19.1: Varanasi: Distribution of Slums and Squatter Settlement, 1989

Table 19.1: Growth of City's Population and Slum Population (1941—91)

Year	Total population	% increase in total population	Slum population	% increase in slum population
1941	2,58,646	—	26,100	—
1951	3,41,923	32.19	48,300	85.05
1961	4,71,258	37.82	69,600	44.09
1971	5,83,856	23.89	86,700	25.57
1981	7,20,755	23.44	1,20,812	39.34
1991	10,26,467	30.77	2,00,000	65.44

Source: Based on Census Records and Municipal Corporation of Varanasi.

Methodology

The basic aim of the present research is to understand the status of civic and sanitary facilities in the context of slum dwellers and its impact on overall environmental condition of the city in general and slum dwellers in particular. It demands, therefore, a methodology which be sensitive and flexible and embraces the views and opinions of slum dwellers towards civic and sanitary facilities. In order to get a fair representation of slum population in Varanasi four slum localities viz. Meerghat, Barigaibi, Lachchmanpur and Durgakund from each category of slums were selected at random. Sample for the survey was chosen by the cluster sampling technique and 100 persons were interviewed from each selected slum for the study. Data were collected on socio-economic, demographic, housing condition, living space, and sanitary facilities through interogating and interviewing the dwellers. Questions were also asked to know their attitude towards shifting to another site of their own choice for having better environmental condition.

Socio-economic Condition of Slum Dwellers

The study revealed that the socio-economic condition of slum dwellers is generally poor in all the four sample slum localities. Literacy was found to be maximum (43.2%) in Meerghat slum. It is because of the fact that most of the residents of this locality have higher income and are employed in services, Barigaibi is characterised by very low level of literacy (26.3%) as most of the dwellers are rural migrants and are working as a daily wage labourers. For having a better idea of their economic conditions income index was calculated (Table 19.2). In the present case income index has been calculated as the modal income of the locality which indicates that majority of the people are earning this amount per annum. Table 19.2 reveals that the maximum income index (Rs. 8,348) was found for the residents of Meerghat while minimum (Rs. 3,474) for slum dwellers of Durgakund. Unemployment was marked maximum (40%) among the youths of Lachchmanpur as most of them are illiterates. So far as the occupational structure is concerned, only 8.6 per cent of the slum working population is engaged in primary activities. Maximum percentage of workers (40.8%) engaged in small scale industries was found in category II slums while minimum (2.5%) in IV category. About 12 per cent of slum workers are engaged in rickshaw pulling. The maximum number of rickshaw pullers was found in slums of third category. They are mostly rural migrants and are characterised by poor economic condition. Bari Maldahia accounts for maximum percentage (78%) of workers engaged in government jobs as most of them are sweepers in municipal corporation. Despite, they live in insanitary conditions.

Table 19.2: Socio-Economic Profile of Slum dwellers of Selected Localities

S.No.	Location	Literacy (in percentage of total respondents)	Income Index (Rs.)	Total No.of households	Population in slum
1.	Meerghat	43.2	8,348	308	1,880
2.	Barigaibi	26.3	6,557	217	1,630
3.	Lachchmanpur	30.8	5,031	237	1,472
4.	Durgakund	39.5	3,474	153	858

Source: Based on Sample Survey, 1989-90.

Housing and Housing Condition

The study reflects that the housing condition is very poor in all the slum categories of the city and are characterised by congestion and overcrowding. Table 19.3 shows that the localities of the inner zone have highest density (1,711 persons/ha). In category III, although the number of persons per house is highest (15 persons/house) but number of persons per room is lowest (4-5). It is because of the spacious size of houses in *abadies*. Although in category I, majority of the houses are big and double or multistoreyed but these are characterised by acute congestion.

There is problem of living space in category I and II as these are situated in old and busy localities of the city, while there is no such problem in category III slum as it lies in the outer zone of the city and residents are mostly owner of the land on which their dwellings were constructed. Housing condition in different categories of slums also varies. Thatched houses were found in lowest number (6.6 %) in first category of slum while maximum (84.9%) is marked in category IV slums. It is because of the fact that the first category of slums are situated in the heart of the city and are mostly pucca. The percentage of mud-brick houses was highest (62.8%) in slums of category III. Majority of the dwellings in all the four types of slum are poorly ventilated. It has been found that only 6.3 per cent of the houses in slum of category I have proper ventilation. In addition to this, a large number of houses in Meerghat slum locality do not receive direct sun rays due to their location in narrow lanes (4-6 feet). Such unhygienic environmental conditions have caused higher incidence of diseases in slums than the non-slum areas (Kumra and Singh, 1992).

Sanitary Facilities

In urban slums not only the housing space is found to be inadequate, denying minimum norms of privacy but there is also found extreme shortage of essential amenities such as water supply and sanitation. Besides, stagnant pools of water and open drains are permanent features. Growth of slums degenerate the entire urban environmental set-up and create pollution. Improper disposal of garbage and stay cattles make the living worse. Table 19.4 depicts the poor status of civic and sanitary facilities in four selected slums of Varanasi.

It is apparent from Table 19.4 that about 75 per cent of the residents of Durgakund slum get their daily water requirement from municipal taps whereas the rest are deprived of any source of water supply. About 60 per cent of the families in Meerghat have piped

Table 19.3: Average Living space in different Categories of Slums

Slum category	No. of houses/ha	No. of persons/house	No. of persons/ha	No. of persons/room
I	118	14.5	1711	7.2
II	102	13.8	1407	6.8
III	37	15.0	555	4.5
IV	189	6.2	676	5.8

Source: Based on Sample Survey, 1989-90.

Table 19.4: Sanitary Facilities in Selected Slums of Varanasi (per 100 dwellings)

S. No.	Localities	Types of lavatory facilities				Type of waste water disposal			Sources of water supply			
		No facility	Sewer	Septic tank	Service	Sewer	Open surface drains	No facility	Municipal taps	Open wells	Hand pumps	None
1.	Meerghat	18	75	Nil	07	69	19	12	60	04	08	28
2.	Barigaibi	76	Nil	15	09	Nil	62	38	52	06	12	30
3.	Lachchmanpur	65	Nil	15	20	Nil	Nil	100	37	16	13	34
4.	Durgakund	82	18	Nil	Nil	18	Nil	82	75	Nil	Nil	25

Source: Based on Sample Survey, 1989-90.

water connections while 28 per cent households do not have any easy access. With reference to water supply, Lachchmanpur suffers badly as only 37 per cent of the surveyed families find access to municipal tap water. The situation is more pathetic when one looks towards the supply or availability of water. Against the requirement of 210 litres per capita per day, the supply is only 147 litres per capita per day (Varanasi Jal Sansthan, 1991). In comparison to this the slum dwellers on an average receive only 50 litres per capita per day of water and sometimes it is only 10 litres. Thus, on account of inadequate water supply, slum dwellers are bound to live in insanitary conditions.

Lavatory facilities were also found to be inadequate in most of the slum localities of the city. Investigations revealed that maximum number of families (75%) having sewered lavatory facilities was found in Meerghat slum as this area is situated in the main city. All other slums selected for present study are devoid of such facility. Further, 15 per cent of the families in Barigaibi and Lachchmanpur slums have toilets with septic tank. It is interesting to note that 82 per cent of the families in Durgakund slum do not find access to lavatories in spite of their employment in municipal corporation, so they are bound to defecate in nearby open spaces. It causes insanitary conditions for the neighbourhood. Excepting Meerghat where 69 per cent of the families have advantage of sewerage facilities, there is a total absence of waste water disposal system in the remaining three slums. Due to lack of waste water disposal system, this worse situation is compounded in the rainy season when the whole area remains waterlogged for a couple of days. It serves breeding ground for mosquitoes causing further deterioration of the living environment.

Attitude of Slum Dwellers towards Shifting to Another Suitable Site

Considering the pathetic situation in slums for human living, a number of schemes for slum improvement have been made but most of them failed as the residents do not responded well. To know their responses and attitudes towards shifting to another suitable site, an intensive survey has been undertaken and the results were tabulated (Table 19.5).

Table 19.5 reflects that only 12 per cent of Hindus and 4 per cent of the Muslim of category I slum have expressed their willingness to move out from this locality provided that better job opportunities and housing facilities are made available to them. In category IV more than 60 per cent respondents are willing to move out to any suitable place in the city. They have recently settled here and their dwellings and jobs are temporary in nature. The rickshaw pullers and casual workers prefer to remain there because they feel that residing away from commercial areas, would deprive them of easy availability of work throughout the year. Investigations revealed that salaried class people in slums of categories I and II would prefer to shift to another suitable site whereas persons employed in skilled jobs will not like to move out of their locality as they are living there from decades.

Table 19.5: Responses of Slum dwellers towards Shifting to Another Site

(in percentage)

Category of slums	Slum locality	No. of slums	Respondents in favour of shifting	
			Hindus	Muslim
I	Meerghat	100	12.0	4.0
II	Barigaibi	100	22.0	11.0
III	Lachchmanpur	100	39.5	23.0
IV	Durgakund	100	68.38	46.60

Source: Based on Sample Survey, 1989-90.

Conclusion

The major finding of this analysis is that the slums of Varanasi are characterised by poor housing condition, congestion, poverty, and are inadequately provided with sanitary facilities. The situation is worst in second category of slums which has been developed haphazardly. The slums of older part, of the city may be rehabitated as the responses of dwellers towards shifting to another site is very encouraging. They want to live in clean environment. The study also highlighted that owing to poor socio-economic condition, the residents of third and fourth category slums, do not prefer to shift from their place of living as their work place is near to their dwellings. The analysis warrants that the process of creating slum should be controlled as their growth deteriorates the urban environmental conditions in general and their own environment in particular.

References

Aruna, C.K. (1985). Spatial and occupational mobility of slum dwellers, in K. Vithal Reddy and R. Ram Mohan Rao (ed). *Urban Crisis in India,* Hyderabad Publishers, p. 148.

Kumra, V.K. and Singh, B.L. (1992). A geographical study of sanitation and health problems in slums of Varanasi, paper submitted for publication in Proceedings of the International Conference on Health and Development, held at University of Ibadan, from 20-25th Jan., 1992.

Singh, Onkar, V.K. Kumra and J. Singh (1988). *India's Urban Environment — Pollution, Perception and Management,* Tara Book Agency, Varanasi, pp. 44—55.

Singh, B.L. (1990). Socio-Economic Condition of Slum Dwellers: A Case Study of Varanasi City, Unpublished Ph.D. Thesis, submitted to Department of Geography, B.H.U., Varanasi.

Varanasi Development Authority, (1989), A Report on Slum Clearance of Varanasi, Varanasi.

20

Interaction Between Population and Urban Environment:

A Case Study of Kharagpur

SUKLA BASU

The study of territorially based systems, of which the urban community is a prime example, is known as *human ecology*. Human ecology refers to a concern with the processes and the form of man's adjustment to environment. The community is a generalised form of that adjustment. Community denotes a territorially localised population that carried on a collective life through a given set of institutions. From a functional stand point the urban community is a mechanism for relating a local population to a larger universe of activities. The unit of observation, therefore, is a population. Environment embraces all of the externalities that impinge on the life chances of a definable population. It includes not only the physical and biotic elements of an occupied area, but also the influences that emanate from other, organised populations in the same and in other areas. The economic opportunities opened up for the modern world provided plentiful, opportunities for cultural contact. Culture is the sum total of human learned behaviour and ways of doing things. Culture is invented, carried on and slowly modified by people living and working in groups. It is a term in anthropology which covers everything from the traditional manner in which people produce, cook or eat their food, the way in which they plan and build their houses and arrange them on the surface of the land, the manner in which men are organised into communities, to the moral or religious values which are found acceptable, on to the habitual methods by means of which satisfaction is gained in respect of the higher qualities of the mind. Thus the emergent cosmopolitan culture is a fusion of elements in varying proportions.

Population Environment: a mutually interacting, dynamic and sequential two way system in which a whole environment affords potentials to a whole population but undergoes change through resource use by human spatial occupancy.

Population—Social Organisation: interacting relationships between size or distribution of a culture group and the culture forms or structures adopted or devised by organised societies.

Organisation refers to the entire system of interdependences among the members of a population which enables the latter to sustain itself as a unit. The parts of such a system— families, clubs, shops, industries, for example— cannot be self-sustaining. They can only survive in a network of supporting relationships. Society exists in and through communication. By means of communication individuals share in a common experience and maintain a common life. Segregations of population take place; first upon the basis of language and culture, and second, upon the basis of race.

In its most visible aspect *urbanization* is a process of increasing territorial scope of organisation. Urbanization is a transformation of society the effects of which penetrate every sphere of personal and collective life. It affects the status of the individual and his opportunities for advancement, it alters the types of social units in which people group themselves, and.it sorts people into new and shifting patterns of stratification.

Study Area

Situated at the crossing of 22°22' and 87°21' the city of Kharagpur covers an area of 123 km^2. It is the largest urban centre of the district of Midnapore. The city has 41,579 occupied residential houses to accommodate 42,064 households and 2,32,575 persons (1981). At present it includes three distinct settlements; the railway settlement, the Indian Institute of Technology and the Kharagpur Municipality. It is 116 km away from Calcutta and is an important railway junction with rail lines deviating in four different directions. It is the headquarters of loco, carriage and wagon departments of the South Eastern Railway.

Kharagpur during the early period

Earlier records reveal that Kharagpur and its adjoining regions were forested. These forests stretched widely engulfing vast areas of Bihar, Orissa, Burdwan and Midnapore. These areas (including Kharagpur) were under full domination of the aboriginals. This is also confirmed by the number of aboriginal derivatives of pre-historic origin. These ethnic groups of population are generally considered to be the autochthons surviving population groups known as Santhals, Mundas, Lodhas, Mahalis and Koras. These groups have their distinctive living standards, socio-cultural patterns and value systems. 'Koupin' an aboriginal dress was worn and the people then lived in huts made of leaves. *Dola* an ordinary form of palanquin, was the only conveyance then. Archaeological evidences show that these vast areas remained untrodden until 1568 A.D. by any civilized kingship, though Bengalees of Aryan culture were found to have come in power as early as 600 A.D.

Kharagpur is situated in a *frontier tract* which occasionally witnessed many migrations, aggressions and conquests which led to cultural transformation. The basic cultural stratum of this region was non-Brahmanical. This tract had to witness Chola invasion, Mohammadan invasion, conquests from southwestern side and that by the Marathas from the west. Last in the chain of incursions were the Europeans. Such invasions and migrations due to various political and economic factors affected the people living in this area most intensively and in their wake, they brought in frequent changes in the social pattern.

In the second half of the nineteenth century Kharagpur was a vast area inhabited by a thin population consisting of cultivators. The establishment of religious centres had attracted some people to settle there. The Nandeswara and Jhareswar temples at Malancha were built in 1719, the material used is of stone. The Khargeswar Mandir at Inda was the product of the labour of the common people and artisans, whoever might be its financier. The town owes its name to this temple of Siva built by Kharga Singh in the seventeenth century. The area came to be known gradually as Curakpur followed by Kharaksur and Kharhgapur and finally Kharagpur. The area had only a few localities along banks of Cosseye river, along the roads leading to old Midnapore town as well as along the Jagannath Path to Puri. The houses, then were of the traditional pattern made of laterite and mud. Owing to the needs of the local people markets come up. Puratan Bazar is once such market which started its retail activities in 1882, dealing with principal essential commodities such as rice, vegetables, fish and meat. This bazar played an important role in the social and economic life of the people, linking the people of the various residential pockets, existed then.

Kharagpur during the British Period

With the advent of the British rule, the nature of the villages changed. Sanction was given to the Bengal Nagpur Railway Company for the construction of a railway line from Midnapore (Kharagpur) *via* Bishenpur to Bijudih a point on the Damuda river opposite the Jharia coalfields. This would, give the railway access to the Palamau and Daltonganj coalfields as indicated in the *History of Railway Projects 1899,* (Simla). A junction was established at Kharagpur near Midnapore and rails were laid from Sini *via* Kharagpur to Kola on Roopnarain river, and from Kharagpur to Balasore. The most vital sections of the railway i.e. Sini-Kharagpur and thence to Kolaghat and Kharagpur-Cuttack were opened for goods as well as passenger traffic in 1898-99. Midnapore, the district headquaters could not be connected a direct line, so a branch from Kharagpur to Midnapore (8 miles) was sanctioned in January 1899 and work commenced in the same year. It was opened for traffic in 1900 upto the bank of the Kosi river. The inauguration of the Bengal Nagpur Railway from 1899-1903 extinguished the major source of Cholera which used to be introduced by the pilgrims to and from Puri along pilgrim roads. The railway benefitted in many other respects as well. By facilitating the disposal of the produce, prices rose and the cultivators were well off. The opening of the railway track rapidly increased trade in forestry goods and wood (*sal*). Particularly the portion on which the present settlement stands now, had an elevation of about 40 feet from the surroundings and was locally known as *Kharagpurer Damdama*. This natural advantage led the then Bengal Nagpur Railway to select the site as a junction point of the rail routes branching off in four different directions.

The Railway Workshop was created in 1904 because of :

(a) nearness to Calcutta Port,
(b) availability of vast open land of hard laterite,

(c) fairly good transport facilities,
(d) water and power resources, and
(e) availability of building material from nearby sources.

Gradually twenty-six villages around the railway junction were brought under the railway settlement. When the process of development of Kharagpur as a railway settlement was accelerated and when the railway tracks were just being laid, cheap labour was imported from the distant southern districts of Andhra Pradesh. With the establishment and further expansion of the railway workshop, fresh waves of immigrants flowed from different parts of Andhra Pradesh and Orissa with the expansion of service sector, business growth and construction activities migratory movements to the city continued.

Table 20.1: Population in Kharagpur City based on place of birth

Place of birth	Total	Male	Female
(A) Born in India:	94.76	49.37	45.39
I. Born in West Bengal	61.90	30.77	31.13
(i) Midnapore district	53.78	27.75	26.03
(ii) Other districts of West Bengal	8.12	3.02	5.10
II. Other states of India:	32.86	18.60	14.26
(i) States adjoining West Bengal	14.08	8.04	6.04
(ii) States other than in (i)	18.78	10.56	8.22
(B) Other Asian countries:	5.18	2.99	2.19
(i) Pakistan	4.85	2.83	2.02
(ii) Burma and Nepal	0.33	0.16	0.17
(C) Countries in Europe and America:	0.06	0.03	0.03

Note: Figures are as percentage of total population of the city.
Source: District Census Handbook, 1951.

Kharagpur during the Post-Independence Period

If we consider the place of birth of the people residing in the city in 1951, it is evident that a little more than half of the population were born within the district. About 46.22 per cent were migrants from different parts of the country and even abroad. A large group of these migrants had the encouragement and support of their relatives and friends in the city who could get them some job and accommodation. One of the strongest forces influencing the development of an ethnic concentration is the operation of such migration chains. *A migration chain* is a process whereby migrants from one specific locality in the place (country/state) of origin are attracted to emigrate by friends or relatives from their own locality who have already emigrated. The links between immigrants and prospective immigrants frequently extend beyond the feedback of information and may include financial support for fares, provision of initial accommodation and even the securing of employment.

In 1971 about 37.67 per cent (60,750 persons) of the total population (1,61,257 persons) were migrants. Migrants may come from rural areas as well as from urban areas as indicated in table 20.2.

Table 20.2: Source of Migrants into the City in 1971

Source of migrants	Total	Rural	Urban	Unclassifiable
(A) Resided it India	86.63	34.11	45.36	7.16
I (a) Within Midnapore District	14.33	9.58	3.86	0.89
(b) Other Districts of West Bengal	19.06	4.50	13.56	1.00
II. Other States of India	53.24	20.03	27.94	5.27
(B) Foreign Countries	13.37	NA	NA	NA

N.B.: Figures are as percentages of total migrant population
Source: District Census Handbook, 1971, Part-XC

More than 50 per cent of the total immigrants in 1971 have come from states other than West Bengal, Andhra Pradesh alone accounting for about 20 per cent and the neighbouring states of Orissa and Bihar accounting for approximately a little over 9 per cent each (Table 20.3).

Most of these immigrants belonged to the working age group of 15-59 years. About 85.48 per cent of the male migrants and 76.03 per cent of the female migrants belonged to this group. The literates outnumbered the illiterates in the case of male migrants. About 29.47 per cent of male migrants and 70.53 per cent of the female migrants

Table 20.3: Immigrants from India other than West Bengal, 1971

Place/State	Total migrants	% of total migrant
Andhra Pradesh	12,000	19.75
Orissa	5,910	9.73
Bihar	5,575	9.18
Uttar Pradesh	2,770	4.56
Madhya Pradesh	1,575	2.59
Tamil Nadu	975	1.60
Maharashtra	956	1.59
Punjab	760	1.25
Others (including union territories)	1095	1.80

were illiterates in 1971. Educated persons seeking employment have come to this city with the hope of seeking a job either in the railways or in the other offices. Another feature noticed was that non-workers (63.61%) outnumbered the workers (36.39%) but people belonging to the working age group were significant in numbers. So the pull towards the city drew the people from far and near resulting in a significant growth of population the intermingling of various cultures and languages giving it a cosmopolitan character. The desire to seek out familiar surroundings in an alien environment is natural. The regional affiliations, mother tongue and previous contact are denominators of social interactions. These factors enabled the migrants to form social groups and maintain ethnic and group identities on the traditional patterns.

The European Institute now known as South Institute was originally set up as a centre of recreation for the Britishers with its stadium, swimming pool and auditorium. Besides there are innumerable clubs in different *paras* formed by the Bengalis (Table 20.4).

Another positive force which voluntarily draws immigrants together is the basis of shared language and culture. Segregations of population take place, first upon the basis of language and of culture and second, upon the basis of race. Obviously Bengalees constitute the largest ethnic group (39.15%) while the rest are represented by people from different parts of the country speaking Telugu (26.04%), Tamil (1.02%), Hindi (17.57%), Oriya (6.3%), Urdu (4.28%), Punjabi

Table 20.4: Social Associations of the Immigrants

	Name of the club/association	Group identity	Location
1.	South Side Society	Telugus	South Block, Railway Area
2.	Modern Association	Telugus	West Block, Railway Area
3.	Rabindra Bharati Association	Telugus	West Block, Railway Area
4.	Rao Brothers' Club	Telugus	Nimpura, Railway Area
5.	Kerala Samajam	Keralites	S.Block, Railway Area
6.	Khalsa Club	Punjabis	North West Block Railway Area
7.	Sat Sangha	Punjabis	" " " "
8.	Harijan Sevak Sangha	Harijans	" " " "

Source: Field Survey

(1.68%), English (0.85 %) and other languages (3.11%) in 1971. The diversity of language and social habits seems to inhibit a closer social relationship between different regional groups. The traditional pull compelled the migrants to revive their lost identities and to safegaurd the group interest of the various sections of workers in the railway workshop or other offices through the formation of various societies. However, employment in the workshop and other departments of the railways and in other occupations has brought the diverse ethno lingual cultural groups to a common meeting ground. This is found to have some impact on the interaction between individual and the social group. Though an individual continues to belong to the group based on traditional pattern i.e. region or kinship or both, the cohesion of this traditional collectivity is gradually weakening. The less the social distance there is between individuals, greater the probability of interaction of some kind. Similarly, the greater the physical proximity between people the more likelihood of interaction of some kind. The increase in transport and communication facilities is responsible for the greater social interaction. The behavioural connection in thought to be a function of opportunity to meet and the decay of distance. Distance decay means simply that, other things being equal, there will be greater amount of interaction (measured in such terms as visits, telephone calls etc.) between people and places close to one another than those further apart. Work places and places of entertainment offer opportunities for social interaction. The location of these activities may be only indirectly related to the residential places of the interacting population. Most of the cinema halls, video centres are located within the railway premises, but this does not hamper the people living in the municipal limits from enjoying such facilities,

Language is one of the strong cements of culture. One thing which divides the Bengali, Bihari, Telugu, Tamil, Oriya, European or Punjabis in Kharagpur is language (Table 20.5). It becomes the tool of group cohesion. Intercommunications is carried on either by a broken form of Hindi or Bengali, or in so far as upper classes are concerned, by means of the English languages. However, the migrants are trying to adopt the language of their new domicile and similarly the Bengalis are interested in learning the language of their co-residents. The data in support of this is available in the 1951 *District Census Handbook* of Midnapore.

Table 20.5: Bilingualism in 1951

A. By subsidiary languages spoken

Mother Tongue Spoken		Subsidiary languages spoken				
		Bengali	Hindi	Urdu	Oriya	Total
1. Bengali	62,883	—	2,086	1,283	1,579	4,948
2. Hindi	23,606	1,765	—	168	64	1,997
3. Urdu	1,738	357	210	—	5	599
4. Oriya	8,995	1,947	1,049	4	—	3,000

B. Non-Bengalis speaking Bengali

Mother Tongue	Total persons	% of Total population	Persons speaking Bengali
1. Telugu	23,839	18.39	589 (2.47)
2. Hindi	23,606	18.21	1765 (7.48)
3. Oriya	8,995	6.94	1,947 (21.64)
4. Urdu	1,738	1.34	357 (20.54)
5. Punjabi	721	0.56	19 (2.64)
6. Tamil	2,321	1.79	105 (4.52)
7. English	862	0.66	119 (13.81)
8. Others	4,664	3.60	1,345 (28.84)

Note : Figures in parenthesis are as percentage of the respective total population.

This Table 20.5 reveals that people are quite keen to learn the language of their neighbours or friends or co-workers. A great section of the Oriyas (21.64%) residing in Kharagpur can speak Bengali. Similarly 20.54 per cent of the Muslims (1,738 persons) can speak the local language.

The railways area gives a much more distinctive picture of the cosmopolitan character of this nodal junction. It is divided into 6 residential blocks — south, southwest, northeast, northwest and lines (i.e. still further west). The linguistic composition of these blocks has been illustrated in Table 20.6.

The rows indicate the number of families speaking a particular language while the colours indicate the blocks linguistic composition. This table clearly indicates the segregation of different groups although the boundary cannot clearly be defined. The Bengalis constitute 30.85 per cent of the total families, closely followed by the Telugus (26.96%). However, a greater number of Bengalis reside in the south, southwest and northeast blocks i.e. near the railway station working in the various offices the Telugus outnumber towards the northwest, west and further west i.e. surrounding the railway workshop.

To meet the needs of the different linguistic groups, separate schools in different media have been set up. Besides Bengali, there are several English (11), Hindi (27), Telugu (8), Oriya (5) medium government recognised schools in the entire city. The impact of the British rule is felt even today. The Church and England missions was established here. Similarly, the Sacred Heart Roman Catholic Church in the sixth avenue at Kharagpur Railway Settlement Area was built in 1916. The European Institute, popularly known as the South Institute still exists. Several other missionary schools have been set up which are quite popular today. Even from great distances like Kalaikunda, Salua, Medinipur, children come here to be educated by buses or vans arranged by the people of these places.

During the Second World War, the Government of India realised the importance of expanding and recognising facilities of technical education in the country. The first of these higher technical institutes to be set up in the east near Calcutta was at Kharagpur in 1950. The Indian Institute of Technology centred round a rural surrounding has cropped up on the same spot of the Hijli Jail. The area including the institute occupies about 1,400 acres of newly cleared land of the once

Table 20.6: Family Composition (based on mother-tongue) within the Railway Settlement Area, 1991-92

Family type	Location (I.O.W. Block)													
	South		Southwest		Northeast		Northwest		West		Lines		Total	
Bengali	693	(58.98)	257	(35.69)	714	(46.33)	955	(22.86)	801	(27.3)	276	(19.27)	3,696	(3085)
Telugu	125	(10.65)	125	(17.36)	176	(11.42)	1219	(29.16)	974	(33.2)	612	(42.74)	3,231	(26.96)
Other South Indian	45	(3.83)	73	(10.14)	73	(4.74)	527	(12.61)	234	(7.98)	93	(6.49)	1,045	(8.72)
Oriya	81	(6.89)	84	(11.67)	173	(11.23)	235	(5.62)	207	(7.05)	96	(6.7)	876	(7.31)
Hindustani	47	(4.0)	66	(9.17)	168	(10.91)	332	(7.94)	195	(6.65)	89	(6.22)	897	(7.49)
Christian (English)	55	(4.68)	44	(6.11)	22	(1.43)	77	(1.84)	29	(0.99)	16	(1.12)	243	(2.03)
Punjabi	22	(1.87)	06	(0.83)	9	(0.58)	47	(1.12)	43	(1.46)	30	(2.09)	157	(1.31)
Muslim	22	(1.87)	07	(0.97)	50	(3.24)	85	(2.03)	33	(1.12)	14	(0.98)	211	(1.76)
Others	85	(7.23)	58	(8.06)	156	(10.12)	703	(16.82)	418	(14.25)	206	(14.39)	1,626	(13.57)
Total Families	1,175	(100.0)	720	(100.0)	1,541	(100.0)	4,180	(100.0)	2,934	(100.0)	1,432	(100.0)	11,982	(100.0)

Note : Figures in parenthesis are percentages of the total number of families of the respective blocks.

Source: Compiled from the data collected from the Quarter Allotment Registers of I.O.W. (Inspector of Works) offices, South Eastern Railway, Kharagpur.

wooded Sonamukhi village and Hijli upland. Dedicated to the service of the nation, the institute has a truly national character as it draws pupils from diverse linguistic, racial and cultural groups.

Religious institutions promote group solidarity, where the orthodox religion of the immigrant group is alien to the host society, religious institutions and practices may encourage territorial as well as cultural unity. Besides temples, churches, *masjids* and *gurdwaras* have been constructed in different parts of the city to meet the needs of Hindus (90.98%), the Muslims (5.33%), the Christian (2.34%), Sikhs (1.07%) and others (0.28%). The percentages are that of the total population in 1971.

Housing Pattern

The east-west railway track divides the railway settlement into two parts. The northern side has mostly Indian styled houses while the southern side has mostly European styled bungalows and quarters with generous spacing and better planning. The quarters are brick built with concrete roof or tiles. They are alloted to the employees on the basis of economy and status ranging from one roomed labour quarters to first class large bungalows of the managerial staff.

Within the I.I.T. complex, buildings are nice samples of modern housing system and allied facilities occupied by the enlightened groups of men attached to the institute. The municipality reflects a diverse housing pattern ranging from first class modern buildings of the richer section to the thatched or mud walled ill-ventilated cottages of the poorer section (Table 20.7).

Urbanisation affects the institution of family, i.e. the joint or multi-generational family type to yield to the nuclear family type. There are no data available to indicate such change. However surveys conducted reveal that the number of nuclear type families exceed the joint families. This is concomitant with the gradual change in attitude towards the joint family. Another possible reason is that those who are in the railway services with alloted family quarters might have brought their wife and children along with them from their native places and set up nuclear families. The growing economic pressure and housing shortage are perhaps responsible for the breaking up of the joint families as well.

Table 20.7: Building Pattern of Kharagpur Municipality (1989-90)

Building Structure		Stories		Rooms	
1. Pucca	6,008 (35.94)	One-Storied	14,742 (88.18)	One Roomed	5,555 (33.23)
2. Kuccha	6,596 (39.45)	Two-Storied	1,810 (10.83)	Two Roomed	6,546 (39.16)
3. Semi Pucca	4,114 (24.61)	Three Storied	149 (0.89)	Three Roomed	2,496 (14.93)
		Four and above	17 (0.1)	Four Roomed	1,258 (7.52)
				Five Roomed	413 (2.47)
				Six Roomed	450 (2.69)
Total	16,718 (100.00)		16,718 (100.00)		16,718 (100.00)

Source: Compiled from Tax Collection Register, Kharagpur Municipality

The flux of commerce brings together people from various cultures and backgrounds, and requires that they learn to live and work together. They become detached from the parochialisms of locality and are led to dwell in a wider realm of experience. Out of necessity they become accustomed to.pursue their needs and interests through relatively complex organisational arrangements. A market centre is thus a mirror of life and economy of the people. It is a focus of various multifunctional human activities. Before 1911 there were only 62 shops in the railway town and these were the growth points of today's Kharagpur city (Table 20.8).

The establishment of the railway workshop followed by its expansion, the Indian Institute of Technology and several other such technical training institutions encouraged the opening up of several, ancilliary industries which in turn recruited managers, supervisors and skilled, semi-skilled and unskilled labourers. Several such engineering industries were located on major transport routes like the National Highway 6, Orissa Trunk Road near Inda and Puratan Bazar and Malancha Road (all within the municipal limits). In 1981, there were 94 registered factories, 1418 unregistered workshops totally employing 7,500 persons (excluding railways). Besides, a large number of electrical, agro based, service industries have been set up.

Table 20.8 : Temporal and Spatial Development of Markets

Location	Population (1981)	No. of shops (Licensed)	Percentage of Total Shops	Names of Markets	Year of establishment
1. Railway	82,100	3400	54.4	Gole Bazar (W.S. + Retail) New Settle-	1905
				ment Market	1942
				Gate Bazar	Not known
				Traffic	"
				Nimpura	"
2. Municipality	1,40,180	2700	43.2	Puratan Bazar	1882
				Kharida Bazar	1912
				Talbagicha	1950
				D.V.C. Settlement and Pirbaba Bazar	1954
				Malancha Market	1955
				Prem Bazar	1970
3. I.I.T.	11,000	150	2.4	Technology Market	1952

Source: District Census Handbook, Midnapore, 1961 and 1981

The basic means of communication within the city are the bicycles and cycle rickshaws. In 1985-86 there were 2,279 registered rickshaws in Kharagpur. However with the increase in income of the household use of automobile vehicles (particularly twowheelers) has increased. Bus services, auto-rickshaws, telephones have increased the interactions and linkages within the society. Several, clubs, theatres, cinema halls, video centres and libraries have brought about cultural interactions. The love of sports of the people of Kharagpur is reflected in the creation of the SERAA Stadium, (South Eastern Railway Athletic Association), the first of its kind in West Bengal. Here many inter-railway football tournaments and a number of games of all Indian attraction are held. Besides, there are several smaller play grounds, a few swimming pools scattered within the city. Hotels and restaurants too play an important roles in social interaction within the city

To cater to the medical needs of the people hospitals (such as the Railway Main Hospital, the State Government Hospital within municipal limits), the Bidhan Roy Hospital (within I.I.T. complex), several dispensaries, nursing homes and clinics have been set up.

Population pressure has resulted in an increase in land values and tenure. Vacant lands and waterbodies are gradually being converted to residential-cum-commercial use. Within the planned settlement of I.I.T. and the railway, slums have come up owing to the practice of allotment of living accommodation according to the level of income and not requiremnt for space. A lower class employee with a large family indulges in overspill of activities outside the house. This temporary encroachment to compensate for deficiency of living space within the house, subsequently becomes a semipermanent extension of the basic dwelling unit.

The slums in the freely developing areas had its beginning in the establishment of railway workshop and immigration of the business populace to the city. Since the formation of the municipality in 1954, the continuing existence of slums is explained by the fact that inhabitants cannot afford good housing and business as private enterprises do not supply at prices which they can afford. Their existences is therefore attributed to landlords, the tenants and the community.

Area of notified slums in Kharagpur (in 1979) was 1.71 km^2. and its population 43,080 (Civic and.other amenities, 1979). Flimsy and dingy dwellings, cramped camps, tortuous paths, little light and no privies open and overflowing drains, rotting garbage and excreta are the main features of slums. The degeneration of the urban environment is attributed to the arrival of the rural poor and their urbanity based on rural values. The people here are generally rickshaw pullers, domestic servants and the owners of roadside squatter restaurants, cycle repairing shops and tailoring shops. The culture of these slums is a result of the synthesis of the culture of the lower class and the culture of poverty.

Other acute problems faced include:

(a) lack of proper drainage system particularly within the municipal limits leading to water logging,

(b) shortage of water supply as every year the subsurface water level goes down due to lack of recharging in the acquifer (due to laterite soil conditions),

(c) lack of proper planned roads within the municipality,

(d) improper and inadequate sanitary system. According to 1971 census of Midnapore we find that in the developed areas of the municipality where there were full facilities 10 persons were using one latrine while in the less developed areas. About 17 persons were using one latrine. Service latrines outnumber the septic tanks,

(e) shortage of land, increase in land value and tenancy rate and hence housing problem.

Conclusion

The socio-anthropological history of the area around Kharagpur is the story of changing life pattern of the nomadic tribal population leading to their conversion into the agricultural peasantry.

After the establishment of railway workshop, Kharagpur grew into a cosmopolitan town and further developed with the establishment of the I.I.T. As there were no skilled mechanics available locally when the workshop was started, it was found necessary to import all the skilled labour and to pay high wages. Consequently, there is a curious miscellany to be found in the works of men from the Punjab, Bombay, the United Provinces, the Central Provinces, Bengal and Madras. As a work place of government and other officials the Kharagpur urban agglomeration is a meeting place. As a business centre it is a place where people from its hinterland go to purchase luxury goods and specialized goods. The assimilation of different cultures which have developed in different periods have kept their impression on different elements of the town. The city has a.definite tendency to thwart the formation of groups on the basis of regional affiliations and cultural traditions. Migrants from different parts of India with different religious and caste indentities are found to be loosing their old ties. People having different social and economic status, religions, caste and linguistic groups live side by side.

Several market centres have cropped up owing to the needs of the local people bringing about social integration. Employment opportunities are availed of at several ancilliary industries, commercial establishments, hospitals, educational institutions besides the railway workshop and other such railway offices, the I.I.T. and the municipality. Functional changes have resulted with the passage of time. With the increase in urbanisation more and more rural areas are being encroached industrial estates have been set up on the outskirts. Increase in population both by natural increase and by migration has resulted in a gradual change in physical, social and cultural environment.

In short the population—urban environmental interaction can be expressed as:

Change in Environment caused by	Resulting in:
(i) Opening of railway lines	Distance-decay, more linkages
(ii) The railway workshop	Employment opportunity—migration—segregation
(iii) I.I.T.	Mixing of different caste, creed, religion—cosmopolitan city
(iv) Development of industrial estates	Employment, more influx, slums
(v) Construction of Roads and opening of bus routes	Inter and intra-urban linkages, increase in vehicular traffic, social and cultural interaction
(vi) Development of market centres	Functional change, growth of commercial establishments, hotels, video centres
(vii) Growth of slums	Crimes, degeneration of the environment, increase in informal activities

References

Bose N.K. (1971), *Culture and Society in India,* Calcutta, p. 15.

Dixit, R.S. (1984) *Market Centres and Their Spatial Development in the Umland of Kanpur*, Lucknow, pp. 13-5.

Hawley, Ames H. (1971), *Urban Society: An Ecolog cal Approach,* New York, pp. 10-5.

Peach, Ceri (ed.) (1975), *Urban Social Segregation,* New York, pp. 10-2, 25-8.

RGI (1951), *District Census Handbook,* Midnapore : 1951, Census Publication.

RGI (1961), *District Census Handbook,* Midnapore, 1961, Census Publication.

RGI (1971), *District Census Handbook,* Midnapore, 1971, Census Publication.

RGI (1981), *District Census Handbook,* Midnapore, 1981, Census Publication.

Spencer, , J.E. and Thomas, W.L. (1973) *Introducing Cultural Geography,* California pp. 6, 21, 41, 191-2.

Trevor, R. Lee. (1977) *Race and Residence: The Concentration and Dispersal of Immigrants in London,* Oxford, pp. 27-8, 44-5.

21

Urban Renewal in Calcutta: A Perspective Study

UDAYAN MAJUMDAR

Inner City; Slum; Outer City

Urban renewal implies the creative use and re-use of the dilapidated quarters of the city–the slums–taken as a whole. Moreover, the idea of urban renewal advanced in this submission involves infusing vitality into areas judged to be in some way, substandard. Especially slum areas, characterised by poor quality of life situation and typically in the inner parts of the city and often home for lower-income families having social, economic and cultural values different from the more privileged in society, living in the outer city.

Squatter settlements, having some characteristics in common with slums, are also parts of the inner city. Nonetheless, this chapter primarily focuses upon slums, since squatter, settlements, having constant flows of new comers constitute a floating population. Further still, squatter settlers are largely unorganized (politically as well as economically) and are also incapable of articulating their demands.

Inner cities, anywhere in the world, have no classifiable boundaries setting them apart from the rest of the city, but they generally include the densely populated, deteriorating older parts of cities, slums and the homes of squatters living by the sides of railway tracks and canals, which usually intertwine with or encircle the centre of cities.

Here, the term inner city is used, not geographically but as an euphemism for slums that have been left behind. Even sections, away from the declining older core of the city, but sharing similar environmental characteristics, are termed inner cities. More accurately, they are people in slums who have been left behind by history and prosperity, when society and the economy moved on, when patterns of work and demography changed.

The inner city, whatever its precise boundaries, is easily recognised by its multitude of environmental problems. Air pollution, a problem for nearly all of the city, lays its pall most heavily over the inner city. Open space, parks and recreational opportunity are high priority for a better environment. Yet in the inner city, they are lacking to a higher degree than in the other parts of the city. Problems of noise, sanitation, lack of piped potable water supply, crime, congestion and general corrosiveness affect nearly all sectors of the larger cities. But overcrowding, ill-ventilated dwellings in a poor state of repair, squalid slums, unemployment, poor street lighting, rats, narrow streets that are unsuitable for motor vehicles, or in many instances, for the volume of pedestrian traffic to which they are subjected, and everpresent litter and garbage are afflictions more typical of the inner city. And these environmental problems interact with social, economic conditions to endanger the survival of the city as well as, sustain a poor "quality of life" situation.

Justice is essentially to be thought of as a principle (or a set of principles) for resolving conflicting claims. One is aware, however, that there is no one generally accepted principle of social justice to which we can appeal (Harvey, 1973).

And this requires a rethinking of the entire renewal endeavour along the lines of social justice as adumbrated by Manuel Castells (1977), Pickvance (1976), Pahl (1975) and Havey (1973) together comprising the subject matter of the sociology of consumption "new urban studies."

Considering that the focus of this paper, is primarily on developing a skeleton concept of "a just distribution justly arrived at" regarding urban renewal (essentially a multi-dimensional concept) rather than upon the validity of the dismal image of Calcutta itself (which in its own way reveals a lot about the city's urban crisis), an attempt has been made to understand the problem of urban renewal with special reference to Calcutta. Particularly with an emphasis on its relevance to the condition of the economically weaker sections of the city who are considered to be the residents of the inner city. Accordingly, this chapter also emphasises on the issue of improving access of inner city residents of Calcutta to basic infrastructure and much required social services—imperative to the improvement of quality of life.

The sociological problem thus, is to suggest a form of social-spatial-economic organisation which maximises the prospects of the least fortunate people–the inner city residents within the given constraints of space and other resources. For, it is known that renewal measures clearly move in a direction which bears little relationship to need or to the condition of the least advantaged people in the slums or of the economically backward people. And this has led to the creation of localized pockets of high unfulfilled needs, such as those now found in Calcutta's slums.

Obviously then, it is necessary to study the entire evolution of the concept of urban renewal and its actualisation, alongwith a field study of inner city residents in Calcutta.

The stimulus for this paper was provided by the fact that, notwithstanding all efforts on the part of the CMPO, as well as that of the CMDA (set up in 1970 through what is known as the swiftest piece of legislation ever made for urban development in the country), along with that of a host of other civic bodies, the Calcutta Metropolitan Area (the largest and the oldest metropolitan centre in the country with an area of 1,350 square kilometres) even today continues to present a dismal picture of urban degeneration (Sivaramkrishnan, 1978). For instance, as the State Planning Board suggests, the dwelling structures within the CMDA are in a rather decripit state 37 per cent the total structures are temporary and made up of non-durable building materials and 32 per cent the housing stock have walls made of durable materials but roofs are non-durable. Further as far as the *Bustee* dwellings are concerned—3,000 in the city of Calcutta—32 per cent of huts occupy less than 1000 square feet land; 78 per cent of huts are shared by more than six families; and 90 per cent of familes live in one room—indicating spatial inequality of the worst kind, leading to a poor quality of life situation. (SPB, 1990). Surely such a general picture of living conditions in the *Bustees* are below human acceptability. And unsanitary living conditions and other related social constraints being predominant in these areas, a vast number of the *Bustee* population is often compelled to turn to antisocial activities mainly due to the manifold social deprivation (SPB, 1990).

Since 1971, a massive programme for *Bustee* Improvements was undertaken by the CMDA for improvement of overall environmental condition along with extension of civic facilities to the *Bustee* dwellers. Such *Bustee* Improvement Programme is still continuing in some form or the other in the *Bustees* within the Calcutta Metropolitan District. Attempts were also made in certain *Bustees* within the creation of new buildings for *bustee* dwellers within the existing *bustees* under the Slum Clearance Programme. However, as the State Planning Board observes not much success has been achieved.

Again, the proliferation of highrises in Calcutta has led to dire consequences in human terms. In many instances poor quality construction and poor quality of services like water-supply, sewerage and drainage, solid waste management, etc. and inadequate maintenance of infrastructural facilities have further compounded the difficulties. And their total collapse can also not be ruled out.

Such highrise buildings have further, concentrated population in a limited area, and this acts adversely on the civic infrastructural facilities.

According to the Human Development Report 1990 in Calcutta, some three million people live in shanty towns and refugee settlements without potable water. They endure serious annual flooding and have no way to dispose off refuse or human wastes. Another 2.5 million live in under serviced older areas. Piped water is available only in the central city and in parts of adjoining municipalities. Sewerage connections are limited to a third of the former colonial core, and the poor maintenance of drains allows periodic clogging and accentuates flooding. Further it has been suggested that Calcutta needs additional resources if Calcutta is to manage itself, solve its own problems and pay its own way (UNDP, 1990).

It has also been suggested that even in the present context, "the lack of infrastructure, particularly communication, the absence of a planning and implementation process and the dearth of a coherent, realistic development policy are at the root of the many problems which plague Calcutta". (Basu, 1991). Though the CMDA has been endeavouring for the past two decades to arrest the decline of this primate city, and though programmes of urban renewal encompassing slum development, strengthening transport and communication network investment in new housing in the metropolis has also been taken up, Ashok Mitra suggests that, most of these urban development bodies/institution, having "mainly hitched their wagons to the star of urban arrangements and aspirations of Western, and, of late, American supercities" are incapable of regenerating Calcutta and cater to the city's myriad problems, particularly in the inner city areas characterised by their range of environmental problems, whatever their precise boundaries. (Mitra, 1979). Now whether it is due to, as Mitra suggests, the mindless emulation of archaic Western concepts and designs of urban renewal, or due to the lack of any coherent, realistic development policy, or due to the elitist views of Calcutta's commercial-civic leadership regarding urban renewal, in reality the various methods employed to tackle the city's problems have been largely ineffective. Perhaps in the case of poor, low income population living in the inner city areas, the reason for the perpetuation of this state of affairs could lie in the fact that the people are often not consulted when the civic authorities are planning schemes of urban renewal. And it could be that, for this reason, the people consider these schemes as schemes of the government, not their schemes. Surely then, this makes the effects of poverty on urban renewal and environmental development truly overwhelming.

However, the environmental problems and human misery associated with the slums in Calcutta have been mentioned *ad nauseum*. Hence, the problem of increasing poverty and declining inner city cannot be swept under the carpet any more if Calcutta is to retain its conviviality, religiosity, and politics–qualities that combine for the good of life (Kohr, 1984). For, it is only by caring for the inner city—home of the bulk of the city's population that everything can be kept in order. It is therefore, time that the city's civic authorities and those concerned with the city's welfare, cease to slur over the plight of the urban poor and returned to appropriate urban renewal policies, firmly grounded on social, historical and economic realities.

The Idea of Urban Renewal : Evolution of Concept and Practice

Cities are as old as human civilizations and their images are a reflection of their culture, polity, economy and society. And attempts at urban renewal, dealing with altering the urban image from time to time, is also quite old. Thus, in order to understand the process of urban renewal, it is necessary to look into the various characteristics of the urban image. Characteristics that extend much deeper than mere physical and spatial representations—founded upon the social, political, economic and cultural ways and aspirations of the people who live in cities, i. e. their social history, behavioural patterns and specific cultural attributes.

Here in India, and in the "chance-directed, chance-erected" city of Calcutta that began as a trading post to cater to the mercantile interests of colonial rulers and is today the natural habitat of 10.2 million modern-day Indians despite its present parlous state, a lot of urban renewal activities had been initiated by the government as well as several quasi-governments and private agencies from time to time. Calcutta's civic authorities created a detailed historical record of such renewal measures which are available for examination.

Prior to India's independence, the British government's urban development and renewal efforts in Calcutta were largely concentrated on the European quarter of the city. As a result of this, the Native quarter grew up as the inner city from the very beginning (SPB, 1990).

However, since the beginning of the 1960s and with the setting up of the Calcutta Metropolitan Planning Organisation in 1960, and further with the formulation of the Basic Development Plan (1966 to 1986), Calcutta's civic authorities, in common with their counterparts in other Third World cities, realised that the time has come to set the house in order. That is, bring order out of chaos, incident to rapid growth, and especially to the influx of people of many regions without common traditions or habits of life. While all Western attempts at urban renewal were essentially directed at wiping out the slums breeding "disease, moral depravity, discontent, and socialism", Calcutta's civic authorities were struggling to combat, since the 1960s, the population explosion and its consequences, caused due to the constant inflow of people into Calcutta (SPB, 1990).

The population explosion occurred as a result of the gradual decay of the rural economy and frequent outbreaks of cholera and other epidemics which actually encouraged the richer and elite sections of the rural population to migrate to the city. And since Second World War, migrants had been moving into the metropolis in waves, first the Indians from Burma, secondly the displaced persons from the erstwhile East Pakistan (now Bangladesh), and now the immigrants from the rural areas of West Bengal and other neighbouring states.

Naturally Calcutta became the most populous city in India

(9,194,018 in 1981) and one of the first ten cities of the world. And today, the urban aggregate has become so overloaded with population that it now seems to be well beyond the capacity of the city's civic and service institutions to square up with the situation. In fact, as lately as in 1966, the BDP observed the situation of the metropolis as "alarming" (CMPO 1966).

Research Findings in Calcutta and Discussions

According to urban historian Agnew *et al.* (1984) when urban growth and urban form in a particular region are seen in cultural context, the basis for understanding their patterns is provided by the region's historical experience, including the changing character of the region's ties to an increasingly integrated political economy.

Accordingly, a perspective study of urban renewal in Calcutta over a rather long period of time, a study of the city's growth beginning from the city's emergence out of the three villages—Chuttanutty, Calcutta and Govindpur, and also the city's changing form from time to time, caused by a multitude of changing structural factors—economic, cultural, social and political, including the city's changing "ties to an increasingly integrated political economy", reveal that, as a result of urban renewal in Calcutta, new residential zones have been developed and the central business district of the city has also been redeveloped and old shophouses replaced with multistorey blocks of flats. They are all architectonic masterpieces, but success, for such it must really be counted, has only been achieved at a cost. It is suspected that many old communities have been broken up and the low income population driven out. Further, much of character that differentiated one *mohalla* (spatial units encompassing the entire gamut of familial and sociocultural relationships among its individual residents, creating, in a sense, a large extended family) from another has been lost.

Again, dozens of new shopping centres have sprung up at strategic locations in Calcutta over the past ten years or so, in several instances, replacing popular pedestrian shopping malls. These shopping centres are mostly air-conditioned mazes given over to the single activity of spending money. And because these new shopping centres, charge high rentals from merchants, only the better sort of shops can open up along their dust-free corridors. Unlike the street, a shopping centre is by its very nature exclusive. Only those who can afford to buy its luxurious merchandise are also allowed to contemplate its waterfalls and glittering mirrors. The majority of the city's population, are as a consequence, deprived of their earlier freely accessible streetshops, and also denied entry into these new shopping centres.

Further, it is a truism that, till date, almost all legitimate efforts at urban renewal in Calcutta, designed to cope with the physical and material needs of the city's residents have only given way to glitzy, image conscious renovations characterised by a dedication to aesthetics over ethics, while life conditions in the inner city slums continue to decline. Small wonder then, Calcutta today, is the centre of conspicuous consumption and of potentially dangerous poverty at the same time. Even by conservative estimates, 45 per cent of the CMA population belongs to the category of economically weaker sections, and one-third of the city's population lives in squalid slums.

Furthermore, field work done in two inner city areas spread across the city, areas that have undergone "improvement" since quite some time—Cossipore, a predominantly Hindu area in the north western most corner of Calcutta; and Kidderpore, a predominantly Muslim area in the south western part of the city—within the Calcutta Municipal Area, which lies within the jurisdiction of the CMDA, in order to assess the impact of urban renewal/ "improvements" in inner city areas, revealed that provisions of essential urban amenities (safe drinking water, underground open sewerage and latrines, lighting, paved streets, community baths, etc.) that account so much for good quality of life, have been largely inadequate and poorly maintained (Foster, 1981). Most inner city residents stated that water supply was grossly inadequate and the source of supply (water taps/tubewells) was not maintained properly. The sanitary latrines are inadequate and that they were not kept in working condition and lacked flushing water. Again residents indicated that the surface drains and underground sewerage being not maintained adequately, led to unhygienic conditions. The roads inside the *bustees* became unusable during the monsoons. Adequate bulbs are not repalaced on electric lamp-posts when they go out of order leaving the narrow streets without lighting for long stretches of time.

The survey, thus, revealed the sheer inadequacy of the urban amenities within *bustees*. It was noted that on an average 40-50 people use a single latrine per day. The figure soars to 100 when any one or more of the latrines become unusable. *Bustee* dwellers pointed out that repeated demands for provision of more latrines have as yet remained unanswered. Similarly, on an average, 35-45 people use a single source of water (tap/tubewell), but this number rises to 100 or even 125 if any one of the supplies go out of order.

Again, it was observed that the occupants made repeated complaints that water supply was inadequate and that more taps/ tubewells should be installed; there were similar complaints about the open surface drains within the compounds which were frequently choked and were not cleaned by the municipal workers. However, more than 80 per cent of respondents opined that conditions relating to urban amenities were better than it used to be before "improvements". 'This obviously indicates that civic amenities provided within *bustees* were greater in number now, but increasing population seem to defeat all renewal effort. The maintenance of civic amenities was very poor, at least 25 per cent facilities did not function for most of the year.

Sadly enough, "improvements" did not touch upon the dilapidated dwelling units or shelter of inner city residents. The area of dwelling units varied from 80 sq ft to 120 sq ft and a few of them had small verandahs in front which serve as kitchens. Most dwelling were *kutcha* (temporary) with roofs of clay tiles or corrugated iron sheets while the walls were mostly of mud with bamboo-split bat tens; some of course had brick soling. Most dwelling units or huts were ill-maintained. The hut owner was reluctant to spent money, considering it uneconomical. Occupants stated that maintenance on their part was impossible because of limited earnings. About 6 to 8 families occupied each hutment.

People's Participation and Responses

Renewal projects were mostly undertaken without the active participation of the residents of inner city areas. This is indeed unfortunate considering the fact that about 70 per cent of respondents expressed eagerness to participate in the renewal planning as well as renewal processes. These respondents also showed considerable willingness to join workshops for acquiring skills in inner city renewal and construction of housing.

The total absence of participatory planning has resulted in total apathy among the inner city residents regarding further development of renewal programmes on their own initiative and the maintenance of the amenities installed.

Regarding their own efforts at improving their quality of life by bringing about complementary urban renewal, nearly 75 per cent respondents opined that they know best, and that if only the government could manipulate land valuation, planning and shelter policies to work, with their aspirations instead of against them, they could well manage their own future.

The *bustee* residents of Cossipore and Kidderpore have mostly come from the rural and semi-rural areas of West Bengal and from the neighbouring state of Bihar. Unemployment is high among youths and literacy level is rather poor.

Nonetheless, with the advent of Calcutta Urban Service Consortium or CUSCON (a voluntary organisation) into these areas youths are being organised towards attaining self-employment. CUSCON has also succeeded to some extent in improving medical care facilities within these *bustees*. The *bustee* residents largely prefer spending their leisure within the slum areas and there is sufficient participation in festivals and sports within *bustees*. The survey brought to the fore that *bustee* residents maintain strong links with their places of origin, notwithstanding their long stay in the *bustees* and also that their primary aspirations is to return to their places of origin and work upon social mobility of any kind whatsoever at their respective native places.

Enquiry into the social backgrounds as well as current social positions of inner-city residents revealed how "a significant proportion of the city's inhabitants were, and are, isolated", rather encapsulated from the social consequences of lack of physical distance, number, density and heterogeneity, by social structures and cultural patterns which they either brought to the city or developed as a result of living in it (Wirth, 1938).

Questions were also asked to inner city residents on, whether they lived in the inner-city by choice, or partly because of necessity, or partly because of tradition, or because they have no other choice.

Based on the understanding that the social and cultural moorings of the inner-city population are best described by the particular type of inner-city residents, they represent according to the classification scheme of sociologist (Gans, 1968):

The cosmopolites—being such residents who remain in cities to seek, advantage of the unique cultural and intellectual benefits, viz., writers, artists and scholars.

The unmarried or childless—being those people who choose to live in cities because of the active night life and varied economic and recreational opportunities.

The ethnic villagers—being people who prefer to live in their own tight knit neighbourhoods. Being constituted of immigrant groups they isolate themselves in such closed communities in order to avoid resentment from well established urban dwellers.

The deprived—being those very poor persons and families having little choice but to live in low-rent, and often rundown neighbourhoods.

The 'trapped' and downward mobile—being those city residents who wish to leave urban centres but cannot because of their limited economic resources and prospects. For example, families who feel that their, neighbourhoods have been "invaded" by non-residential land users or unfamiliar immigrant groups often desire to live elsewhere. And, the "downward mobiles" are those people who once held higher social positions but who are compelled to live in less prestigious neighbourhoods owing to loss of a job, death of a wage earner, or old age.

Among the inner city residents in Cossipore and Kidderpore, the majority (approx. 65%) would perhaps belong to Gans' category of "ethnic villagers." The rest seemed to live in the inner city as a result of being "deprived" (approx. 27%) and "trapped" (approx. 8%) Further, the enquiry into the social, cultural and economic way of life of the inner-city residents revealed that in their present situation, these urban dwellers must silently accept the dilapidated housing and blighted environment to which the housing market relegates them, as capital has been consistently transferred to other sectors or out to the much more profitable private building market in the outer city, although among the inner city residents are some for whom the *bustee* is a hiding place, or a temporary stop-over to save money for investment at home in rural areas.

The field work sought to establish the extent to which Calcutta's civic authorities have been successful in doing inner city renewal and also that merely being at the receiving end does not necessarily lead to higher quality of life standards. A thorough study of the story of urban renewal in Calcutta from secondary source materials, ranging from the social anthropological through the purely historical, to the largely statistical, both qualitative as well as quantitative, nonetheless critical, revealed Calcutta's true urban ethos. The cultural values about the nature of the city and urban life are shared by a large number of the city's residents, especially Calcutta's commercial and civic leaders, who always used the renewal process for the city to define and control the image of Calcutta during the periods of rapid social transformations. Surprisingly, and, perhaps happily, the sample of inner city residents showed high awareness of their own problems and were able to make almost specific suggestions regarding inner city renewal with relation to the wider market economy.

Inner city residents prefer gradual reform opposed to any drastic approach to renewal. Further, they advocate creation of social safety net for inner city low income families and others living on a pittance in order to beat rent rise related problems that usually arise with redevelopment and renewal. Of course, the government has to take care of funding such safety net programmes.

The residents of inner city urge allotment of home ownership. The government, they feel, should sell complete dwelling units with land and where necessary, even give away free-of-charge, such dwelling units. This they feel, could be a major help to them and a possible way out in the kind of shortage economy like ours. Of course, at the same time, all residents opine that owning poor quality of housing is more like incurring a debt than acquiring an asset. Maintenance costs being rather high and very few inner city residents being actually able to afford it. A striking feature revealed is that dwelling accommodations within the inner city are available at very low rents. Residents feel that this is so mainly to accommodate low incomes and perhaps for ideological reasons too. Infact, the rents never even cover minimum maintenance costs of the existing dwelling units. And this has led to the dilapidated situation of huts. Certainly major renovations are urgently required.

The majority of those who responded, suggested that urban renewal of any kind whatsoever cannot be imposed from above. For impositions only lead to social justice and individuality being hampered. Moreover, respondents also largely regarded the civic authorities decision making process as grossly unfair, arbitrary, or even plainly mysterious at times. Very few respondent regarded the ongoing renewal process as fair and straightforward (these respondents had been successful in procuring job placements with the government's assistance). This vindicates Dunleavy's argument that "people's political alignments are closely connected to their relationships to the state" (Dunleavy, 1979). The majority of inner city residents claiming the renewal process as unfair, drew particular attention to what they regarded as unjust distribution of materials and services. Further, they said that policies were implemented in different ways in different areas. Of course, the fairness of these somewhat jaundiced comments and views is not the issue. Although civic authorities have been endeavouring for quite some time to improve the overall quality of life of inner city residents in Calcutta, there exists a wide hiatus between the actual requirements and aspiration of inner city residents, and the civic authorities actions and intentions. In brief, the civic authorities are indifferent to the inner-city residents' priorities. and rarely if ever manage to deliver the goods.

Thus, what Calcutta needs urgently is to follow renewal policies that are based on and also promote the principle of social justice rather than emphasize on piecemeal physical improvements of select areas only—catering to the more better-off population in the city. And for this it is imperative to follow examples of inner city renewal set by other cities sharing similar problems and life conditions as Calcutta—as for instance with regard to rent control, pollution control, improvement of existing housing stock, etc. Especially since, it is a fact that in matters of urban development, design and renewal, Calcutta like most other Indian metropolises has so far followed the footsteps of advanced industrialised nations. Much of the urban change that has taken place in Calcutta over the last few decades which has drawn

Calcutta closer to the same forces that presently shape the destinies of other Indian cities as well as cities in other countries. In fact, with Calcutta's adoption of contemporary western styles of building designs, ekistics, urban planning and associated consumption values and patterns, today it has emerged as a giant metropolis of a truly global style.

Calcutta's central business district is dominated by skyscrapers, is inhabited by international corporations and banks, whose employees wear similar clothes and grab fast food during lunch breaks—like central business districts in any other metropolises in India and abroad. Accordingly Calcutta now, also shares many more common urban problems than ever before. Problems stemming from increasingly similar–and interwoven–economic and administrative systems. And this implies that Calcutta now also has to share solutions (success stories) to her urban problems more than ever before.

An attempt must be made to share success stories with other mega cities of the world and also to propose new found prospects for the forging of new partnerships for innovation which include the public, private and voluntary sectors' working in close co-operation with the social science community. The sharing of success stories becomes important as Perlman rightly says, for avoiding replication of mistakes, or avoiding costly invention of the wheel. Moreover, the sharing of success stories also helps in legitimising innovation, and providing the much needed credibility for its proponents (Perlman, 1986). For instance, even while the West Bengal government is contemplating introducing a partial ban on the movement of cars one day a week to ease traffic congestion and to reduce petrol consumption, in Mexico City, attempts by the municipality to check, air pollution, automobile fuel consumption and traffic congestion by a programme called *Hoy No Circula* (day without a car) has already proved successful.

Finding that the main source of atmospheric pollution is automobiles, the municipality in this most populous city in the world with the dubious distinction of having the most polluted air in Latin America, has banned use of private cars from 5 am to 10 pm one work-day a week. The results have been dramatic. Within a year of launching this programme, Mexico City has seen a 23 per cent reduction in total pollutant emissions. Over-all, air quality indexes have also improved from 10 to 15 per cent.

Further what really lends credibility to the West Bengal government's plans along similar lines is that, after the introduction of the *Hoy No Circula* programme, traffic circulation in Mexico City has improved considerably and automobile fuel consumption in the city has also dropped. Most importantly, this programme has received a 75 to 90 per cent approval from the public.

Of course, for such programmes to actually become effective, public transporation in the city needs upgrading. In this aspect Calcutta can perhaps follow the example set by the Regional Transit Company in the Munster, Germany Metropolitan Region. In the Munster, the new bus fleet operated by the RTC has actually made hour-long bus rides to and from work enjoyable. The buses have soft reclining seats with earphones that offer two radio programmes and two taped music programmes. Commuters are also provided with daily newspapers and magazines. On request, bus drivers even call to have taxis meet passengers at bus stops.

Although in India's and Calcutta's context this might seem apparently a luxury expenditure, this service, as seen in case of Munster, can actually discourage. the use of privately owned automobiles for commuting purposes. It can also bring about a positive impact on traffic, pollution and lifestyle.

Surely, such successful renewal programmes that also help revive the economy in times of crisis can be profitably adopted by metropolitan authorities in India on the basis of inductive reasoning especially in Calcutta, currently plagued with problems of air pollution, automobile traffic congestion and poor public transport system. Urban experience worldwide indicates clearly that urban problems of these kinds are by their very nature, more global than local. Of course, one is aware that, no particular city with its specific standards and renewal model can lay claim to general validity, as, in the various countries of the world, cities are determined, by differing historical backgrounds. While there is an apparent similarity in the explosive population growth in many cities of the world, nevertheless, experience has taught us that ways of coping with cities differ, and are characterised by the effects of social and cultural conditions specific to the region combined with international influences. Nonetheless, experience with urban renewal worldwide has also shown that dovetailing success stories regarding the problem of urban renewal does help in improving access of inner city residents to basic infrastructure and much needed social services. The idea being to bring about a new approach to social-spatial-economic organisation that maximizes the prospects of the least fortunate people. This has to be drawn along the lines of social justice in distribution as proposed by the 'new urban studies', found in the early works of David Harvey, R.E. Pahl, C.G.H. Pickvance and Manuel Castells. The inner city residents must be provided with collective consumption, mediated by the state. Surely the provision of such collective consumption could lead to conflict of interests and values, especially since state intervention in this issue highly politicises. Nonetheless, such conflict of interests and values can be mediated by careful analysis and intellegent choices, once the nature of Calcutta's local culture and experience replete with traditions of powerful social movements, are recognised.

References

Agnew, J.A., J. Mercer & D. Sopher (1984), *The City in Cultural Context,* Allen & Unwin, Boston.

Banerji, H. (1966), *The Socio-economic Plan Frame,* Town & Country Planning Department, Calcutta Metropolitan Development Organisation, Calcutta.

Basu, Sreemay (1991), Urban building : System constraints in Calcutta, *Economic and Political Weekly,* vol. XXVI, Nos. 31 & 32.

Castells, Manuel (1977), *The Urban Question—A Marxist Approach,* Edward Arnold, London.

C.M.D.A. (1986), *Development Need of the Metropolitan Area,* Calcutta Metropolitan Development Authority, Calcutta.

C.M.P.O. (1967), *Traffic and Transportation Plan for the Calcutta Metropolitan District, 1966 to 1986,* Planning Department, Calcutta Metropolitan Development Organisation.

C.M.P.O. (1964, Revised 1967), *Bustee Improvement Programme—Calcutta and Howrah,* Planning Department, Calcutta Metropolitan Planning Organisation.

C.M.P.O. (1966), *Basic Development Plan for the Calcutta Metropolitan District, 1966-1986,* Calcutta Metropolitan Planning Organisation, Government of West Bengal, Calcutta.

C.M.P.O. (1967), *Interim Report—Housing, Calcutta Metropolitan District,* Department of Town & Country Planning, Calcutta Metropolitan Development Organisation.

C.M.P.O. (1971), *A Memorandom on a Perspective Plan of Calcutta Metropolitan District and West Bengal, 1971 to 1989,* Planning Department, Calcutta Metropolitan Development Organisation.

C.M.P.O. (1971), *A Note on the Estimation of Employment Potential During the*

Fourth Five Year Plan Period based on CMD Development Schemes, Planning Department, Calcutta Metropolitan Development Organisation.

Dunleavy, P. (1979), The urban bases of political alignment, *British Journal of Political Science,* vol. 9, pp. 409-43.

Foster, Margaret (1981), A Study of Some Consequences of the Calcutta *Bustee* Improvement Programme, Unpublished MA dissertation submitted to the Institute of Planning Studies, University of Nottingham.

Gans, Herbert J. (1980), Urbanism and suburbanism as ways of life, in Robert Bocock, Peter Hamilton, Kenneth Thompson & Alan Waton (eds), *An Introduction to Sociology—A Reader,* Fontona paperbacks in association with Open University Press, pp. 397-421.

Harvey, David (I 973), *Social Justice and the City,* Edward Arnold, London

Phal, R.E. (1975), *Whose City? And Further Essays on Urban Society,* Penguin Books (UK).

Jagannathathan, Vijay N. (1987), *Calcutta basic development plan : In retrospection Economic aspects of Cǎlcutta's BDP,* Calcutta Metropolitan Development Authority, Calcutta and National Institute of Urban Affairs, New Delhi.

Kapoor, R.M. (I 982), Issues in metropolitan planning in India: Urban renewal—The Calcutta Context. Paper presented in the seminar on Land in Metropolitan Development.

Mitra, Ashok (1979), Where do we go from here ? The problems of Calcutta Metropolitan region, *Nagarlok,* vol. XI, No. 2, April-June.

Mukherjee, S. (1976), Slum improvement in Calcutta, *Nagarlok,* vol. VIII, No. 1, January - March.

Perlman, Janice E. (1 986), Mega-cities and innovative technologies, paper presented in the XI World Congress of the International Sociological Association, New Delhi, August 18-24, 1986.

Pickvance, C.G. (ed) (1976), *Urban Sociology—Critical Essays,* Tavistock Publications, London.

Roy, Dilip & Jai Sen (1980), Calcutta's *bustee* improvement programme, *Science Today,* April.

Sivaramakrishnan, K.C. (1978), *Indian Urban Scene,* Indian Institute of Advance Study, Shimla.

S.T.B. (1973), *Comprehensive Area Development Programme (CADP)—A New Stretegy for Development,* State Planning Board, Government of West Bengal, Calcutta.

S.T.B. (1990), *A Perspective Plan for Calcutta 2011, Development and Planning Department,* State Planning Board, Government of West Bengal, Calcutta.

U.N.D.P. (1990), *Human Development Report,* United Nations Development Programme, Oxford University Press, New York.

Wirth, Louis (1938), Urbanism as a way of life, in Paul Hatt & Albert J. Reiss Jr. (1957), *Cities and Society,* The Free Press, Glencoe. pp. 46-64.

22

Temporal Variations in Residents Perception of A Planned City: Gandhinagar

N.G. DIKSHIT

Perception of a city is a process which can be studied over a period only. It is important specially in the case of a planned city, where various amenities and infrastructural facilities may be inadequate in the initial phase of its development and expansion. Such a situation may generate feelings of dissatisfaction and unhappiness among the early settlers of the city. They tend to then compare new and old towns with reference to the amenities, social environment, infrastructural facilities etc. Residents in the new city may alter their opinion (or impression) i.e. perception, about the new city over a period as a result of manifold development there. This study is carried out with this background.

The Study Area

Gandhinagar, the study area, is the newly planned city and also capital, for the State of Gujarat, created following an administrative decision. It is the third planned capital city for a State in India. The city started functioning since 1970, when the entire administrative complex of Gujarat state was shifted here from Ahmedabad alongwith the government employees. These early residents of Gandhinagar were the selected residents as they were required to shift to Gandhinagar from Ahmedabad or other towns in Gujarat on account of their transfer.

The city is located at about 35 km in the northeast of Ahmedabad along the River Sabarmati. It is planned to have a grid pattern of roads and housing blocks called sectors. The city is divided into 30 sectors formed by rectangular inter-sections of roads. These roads are numbered from 1 to 7 running almost west to east and are labelled alphabetically as K, Kh, G, Gh, Ch, Chh and J, running almost north to south.

Each sector contains different types of government owned residences, revealing a hierarchy of income of residents. Private residences appeared only after 1980, hence only government-owned residences are considered for this study. Each developed sector is provided with the basic facilities and associated infrastructure. Out of 30 sectors planned, eighteen sectors have been developed by now (1990). The city is designed to develop a multi-nuclei structure of urban development.

Objectives

This study is undertaken with following objectives. The residential accommodations, provided to the early settlers in Gandhinagar, were owned by State Government and were based on the income groups of the residents. At present, there are eight types of government owned residences available to government employees only. These residences are planned and constructed in different sectors during seperate time phases.

Most of the early migrants-residents here were used to stay in a housing pattern which may have been based on social parameters like caste, language, religion, region or even occupations. People have lived in such a social environment over a number of years. Here, the immediate or near neighbours belong to a homogeneous social group formed by the social parameters. But, in Gandhinagar, the government owned housing pattern is based on the income hierarchy only, where the immediate or near neighbours may not belong to the same caste, religion, region, language. Such a situation affects the pattern of interactions by residents, who may prefer to interact longer distances frequently in order to maintain their traditional behaviour of interactions with people belonging to their own communities. Many of these residents may find it difficult to adjust to the new environment created on the basis of economic parameter—like income. This paper attempts to study whether the residents have perceived positively the non-traditional pattern of housing based on income hierarchy.

Amenities from an integral part of human life in modern era.

With various innovations and technological development, changes have occurred in the types of amenities. Life style of people also changes over a period. Larger and already developed cities offer a wide range of amenities, services and infrastructure, which are well organised. Residents in such large cities are very much used to such modern life style, which becomes almost a part of them.

In Gandhinagar, amenities and infrastructural facilities were not available adequately during its initial phase of development, i.e. between 1970-75. Government employees who were then transferred to Gandhinagar as early settlers, were expected to acclamatize to the prevailing conditions. From the beginning, they had to depend on bigger towns nearby, like Kalol, Mahesana and Ahmedabad for almost all of their occasional needs. Even their daily requirements were also not available adequately and consistently. Frequent shortages created a lot of stress and strain on them. As a result, they interacted with Ahmedabad and other towns nearby frequently. There was no scope for luxury purchases in Gandhinagar. Apart from shopping, residents in Gandhinagar also depended heavily on Ahmedabad for various services like technical and higher education, specialised medical treatment and recreation.

With the passage of time, many facilities are made available in Gandhinagar. This has changed the pattern of interactions by residents, from occasional long distance interactions to other towns to frequent short distance interactions within Gandhinagar. As a result, their perception about. Gandhinagar is likely to have changed. This study attempts to know as to what extent, the residents in Gandhinagar have adjusted to the prevailing conditions over a period of two decades from 1970 to 1990.

Thus following are the specific objectives of the study:

1. To study the impact of income based hierarchy of housing pattern over the traditional housing system based on social parameters.
2. To examine the changes in the perception, on the whole, about the town by residents over a period from 1970 to 1990.

Data-Base and Methodology

This study is based on primary data. A field survey was carried out and 250 households were covered. This sample survey has covered about 2.5 per cent of population of Gandhinagar as per Census of 1981. Primary data is grouped into various cross tables as per the requirements for the purpose of analysis.

Respondents are classified into five groups, each group indicating an independent parameter. Each group is again sub-divided on the basis of few more parameters. In all, 27 parameters have been considered as follows. Details are given in Table 22.1.

Major Groups of Respondents	No. of Parameters
1. Duration of Stay in Gandhinagar	04
2. Level of Income	06
3. Level of Education	06
4. Occupation	06
5. Age Group	05
Total:	27

Respondents were asked to give their opinion about the city with reference to the amenities available in Gandhinagar at two points of time: (a) at the time of their arrival in Gandhinagar and (b) at present. In all, 20 items were selected for collecting respondents' perception, along with the qualitative ranking such as good, fair and poor (Table 22.2).

The following hypothesis are formed to arrive at a conclusion. They are:

Hypothesis I: Residents like to interact more with their best friends in their own communities, irrespective of income-based housing and constraint of distance.

Hypothesis II: Perception by residents about Gandhinagar has changed depending on the period of stay in Gandhinagar, socio-economic status of the respondents and availability of goods and services in Gandhinagar.

Table 22.1: Overall Impression about Gandhinagar

Groups		Image of Gandhinagar is			
		Improved		Deteriorated	
		Respn	%	Respn	%
A: Age-Group					
1. 20–30	Years	30	88.2	4	11.8
2. 31–40	Years	68	94.4	4	5.6
3. 41–50	Years	97	97.0	3	3.6
4. 51–60	Years	40	97.6	1	2.4
5. 60+	Years	2	66.7	1	33.3
B: Income Groups					
1 Below 1000	P.M.	6	100.0	—	—
2 1001–1500	"	30	90.9	3	9.1
3. 1501–2000	"	57	93.4	4	6.6
4. 2001–2500	"	57	95.0	3	5.0
5. 2501–3000	"	49	96.1	2	3.9
6. 3000+		38	97.4	1	2.6
C: Education					
1. Primary		11	100.0	—	—
2. Secondary		60	90.9	6	9.1
3. Graduate		117	95.1	6	4.9
4. Post-Graduate		45	97.8	1	2.2
5. Ph.D.		3	100.0	—	—
6. Others		1	100.0	—	—
D: Occupations					
1. Below clerks		10	100.0	—	—
2. Clerk/Head clerk		91	93.8	6	6.2
3. Officer		78	97.5	2	2.5
4. Sectional Head		25	100.0	—	—
5. Teacher/Professor		16	80.0	4	20.0
6. Others		16	94.1	1	5.9
E: Period of Stay					
1. Less than 1 year		9	90.0	1	10.0
2. 1–5 Years		60	90.9	6	9.1
3. 6–10 Years		78	95.1	4	4.9
4. 10 years and more		90	97.8	2	2.2

Note :Figures indicate percentage of respondents.
Source: Field Work by the author.

Table 22.2: Variations in Perception of Gandhinagar — 1970-90

Particular	Good	Fair	Poor
1. Daily Needs	152	– 20	– 92
2. Grocery Shops	240	– 15	– 88
3. Provision Goods	142	18	– 97
4. Medical Facilities	144	– 56	– 82
5. Social Celebrations	80	94	– 32
6. Religious Celebrations	100	142	– 30
7. National Celebrations	35	142	– 21
8. Hotels/Eating Houses	140	207	– 78
9. Cinema/Shows	117	73	– 15
10. Exhibitions	143	95	– 69
11. Traffic Facilities	86	– 04	– 56
12. Traffic Problems	193	Nil	– 20
13. Noise Pollution	42	21	– 10
14. Relations with Neighbours	95	37	– 84
15. Cleanliness nearby	42	– 31	– 60
16. Water Supply	11	– 19	– 121
17. Power Supply	28	– 20	– 11
18. Banking Facilities	67	35	– 91
19. Postal Services	80	83	– 79
20. Educational Facilities	70	22	– 85

Note : Figures indicate percentages of respondents.
Data Source: Field Work by Author

Findings and Conclusion

The following are the findings based on the testing of the hypothesis.

Hypothesis I: "Residents like to interact more with their best friends within their own communities, irrespective of their income-based housing and constraint of distance."

Generally, residents interact considering the socio-cultural factors such as caste, religion, language, position in job, regionalism etc. Geographical distance may not prove to be a barrier occasionally.

Three basic factors decide the community. They are : (a) mother tongue, (b) religion, and (c) caste. This hypothesis is tested through these three parameters. Data are collected for the *best friend* of respondents for two locations, namely (i) within Gandhinagar, and (ii) beyond Gandhinagar. Best friends are further.classified into three categories, namely: (a) colleagues, (b) neighbours, and (c) people of same caste. Table 22.3 shows a detailed break-up of the best friends of the respondents. Findings are as follows:

About 206 respondents had colleagues as their best friends. Of them, 86 per cent of respondents had colleagues speaking the same language within Gandhinagar. Almost identical situation is noticed in the other location i.e. beyond Gandhinagar, where 83 per cent of respondents interact with their *best friends* colleagues having their own mother tongue. Thus, the data reveals that a majority of respondents like to interact more with those colleagues who are their best friends and also speak their own mother tongue.

About 94 respondents have preferred neighbours as their best friends within Gandhinagar. Of these, 87 per cent respondents interact more with those neighbours whose mother tongue is the same as their own. In beyond Gandhinagar locations, the difference is more. Here, 90 per cent respondents have preferred those ex-neighbours as best friends whose mother tongue is the same as their own. It is evident from these figures that the majority of respondents prefer to interact more with those neighbours who speak their own language as mother tongue.

Nearly 183 respondents have *people of same caste* as their best friend within Gandhinagar. Out of these, 99 per cent respondents interact with those best friends who speak their own mother tongue. Similarly, in other location i.e. beyond Gandhinagar also, 99 per cent respondents interact more with those best friends who speak the same mother tongue as their own.

Almost similar observations have been recorded on the interactions based on other two parameters namely religion and the caste of respondents. Interactions are found to be intense among respondents observing same religion. Caste factor is also found to be dominant. Thus, overall finding is that the nature of interactions by respondents is positive to mother tongue, religion and caste. This part of the hypothesis is tested by applying the chi-square test. Result shows that the hypothesis is accepted at 95 per cent level of confidence.

Table 22.3 : Best Friends of Respondents

Locations	Mother Tongue			Religion			Caste		
	Same*	Diff*	Total**	Same*	Diff.*	Total**	Same*	Diff.*	Total**
A. Best Friends within Gandhinagar									
1. Colleagues	86	14	206	85	15	206	24	76	206
2. Neighbours	87	13	94	87	13	94	25	75	94
3. People from same caste	99	1	193	98	2	193	98	2	193
B. Best Friends Beyond Gandhinagar									
1. Colleagues	83	17	107	89	11	107	15	85	107
2. Neighbours	90	10	148	79	21	148	15	85	148
3. People from same caste	99	1	188	99	1	188	98	2	188

*Percentages ** Absolute Figures

Data Source: Field Work by Author

Distance does not seem to be a major problem to the respondents. Table 22.4 gives information about the distance covered by respondents to interact with people of their own caste and the communities. It can be seen that 75 per cent of respondents have preferred to interact with people of their own caste residing in distant sectors. They are travelling a distance of even seven kilometres or more for the purpose. About 23 per cent respondents interact with their caste men residing within one km., but 60 per cent of respondents travel a distance between 4 to 7 kms. or more for caste-based activities. The immediate neighbours are not preferred. This fact indirectly conveys that the respondents have either not preferred or not adjusted to the income-based housing system of this planned city. The trend also suggests that the social hierarchy, decided by castes in India, still has a stronger hold over people and the traditional ways of interactions still persist. This is an important parameter conveying the perception of housing system preferred by residents.

Table 22.4: Distance covered for interactions with People of Same Caste

(A) Locations:	1 Same Sector :	157 Respondents	25 per cent
	2 Different Sector:	468 Respondents	75 per cent
(B) Distance Covered			
Distance		No. of Rspn.	Percentage
1.	Less than 1 km.	166	23
2.	1-3 kms.	123	17
3.	4-5 kms.	192	26
4.	6-7 kms.	113	16
5.	More than 7 kms.	134	18
(including other towns)			
	Total:	728	100

Data Source: Field Work by the Author

Hypothesis II: Perception by residents about Gandhinagar has changed depending on the period of stay in Gandhinagar, socio-economic status of the respondents and availability of goods and the services in Gandhinagar.

Perception about a city changes over with passing of time. Urban centres grow in size and, also grow the amenities and infrastructural facilities in the city. Over a period of time needs of people also change. Residents may also develop some expectations about their requirements. Besides, impression about any aspect is always subjective.

Gandhinagar has developed very much during last two decades. There has been an increase in the amenities and infrastructural facilities along with the spatial growth of the city. Data on the likes and dislikes of residents about the city are grouped in five major sections. They are: (a) age-groups (b) income groups, (iii) level of education, (d) occupations, and (e) period of stay in Gandhinagar. Table 22.1 gives overall impressions of respondents classified into these five groups.

It is evident from these tables that more than 90 per cent of respondents from among all classifications, feel that the overall situation over a period of time in Gandhinagar is good or satisfactory. Respondents in six locations feel that there has been more than 100 per cent improvement in the conditions in Gandhinagar over last twenty years, i.e. from 1970 to 1990. Yet, there are two locations where an adverse opinion is also expressed. About 12 per cent of the respondents, belonging to the age-group of 20-30 years. have commented an overall deterioration. Similarly, 33 per cent respondents from above 60 age group also have supported the deterioration. Thus, a section of highest and lowest age-groups shares a feelings of deterioration in the conditions in Gandhinagar over certain period.

These negative feelings can be explained as follows:

Those respondents belonging to lower age group are likely to have come to Gandhinagar recently and were residing in well developed cities like Ahmedabad, Vadodara or Surat, where goods and services were available easily. Lesser development in Gandhinagar has, thus, generated the feelings of dissatisfaction towards the city.

Same reason can be attributed for respondents above 60 years partially. Few respondents belonging to this age-group also might have come from well developed cities. Their expectations may be more compared to what Gandhinagar can offer. This is true specially for the case of few services like education, specialised medical treatment and modern ways of entertainment. This has obviously reflected on their impression about their perception of Gandhinagar.

Changes in the perception is also measured based on respondents' impressions about availability of goods and services. As mentioned earlier, their impressions were asked for over two points of time (a) at the time of their arrival in Gandhinagar, and (b) at present. Respondents were asked to categorise their impressions into *good, fair* and *poor* rankings. Table 22.2 gives the total variations for 20 parameters. It is quite evident from the data that a great majority of respondents have opted for *good* ranking for almost all parameters and there is a substantial reduction in number of respondents who opine overall that the conditions in Gandhinagar are *poor.*

On the whole, respondents have perceived this city and its environment positively irrespective of their classification in various groups. This is an evident reading from the data. The study raises an important issue. Respondents are reported to the travelling a longer distance in order to meet their best friends and relatives. They undertake more interactions of longer distance ignoring cost of transport and time consumed for such interactions. Immediate or near neighbours are not considered important. Level of income, different housing categories etc. have not affected these interactions. So the income based housing system is not preferred. The issue therefore raised is how far is the western system of income based housing suitable to a traditional Indian society.

Acknowledgement

Author wishes to acknowledge the valuable suggestions offerred by Dr. (Mrs.) Anjana Desai, Head, Department of Geography, Gujarat University, Ahmedabad.

References

Ayeni, M.A.O. (1975) : Some determinents of the propensity of interactions in an urban system : A case study of Jos (Nigeria): *Nigerian Geographical Journal,* 18, 2.

Bouce, D.E. (1965) : The effect of directions and length of personal trips on the urban travel pattern, *Journal of Regional Science,* 6, pp. 65-80.

Carrol, J. Douglas (1955) : Spatial interactions and the urban metropolitan description: *Proc. of Regional Science Association,* Vol. I.

Clark, P.J. and Evans, E.T. (1954): Distance to nearest neighbours as a measure of spatial relationship in population: *Ecology,* XXXV, pp. 445-53.

Desai, Anjana P. (1980) : The environmental perception of an urban landscape: The case of Ahmedabad: *Ekistics,* Vol. 47, No. 283.

D'souza, Victor S. (1978): *Social structure of a Planned City*: Chandigarh : Orient Longmans, New Delhi

23

Possible Use of Old Garbage Dumped in Dhapa-Bantala Area

S.K. THAKURTA, N. SOM and P. NAG

Since inception of the system of garbage or solid waste disposal of the city with the opening of a Scavenger's office in June 1785, in order to clear the forest around Calcutta the method followed was simple dumping. Even now, this traditional system of dumping is continuing at Dhapa-Bantala area in the eastern fringe of Calcutta. At present the total maximum quantity of garbage or solid waste generated in the city was around 2700 metric tonnes (100 wards of CMC) and another 700 metric tonnes in the added areas (wards 101-141 of CMC), giving a total of about 3,400 metric tonnes. Generation of solid waste is always increasing with the increase in population, amenities and standard of living owing to both increase in income level and technological advancements and of course expansion of the city limits from time to time.

During the past, a number of investigations and studies have been produced by several committees and commissions in Calcutta regarding the disposal of solid wastes with systematic and effective management (Annexture I). In all these reports, bunches of recommendations were made, which could not be adopted properly in time. As a result, a number of side effects were prominent producing severe problems. At the same time, none of the above reports were comprehensive so that they could be fruitfully utilized. A few stop gap measures could only be considered whenever necessary without any systematic and effective solid waste management system being practiced for which a lot of amount have continued to be spent every year. It has been seen that all the committees and commissions so far set up have confined themselves to the programme and improvement strategy of solid waste management system to be adopted for all the stages of activity, viz. collection, storage, transportation and disposal.

But no proposal for removal of existing hillocks at Dhapa-Bantala area and its utilization in scientific manner could be traceable.

The recommendations made in all the reports were based on the only available analysis conducted by the Central Public Health Engineering Research Institute (CPHERI), Nagpur in 1970 for determination of both the physical and chemical characteristics of garbage or solid waste of Calcutta. For this purpose several samples were collected from garbage vats in different wards during the month of March to April 1970. Each of the report thus arrive almost the same conclusions and recommendations with slight modifications here and there, without having any thought for application of modern and low cost techniques available or possible.

It would be evident that change in habits of people in using gas and kerosene in Calcutta as the cooking fuel instead of coal have reduced the solid waste generation. The percentage share of ash and ignited coal content has come down from more than 40 per cent (along with each content) to almost nil. However, the bulk of coconut shells which was previously occupying a large percentage of the total volume (constituting about 5 per cent of the total weight) or the solid wastes in 1970 could be almost minimized disposing them at the generation site along with the involvement of non-governmental organisations. Over and above, quality of the solid wastes in Calcutta could also changed considering due to revolutionary change in packaging materials. By now therefore both the physical and chemical composition of the solid wastes at the generation points as well as at the disposal site have continuously undergoing drastic changes, both in terms of weight and volume.

Depending on the chemical analysis of 1970, in the past sanitary land filling was recommended by almost all the committees or commissions even though historically the sanitary land filling could never be practiced for solid wastes disposal of the city. As a result, such recommendations always remained to be a theoretical exercise indicating increasing requirement of land beyond or at the periphery of the city. The sporadic attempts of incineration or to convert a portion of solid wastes (not more than 25%) to compost by the composite plant at Bantala of the Agro Industrial Corporation could hardly produce any desired results, reasons of which still remains to be unknown.

Probably for the first time the Perspective Plan of Calcutta: 2011 prepared by the State Planning Board has rightly emphasized the need for change in the traditional outlook of solid waste management system for the city providing specific suggestions towards achieving the goal. In view of this, a sample chemical analysis or the dumped solid waste at Dhapa has been carried out which has indicated ample possibility for using the same as bio-fertilizer and soil conditioner. Using these wastes, agricultural out put could also be enhanced particularly when the cost of chemical fertilizer is becoming costly and scare day by day. Furthermore, same area may also be utilized again and again for dumping such wastes if the present deposition is made in the manner suggested and that in turn will curtail necessary acquisition of further land for future dumping.

With these aims and objectives, a short study has been made here. The attempt is to give a new direction towards a viable and lesser cost solution in the solid management system for Calcutta by utilizing the solid waste which has been dumped in the Dhapa-Bantala area during the past decade and become bop-fertilizer in a natural process.

Garbage Dumped at Dhapa-Bantala Area

Main garbage disposal site is existing on the eastern fringe of Calcutta, and the Eastern Metropolitan (EM) Bye-pass virtually runs across the Dhapa-Bantala dumping ground for a stretch of about 1.5 km starting from Metropolitan Housing Cooperative on the north and Ambedkar bridge at the southwestern site. This is bound by Huges road on the western and eastern side being continued a few kilometre from the E.M. Bye-pass. The dumped garbage forming a hillock by the eastern side of Ambedkar bridge is composed entirely of city garbage and deposited over a period of last 20—25 years. Approximate size of the dump is 1,500 × 600 feet and thus having a cubic content of about 180 lakh cubic feet. It has been known from a physical inspection with one of the operational function of solid waste disposal system over couple of decades that this particular site has been declared abandoned a few years from now.

Solid wastes deposited in the past have presently taken the texture of solid earth and blackish-brown in colour. Some of the deposits are almost like good earth in appearance. Heterogeneous composition of such wastes may be identified very clearly. During inspection of the entire area, it has been also been found that plants like parthenium also has been grown in some areas and some of the remaining areas have been cultivated to produce vegetables. Quantity of production of such vegetables has been found to be very high. Some of the area have however been fenced by the forest department as found from the display boards fixed there. The area fenced by the forest department is mostly abutting the E.M. Bye-pass.

Nine samples have been collected from different points as identified by the said Calcutta Municipal Corporation (CMC) officials, the deposits ranges from one and half to twenty-five years old. The samples were collected from site on 8th January, 1992 and packed properly for immediate transmission to the soil-testing laboratory of Agricultural Department, Government of West Bengal on the same day. Soil samples were subsequently tested with the objective of assessing the manure value of he samples collected from the dumping sites so that the possibility of using as bio-fertilizer and marketing the dumped solid wastes could be explored (Annexure II). Result of the soil test of January 1992 is mentioned in Table 23.1.

Table 23.1: Soil Test Result, 1992

No.	Years of deposition	PH	Ec Mmho./ sq. cm.	Percentage carbon	Agr. P_2O_5 (kg/ha)	Avg K_2O (kg/h)
2894	6m+	7.0(N)	0.65(N)	1.2(VH)	36(L)	633(VH)
2889	1y+	6.7(N)	1.9(C)	0.42(ML)	52(ML)	633(VH)
2895	2y+	6.8(N)	2.5(Inj)	1.9 (VH)	52(ML)	633(VH)
2891	2&1/2y+	7.0(N)	1.3(N)	0.46(ML)	46(ML)	633(VH)
2892	5y+	7.4(N)	0.63(N)	1.46(VH)	38(L)	633(VH)
2893	6y+	7.2(N)	0.32(N)	2.1 (VH)	46(ML)	503(VH)
2888	10y+	6.5(N)	0.83(N)	0.36(L)	42(L)	633(VH)
2890	15y+	7.4(N)	0.68(N)	0.18(VL)	40(L)	299(M)
2896	15y+	7.9(N)	0.37(N)	1.6(VH)	72(H)	633(VH)

Note : N-Normal; VH—Very high; M—Medium ML—Medium low; L—Low; C—Critical for generation; Inj-Injurious to plants.

Interpretation of Results

The deposited municipal waste is required to be removed from the site to make a room for future dumping and also for its gainful use. This will avoid further land acquisition for dumping of solid wastes as recommended by all the committees or commissions to look into the related aspects. In view of the recommendations of Perspective Plan for Calcutta, 2011, the results, as available, have been examined by an expert of the Agriculture Directorate, Government of West Bengal. The preliminary findings are as follows:

(a) Salinity level in the garbage is almost normal;
(b) Nitrogen is not too low;
(c) Phosphorous is at lower level. But it can be used along with the mixture of phosphate tock dust sufficiently available in this state;
(d) Potassium exists at high level;
(e) The garbage can gainfully used (minus the metallic components and weeds) for improvement of agricultural lands from production point of view in 24-Parganas district; and
(f) It can be transported to distant areas also for use of farmers if it is processed and packed at site simultaneously. Small packets may be in high demand in urban gardening also.

The above analysis however does not indicate the presence of any bacteria and/or microorganism harmful to the environment. Some more detailed may further be required both at collection and disposal points, round the year which will indicate the locational and seasonal variations. Further, expert opinion from soil chemists and micro-biologists may be necessary to ascertain the other constituents. However, there is a positive indication for this deposit to be used as a soil conditioner and/or bio-fertilizer. There is also a wide possibility to enrich the deposit with additional chemical fertilizer in required quantities in case further analysis reveal any shortage of any of the required ingredients. Moreover, measures can also be taken to neutralize the adverse effects of any microorganism and/or harmful bacteria if at all found present in the deposited garbage.

Processing of the Old Garbage

Methodology to be adopted for processing of the old dump should be very simple and cost effective. With this objective the best suitable method is proposed to be the installation of simple smaller grinding machines similar to those used in *Soorki* mills for grinding and powdering bricks and brick bats from *Soorki*. As indicated earlier, the physical characteristics of the deposited garbage is almost in the form of soil and earth with different other materials which may be easily processed in the grinding mills. If necessary, some sorting arrangements can also be made for separation of the undesirable ingredients. These may be some possibilities for presence of metals (ferrous) which can also be separated with the help of magnetic separators if at all needed. According to the requirements, a number of such grinding mills may be installed within the existing dumping sites of varying capacities in different scales. This will also create employment opportunities for unemployed youths. Sufficient land is also available nearby for necessary installation, storage and packaging of the processed materials. Details of the land area required for processing along with the equipment, machines, power supply and other accessories with cost estimates for installation, operation and maintenance can be worked out. Implementation of this process may fetch following multiple benefits :

(a) The existing hillock creating hazard and eye sore may be removed;
(b) New disposal site in an organized manner may be developed on the vacant space or land, if available;
(c) Gainful use of the bio-fertilizer for farming and urban gardening since CMC, CMDA and forest department are presently engaged also in promoting greenaries in the urban and fringe areas. These fertilizers may be used by these organisations to create interests amongst the cultivators and farmers of nearby areas. There are enough potentials for marketing of this bio-fertilizer particularly when the chemical fertilizers are becoming costly and scarce may be through the Agricultural Marketing Directorate, Government of West Bengal;
(d) Employment opportunities are likely to be created in the process if this project is allowed to run forming co-operatives, and other NGOs who can probably be easily attracted for the easy remunerative operation of the suggested system;
(e) Calcutta Municipal Corporation can be released of procuring

further lands for its disposal site. Also haulage distance will be less if this area can be utilized for dumping solid wastes and thereby cost of fuel required for transportation will be less;

(f) Efficient management of the disposal sites may be ensured if this process is repeated in other similar sites under the control of CMC;

(g) There is a demand of fertilizer in urban areas to those who maintain gardens within their premises. This area of demand may also be properly utilized with adequate publicity;

(h) Revenue income of CMC may be increased through sale of this bio-fertilizer may be through effective involvement of NGOs; and

(i) This area may be properly demarcated with a green belt all around and the width of such green belt should not be less than 100 feet. The greenaries or trees will thus hide the garbage dumping area as also produce oxygen for purifying the environment.

Conclusion

As mentioned above, several committees have reviewed the existing system of refuge collection, transportation and disposal practices in Calcutta. It has been felt that the present system is inadequate owing to lack of infrastructure, hence overstrained. The mode of operation in land disposal sites leaves much to be desired as it does not meet the long term landuse planning requirements. Operational control against environmental pollution and insect propagation is inadequate. As per the estimates considering the increasing population and increase of per capita waste generation, an arrangement has to be made for waste management of at least 2,000 metric tonnes per day for next 25 years or so. In addition, an area of 660 ha of land would be needed by 2008 AD considering the average depth of fill would be 2m. In terms of both cost and environmental protection, sanitary land filling and composting emerge as the most suitable solution. However, acute land shortage in Calcutta will force to adopt distant sites. Considering these aspects, it has been agreed that in next two decades nearly 75 per cent of the city's refuge would be disposed by this method. Further, about 50 per cent of the city's garbage can be utilized for composting. However, this can not be practiced in the city due to non-availability of land.

The wet and marshy lands are available in the eastern fringe of the city in Kalikapur, Barakhola, Jagatpota, Mukundpur, Chalenda, Choubhanga, Birinchitala and in western Dhapa-Manpur area. But such solution is not acceptable, as it will aggravate the already critical drainage situation for the core areas of the city along with other environmental and ecological factors. However, selective use of part of these wet lands for garbage disposal may be considered.

Perhaps a review of various alternative methods of disposal of urban solid wastes being practiced in the country and abroad is desirable. How such wastes can be used for the generation of energy? Which are the methods suitable for Calcutta? Long-term perspective and technological options have to be considered. A serious thought in this direction may show the way of making best use of this pollutant in other Indian cities as well.

References

Das Gupta, S. P. (1986), The structure of Calcutta: Morphology of a congested city, in Jean Racine (ed), *Calcutta 1981: The City, its Crisis and the Debate on Urban Planning and Development,* Concept Publishing Company, New Delhi, pp. 131-50.

Kundu, A. K & P. Nag (1996), *An Atlas of the City of Calcutta and its Environs,* National Atlas & Thematic Mapping Organisation, Calcutta.

Roy, Kalayan (1997), Evolution of approaches of urban development in the CMA, in Jayati Hazra (ed), *Dimensions of Human Geography*, Rawat Publications, Jaipur & New Delhi, pp. 92-103.

Singh, J. (1986), Population, pollution and the urban environment, in H.H. Singh, P. Nag, V. K. Kumra, S. K Singh & H. Prasad (eds.), *Geography and Environment: Issues and Challenges,* Concept Publishing Company, New Delhi, pp. 139-55.

Singh, R. B. & Bhuwan Kumar (1996), Green house gases and atmospheric quality in mega cities of India, in S. B. Singh (ed), *New Perspectives in Urban Geography,* M.D. Publication, New Delhi pp. 153-8.

Sen, P. K. (1994), Growth and changes of the Hooghly conurbation belt, in H. H. Singh *et al.* (eds), *Patterns of Urban Change in India,* Indrasini Devi Publications, Varanasi, pp. 73-84.

Sita, K. & D. Mukherji (1997), The city core of Calcutta, in Jayati Hazra (ed), *Dimensions of Human Geography,* Rawat Publications, Jaipur & New Delhi, pp. 104-20.

ANNEXURE-I

COMMISSIONS AND COMMITTEES ON SOLID WASTES

Year	By	Details
1959	P.N. Chatterjee	Chairman, Standing Works Committee, CMC
1961	P.C. Majumdar	Commissioner, CMC
1962	P.C. Bose	Executive Director & Chief, CMPO
	J.L. Vincenz	Director, Public Works, County of San Diego, California, US
1962	P.K. Sen	Dy. Inspector General of Police (Transport) Suggested for the improvement of collection, removal and disposal of garbage under then existing system
1963	Jean L. Vincenz	Consultant, Ford Foundation Advisory Team, CMPO—Report on the modernization of refuge handling in Calcutta
1970	CPHERI, Nagpur	Feasibility study for alternate methods of garbage-disposal for Calcutta City.
1972	R.N. Chopra	Additional Secretary, Ministry of Home Affairs, Government of India—Report on garbage collection transportation and disposal. Published by Department of Health, Government of India.
1972	S.S. Yechury	Controller of Transport, Ministry of Health & Family Planning, Government of India—Report on the garbage disposal schemes of Calcutta Metropolitan Area
1974	Bhupesha Saha	Gujarat Agro Industries Corporation, Ahmedabad-Feasibility report on compost manure plant for the city of Calcutta.
1975	WHO	Solid Waste Management, City of Calcutta
1975	World Bank	Consultant's report
1977	World Bank	Consultant's post appraisal report
1976-78	AIIHPH, Calcutta	Collection and disposal of refuge and night soil in Municipal Towns in Calcutta Metropolitan Districts.
1979	Frank- Flyiutoff	World Bank Project Manual on solid waste management, Second Calcutta Urban Development Project

Year	By	Details
1982	A.K. Bose	Commissioner, CMC—Solid waste management for the city of Calcutta—Problems and alternate strategies for disposal
1982	J.N. Talukder	Report of the CMC Enquiry Committee, Vols. I & II (covering water supply, drainage, conservation service, motor vehicles and administration), published by Local Self-government & Panchayat Department Government of West Bengal.
1987	Task Force	Report on soild waste management indicating CMDA possible options prepared during the preparation of perspective plan
1991	Dibyendu	Dy. Chief Engineer (Conservancy), CMC—Chowdhury Project note on solid waste management services under Eighth Five Year Plan for CMC

Note : AIIHPH — All India Institute for Hygiene and Public Health.
CMC — Calcutta Municipal Corporation.
CMDA — Calcutta Metropolitan Development Authority.
CMPO — Calcutta Metropolitan Planning Organisation.
CPHERI — Central Public Health & Environmental Research Institute.
WHO — World Health Organisation

ANNEXURE-II

GENERATION OF MUNICIPAL SOLID WASTE WITHIN CMC

Source	CMC wards 1-100	Jadavpur	S.S. Units	C.R. Units
Residential houses	1250.00	64.49	91.35	45.24
Markets	270.00	13.50	27.00	40.50
Hospitals	42.61	8.52	4.26	2.13
Slaughter houses	4.50	0.90	0.90	1.35
Stable	3.43	0.69	—	1.02
Khatals	17.15	3.44	1.72	5.16
Hotels	49.40	9.88	12.35	4.94
Marriage houses	21.00	4.20	5.25	2.10
Halls	18.67	3.74	3.74	1.87
Office etc.	25.72	5.14	2.57	1.29
Petrol pumps	26.00	1.30	1.30	1.30
Drain silt	30.64	6.12	9.18	15.32
Building activities	55.71	2.29	5.57	2.29
Hawker	95.56	4.78	9.56	4.78
Trader	27.00	1.35	2.70	2.70
Shops	118.29	5.91	10.83	11.83
Sweeping	85.20	17.04	21.30	25.56
F.P. dwellers	0.05	—	—	—
Parks	2.00	0.10	0.10	—
Misc.	12.00	2.40	3.00	6.00
Total	2,154.83	155.79	212.68	175.38
Winter & Summer+15%	2,478.05	179.16	244.59	201.69
Spl. Season+25%	2,693.53	194.74	>265.86	219.23

24

Impact of Mining on Environment in Raniganj Coalfield, West Bengal

SMITA SENGUPTA

The environmental aspects of human habitat varies over space in the context of man-environment inter-relationship. Whenever, there is a symbiotic relationship between man and environment it is called ecological balance, but when environment can not support human needs and aspiration then it is called 'ecological imbalance'. The over exploitation of environment causes pollution. Environmental pollution is the product of both high development and under development. So, ecological environmental quality is not necessarily good in low level of development. Natural hazards like cyclones, floods and droughts or large scale soil erosion create physical environmental problems. Precautionary measures can be taken against natural hazards by adopting modern technology. Though human beings may not be able to eradicate natural hazards but their intensity can be reduced by adopting technological measures.

Environmental problems resulted for economic development are also multifarious. The accelerated pace of economic development involved on unplanned and excessive exploitation of renewable and non-renewable resources in many parts of the world. There has been a corresponding increase in man's interference in the ecological balance, beyond the tolerance limit of the natural regenerative processes, seriously imparing the life support mechanism of the ecosystem. Environmental pollution problems are becoming more and more serious due to rapid but unplanned urbanization and industrialization in developing countries. Air, water and land are the most important components of the environment or which the very survival of human life is dependent. So, the human life will be affected adversely by the pollution of these environmental components.

Mining of minerals is a process of fighting against nature for the benefit of mankind. The process of mining either by open excavation method or by underground mining technology create impacts on men, site and adjoining area and properties. Some of the negative impacts, sometimes pose threat to living being inhabiting in that particular locality. In the coal mining industry the water table is reduced due to the extraction of water by underground methods. The water retention capacity of the soil being poor, the land do not support vegetation and thereby causes damages to the elements of environment. The gaseous and the particulate pollutants arising from the production activities or the auxiliary operations affect the air and water. The top soil is completely damaged if it is not separately mined in the beginning with a view to achieve reclamation. The loss of vegetation and productive capacity will be permanent if the land is not restored to its original position. Noise, vibrations and air blast are some of the problems that effect the environment. Mine fires profusely released gaseous pollutants. The unburnt hydrocarbons from smog cause damage to respiratory system of human beings.

In the present paper attempts have been made to focus attention on the impact of coal mining on various environmental elements in Raniganj Coalfield, West Bengal, India.

Raniganj Coalfield

The Raniganj coalfield accommodates the western part of Asansol-Durgapur Industrial complex of Eastern India. It is an integral part of Damodar Valley resource region. The present study area occupies the whole Asansol-Durgapur industrial complex. Usually environmental impacts associated with mining on such a gigantic scale do not remain as a local feature but trigger of a set of chain reactions affecting the environmental equilibrium of a very wide region. Hence, the effects of adverse impacts are likely to reach the Asansol-Durgapur belt as well.

The Raniganj coalfield has a long history of mining. Gee (1932) gives a comprehensive account of development of this coal field till the thirties of the present century. The earliest attempt on record of systematic working of coal was in 1774. However, profitable exploitation started only in 1814 when a mine was opened in Egra.

Since then mining has undergone considerable changes over the years. At the time of nationalisation in 1971, there were 200 working mines with an annual production of 18.44 million tonnes. At present there are 124 operating mines under two division of Eastern Coal Fields Ltd.

Both surface and underground methods of mining are in practice here since the commencement of mining operation. In the past production of coal was mainly obtained from surface working quarries or inclines located on or near the outcrops of coal seams. The quarries were manually operated and were usually very shallow. The underground mines which were generally shallow had vertical shafts upto 30m to 48m. The depths of the pits were increased due to the discoveries of deeper seams. The 'long wall' method of mining was also in practice in some of the mines of earlier days. In those days the environmental protection for men and materials were very inadequate and hence there were number of accidents causing loss of life or property. Moreover, as a result of unplanned depillaring in the underground mines, vast areas overlying the old abandoned mines have either remained unsupported or collapsed to form prominent subsidence zone (Fig. 24.1).

The present day coal mining operation in this field is more or less in the same line with significant changes in the scale of operation due to introduction of mechanised mining. In the manual quarries all operations like cutting of over burden or coal seam, lifting of coal from quarry to coal dumps, loading in trucks etc. are done manually. Sometimes steam haulded trolleys are used to lift the coal from quarry face to the surface. 'Broad and Pillar' method is still the most common method of underground mining, but the only changes are in respect of sizes of pillars and galleries which are of larger dimensions. 'Long wall' method of underground mining is adopted only locally (Fig. 24.2)

At the time of the nationalisation of coal mines in 1970-71 the production from this coalfield was 20 million tonnes of which 90 per cent was from the underground mines and 10 per cent from the manual quarries. Initially a higher rate of production was achieved through optimisation of production from the existing mines. However, due to exhaustion of the reserve quite a large number of the mines were closed down resulting in a drop in reduction by about 6 million tonnes. With

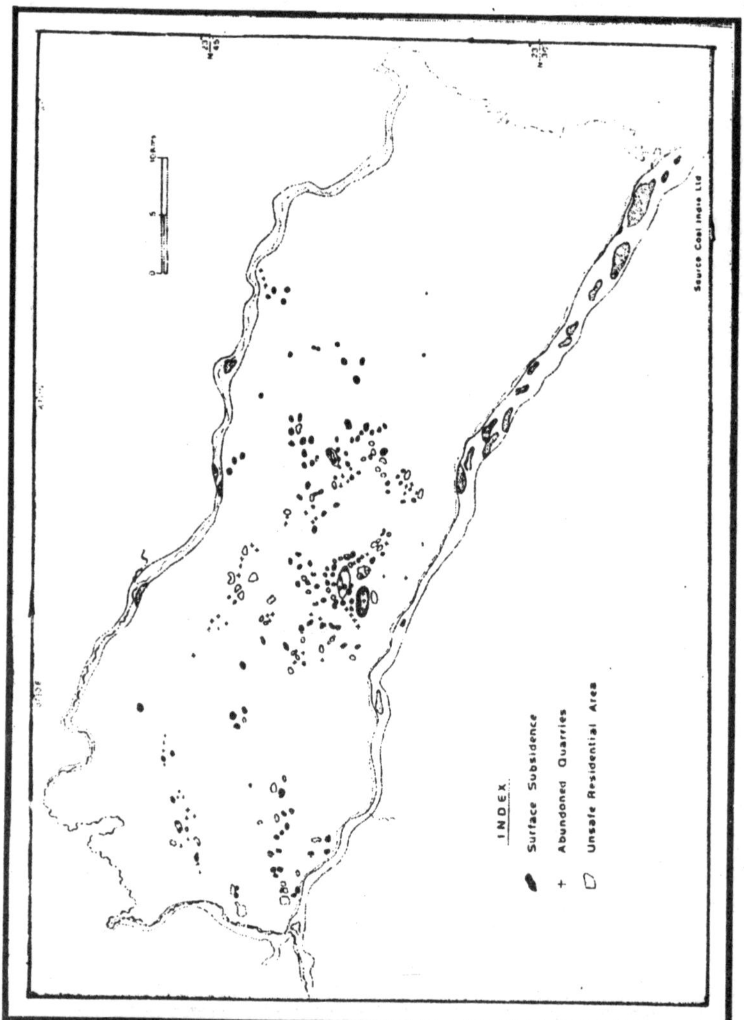

Fig. 24.1 : Mining Hazards

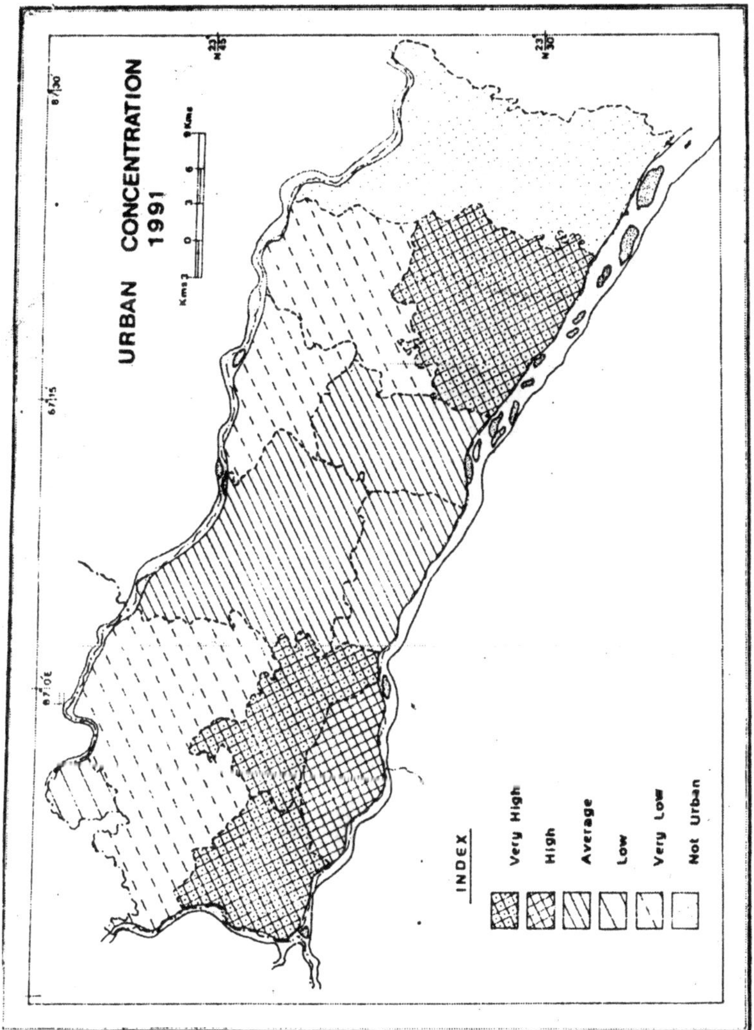

Fig. 24.2 : Urban Concentration

the opening up of the several deep and fully mechanised open cast projects the production again rose to about 25 million tonnes in 1985-86. There is a projection of production of 63.89 million tonnes in 2000 AD. There will be sharp increase in the production from open cast mines from about 20 per cent of total production at present to about 30 per cent within next 10 years. So, this will create a significant impact on environment in this region. In this study Eastern, Northern and Central sector of Raniganj Coalfield along with Durgapur Industrial region have been studied.

Environmental Impacts

The long continued mining operation have resulted in significant changes in the natural environment in Raniganj Coalfield region (Table 24.1). This region is characterised by 143 surface subsidence (depressions ranging 0.3 to 2m) spots covering an area of 50.94 km^2. The localities which are badly affected by subsidence are Hakman-bagan, Babuisol, Ukhra Nimcha, Chapui, Barira, Phatepur, Shyampur etc. (Table 24.2). In addition to this, some important towns and localities like Barakar, Raniganj, Jamuria Bazar, Sanctoria etc. have been declared unsafe by the Director General of Mines Safety as these are located on waterlogged goafs and may be affected by subsidence any time.

Large scale land sacrifice with development of spoil dumps are the most significant impacts of open cast mining in this coalfield (Table 24.3). The total area occupied by abandoned quarries and associated spoil dumps are nearly 20.00 sq km. The height of the dumps range from 5m to 25m and gradient of the slopes vary from 10° to 75°. In old older dumps the outer surfaces are covered with thick bushes and grass. Except for the construction of a play ground on an old spoil dump area at Rajpura colliery no attempt has so far been made for reclamation of the 20 sq km of derelict land. The spoil dumps occupy approximately 50 per cent of the total derelict area. Calculated on the basis of an average height 10m for spoil dumps, there is already an accumulation of approximately 100 million cubic m of mine waste material in the abandoned mines.

There has been considerable changes in the landuse pattern in this region. The area occupied by coal mines and ancilliary industries have increased from about 45 sq km in 1923-24 to 175 sq km at present.

Table 24.1: Details of damaged lands in Raniganj Coalfield

Sector	Area	Area covered by quarries and spoil dumps abandoned several years ago (in sq. km.)	Total area in Col. 3 sector-wise	Area covered by recently abandoned quarries and spoil dumps (in sq. km.)	Total area Col. 5 sector-wise	Area covered by recently active quarries and spoil dumps (in sq. km.)	Total area Col. 8 sector wise
Eastern Sector (Between Ukhra and Asansol)	1. Kumarkhola-Kumardihi-Bankola	0.63	5.60	0.48	2.60	0.98	2.88
	Purushattampur area	3.17		0.82		1.71	
	2. Bahula-Kenda-Parashkol-Dhandradihi-Parasla-Baktarnagar area						
	3. Nimcha-Harbhanga-Kankardanga	1.10		1.00		0.19	
	Searsol-Mahabir area	0.70		0.30			
	4. Sripur-Ratibati-Chapui area						
Central and Northern Sector (Between Asansol and Barakar)	1. Ramnagar-New Damaguria-Dendua area	0.55	4.96	0.95	1.15	0.28	2.19
	2. Benjimari Shyamdih area	1.15		0.20		0.57	
	3. Shyamdih Itapara	1.25		—		1.30	
	4. Gauyrangdih-Chumlio-Dishermohan area						

(*Source*: CMPDI)

Table 24.2 : Details of subsided lands in Raniganj Coalfield

Sector	Area	Nature of damaged lands
Eastern Sector	Babuisol (23° 35'50":87°10'30")	0.5 sq km subsided lands cave circular shaped of diameter 8-9m and depth 2-3m. Tension cracks 5-6m long present.
	Palashban (23°35'50":87°10'00")	
	Ukhra (23°38'30":87°14'30")	Small isolated shallow depressions formed by subsidence over recent mines found
	Kankardanga (23°36'30":87°06'00")	Wide depressed land (4 sq km. approx.) accompanied by old mine fires present.
	Egara (23°36'10":87°06'10")	
	Nimcha (23°39'10": 87°04'30")	Subsided area, measuring about 8 sq km. accompanied with several shallow abandoned quarries, low spoil dumps, old mine openings and installations present. Old mine fires present at few places.
	Charanpur	9 sq km. of subsided land, manifested by depressed lands, accompanied by old mine structures and colliery bustees present.
	Sripur	Cultivation, although present is rather rare
	Chapni (23°89'10":87002' 50")	3.25 sq km of subsided lands, presents. Several cave-in-having diameter 2-3m are present.
North and Central Sector	Noria Khal	
	Barrira (23°45'20":86°52'10")	0.25 sq km. subsided lands present with several small cave in and tension cracks.
	Tentulhjora (23° 45'00":86°51'10")	
	Kankrasol (23° 38'30": 87°13'50")	2 sq km of subsided lands accompanied by a number of shallow abandoned quarries present cave in and tension cracks present.
North and Central Sector	Hakmanbagan (23°42'30": 86°54'00")	Isolated patches of subsided lands (measuring a total land of 0.44 sq km) are present in all these places.
	Bartaria	
	Phatepur	
	Radhanagar	
	Sodepur	

(*Source* : CMPDI)

Table 24.3: Quarry Areas

Sector	Area	Nature of derelict lands
Eastern Sector (Between Ukhra in East and Asansol in West)	A. Kuardihi (23°40'00": 87°16'00")	Some recently abandoned waterlogged quarries with low unstable spoil dump
	B. Kumarkhola (23°41'30" : 87°12'30")	One recently abandoned quarry mostly dry, Mine waste materials available in plenty in spoil tips.
	C. Bahula (23°39'30": 87°11'50") Parashkol (23°38'30": 87°11'30")	A chain of old abandoned quarries and associated stable and unstable dumps extending over an area of 5 km x 0.5 km. Quarries are mostly dry and covered by bushes.
	D. Baktarnagar (22°35'50": 87°10'00")	One big and one small old abandoned quarry with stable spoil dumps. The bigger one is fully waterlogged the recently abandoned one is partly waterlogged with partly stable dumps.
Eastern Sector	E. Chapui (23°39'10": 87°02'50")	Some old abandoned quarries totaly dry, with stable spoil dumps, areas surrounded by agricultural field.
	F. Ratibati (23°39'00": 87°02'00")	One big recently abandoned quarry with unstable spoil dumps with the number of water gullies intense soil erosion.
Northern and Central Sector	G. Banbirdih (23°47'00": 86°53'30") Alkusa (23°46'50":86°54'00") Bhagran (23°47'10": 86°52'00")	Old abandoned quarries are present in these areas. Some depressions at Banbirdih and Alkusa are fully waterlogged with low stable spoil dumps in surrounding areas. Some dry or partly waterlogged quarries are also present. There is one recently abandoned waterlogged quarry with high unstable dumps at Bhagram.

(Contd.)

Table 24.3: *(Contd.)*

Sector	Area	Nature of Derelict lands
	H. Dabar (23°44'00": 86°45'20")	Dry rocky quarry depression with unstable spoil dumps at some adjacent active quarries.
	I. Pahargora 23°47'30": 86°56'30") Amdiha (23°47'30": 86°57'30") Itapara (23° 47' 10" : 86° 58' 50")	Several abandoned quarry depressions and associated stable spoil dumps, depressions mostly dry with only a few water logged.
North and Central Sector (Between Asansol on East and Barakar on West)	J. Gaurangdih (23°49'10":86°58'50") Churulia (23°47'00":87°02'30") Dishermohan area (23°46'30":87°06'30")	A chain of old abandoned quarry depressions, mostly dry associated with number of stable/partly unstable spoil dumps.
	K. Ramnagar (23°45'15":86°50'00") Chaptaria (23°45'50":86°52'30") New Damaguria (23°46'10" : 86°50'50") Dendua area (23°46' 50": 86°52'10")	Several old quarries mostly dry with associated spoil dumps. However one quarry at Chaptaria is fully water logged and surrounded by stable spoil dumps.

Source : (CMPDI)

A vast stretch of forest and cultivated lands have been lost due to mining activity.

Coal by its very composition is liable to spontaneous combustion. 'This susceptibility to spontaneous combustion leads to heating in the coal seams, sometimes leading to active fires. In Raniganj coalfield old mine fire in abandoned underground mines exist at many places like Nimcha, Kankar danga and Belbad etc. Mine fires spread to adjoining areas adding to the losses and also prevent economical exploitation of coal seams in the vicinity.

Many nala courses have been diverted into artificially cut channels to facilitate the extraction of coal by open cast at many places like Hariajam, Nirsa, Mandman, Chanashyar quarries etc. A part of the Nunia nala has been partially blocked by the adjacent old dumps near the village Pahargora. Large point bars have formed within the Singaran nala course due to the deposition of spoil materials.

The extraction of sand from the Damodar river for sand storing of the abandoned mines has resulted in significant chances in the depth of the river bed at some place. There has been a sharp fall in the depth and gradient of the river bed between Raniganj and Waria, where the extraction of sand has been maximum.

The share of opencast mining in the total production is always much less than the underground mines but the damages caused to natural environment are always more in open cast mining. The production ratio between underground and open cast mining will be gradually reduced from 5:1 to 2:5 due to economic reason. Hence the damage potential of natural landscape will increase very fast unless reclamation efforts are taken simultaneously with the mining operations.

Conclusion

The impact of mining on the natural landscape and consequent degradation of valuable land resource in this belt is quite significant. The long continued mining also have wide ranging impacts on the water and air quality affecting the plant and animal life. It is necessary to have a time bound comprehensive environmental management plan for Raniganj Coalfield. Hence, there is great practical importance of socio-psychological thinking by environment concious planners and managers. The effect of a project on people and people's responses may be direct and immediate or remote and alternated.

In general, people are aware of the delectericus effects of mining and coal conversion processes, but with proper planning and adequate precaution these can be kept to a minimum or in some cases bad effects can be altogether avoided. The local people strongly voice for reclamation of land, restoration of original landform and rehabilitation of disturbed population.

The present scale of activity in this mining belt has significant impact on natural landscape. Due to lack of realisation of the degradation of the valuable land resource little effort has so far been made for rectification. This being the most important coal field in the Eastern India. The coal mining industry will continue to grow very fast. The vastly enlarged activity, particularly in the opencast sector, is likely to cause more serious degradation of the natural environment. So, environmental safeguards are required through implementation of appropriate environmental management plans. This should be achieved as early as possible because not only this mining belt but the adjoining area is also facing critical environmental problems due to unplanned urban and industrial growth.

Acknowledgement

The author is thankful to C.S.I.R. for funding her project. She is also thankful to the official of Coal India, CMPDI, CMRS for their help and co-operation.

References

Bose, L.K. (1989): Scenario of habitation *vis-a-vis* mining in Raniganj Coalfield-past, present and future, *Journal of Mines, Metals and Fuels,* Vol. 37, No. 8.

Briggs, H (1929): *Mining Subsidence,* Edward Arnold and Co., NewYork.

Gee, E.R. (1932): *The Geology and Coal Resources of the Raniganj Coalfield.* Mem. No. 61, Geological Survey of India, Calcutta.

Khoshos, T.N (1984): Land reclamation in open cast mines, Twenty-seventh Holland Memorial Lecture of Mining Geological and Metallurgical Institute of India, Calcutta.

Malhotra, K.K. (1990): Environment and Ecology—Relevance to Coal Mining, *MINETECH,* Vol. 15, No. 5.

Mukherjee, S.K. *et. al.* (1988): Critical analysis of some occurrence of underground fire in Raniganj Coalfield, *Journal of Mines, Metals and Fuels,* Vol. 36, No. 6.

Murr, R.E. (ed) (1979): *Environmental Impact Assessment: Principles and Procedures,* John Wiley and Sons, New York.

Ray, U.K. Dutta, A. and Dutta, R.K. (1986): Geological report pertaining to impact on environment due to coal mining and coal based industries in parts of the Raniganj Coalfield, Districts, Barddhaman and Birbhum, West Bengal and Dhanbad, Bihar, Geological Survey of India.

Saxena, N.C. (1991): Subsidence Management in Jharia Coalfield, *MINETECH,* Vol. 12, No. 6.

Sengupta, S. (1991): Land Subsidence—An Environmental Hazards in Mining Region, *ILEE,* Vol. 12, No. 2.

25

Wide Scale Industrialisation and Concomitant Environmental Pollution : A Case Study of Damodar Valley

SWAPNA BASU

The modern world is passing through a rapid phase of industrialisation which is considered to be the prime tool in bringing about economic development and prosperity of a nation. The strong economic base of a region may largely be measured by its industrial potentiality only. This industrial development supported by efficient scientific innovations have indeed given a new dimension to the standard of living of the modern man, but it has also brought in its wake a number of serious problems which are slowly but seriously threatening the very existence of human society. The problems are many fold, chief among them being the disposal of industrial waste water which contaminates the chief sources of potable as well as water meeting other requirements. The present paper deals with the problem of water pollution resulting from large scale industrialisation in the Damodar Valley region.

Preamble

The Damodar Valley region has made tremendous industrial development during last four decades and has emerged as one of the most lucrative industrial sectors of the country. A huge industrial complex has gradually turned up along the bank of the Damodar, stretching from Bokaro in the west to Uluberia in the east, covering a stretch of 50,000 sq km and serving a population of twenty-five millions. This has given rise to spectacular improvement in the national economy and has contributed largely to the improvement of the economic standard of the people concerned.

This phenomenal industrial growth however has brought in its sequel a number of adverse effects, which are seriously affecting the health and hygiene of the valley by way of disposing off industrial pollutants into the Damodar river. The present paper attempts to bring forth the degree of pollution in the Damodar resulting from contamination by industrial effluences and its bearing on the social status of the valley. Different reports and data collected from authentic sources have been consulted during the preparation of the paper, which have subsequently been verified in the field study.

Environmental Issue

Rapid industrialisation in the Damodar Valley has aggravated serious waste water disposal problems in the valley. The growing concentration of people and industrial units together have been creating environmental as well as public health problems which are gradually increasing in dimension every day. The water pollution in the Damodar Valley has become a talk of the day and both people of the locality and the administration are getting more and more aware regarding its consequences.

Reasons Causing Pollution

Some physical as well as man-made factors are jointly responsible for the present state of the river. They may be summarised as follows:

(a) *Rain-fed nature of the river:* The Damodar is a rain-fed river and its natural flushing capacity decreases with the decreasing volume of water. As such, the river is more prone to pollution in its lean period during summer and winter than it is in the rainy season.

(b) *Decreasing capacity in natural flushing:* The river being dammed at five different points upstream has lost much of its original volume and discharge capacity. As such many of the pollutant loads are suspended within the water, thereby deteriorating the water quality.

(c) *The reservoirs collecting heavy pollutant loads:* The stagnant water in the reservoirs are the gathering-ground of toxic pollutants from the industrial effluences upstream. These are again released through distributary canals to the crop fields as also to the dwelling localities thereby aggravating the problem.

(d) *Disposal of untreated waste water:* Most of the industrial establishments on the bank of the Damodar treat the river as a sewage for disposing off their effluences. These are the chief miscreants for deteriorating the water quality of the river, which has become a great concern these days.

(e) *Industrial disaster:* To add to the injury, there often happen industrial disasters which deteriorate the water state to a perilous level. For example the sudden oil spill in the Bokaro Steel Plant area in 1980 caused a havoc in the surrounding locality affecting both the human and cattle population.

Source of Pollutants

The pollutants can be put in three groups, based on their origin :

(a) *Industrial pollutants:* The Damodar runs through a coal mine belt in its upper stretch and receives the pollution load like fly ash from Chandrapura, Bokaro and Patratu thermal power stations, coal fines, shales and dirts from coal washeries and large scale over burden dumps, chemical effluences from Bokaro Steel Plants, Sindhri fertilizer and cement factory, lead smelting, Indian Explosive Ltd. at Gomia etc. In the middle valley it gets pollution load from Chitteranjan Locomotive, Raniganj Paper Mills, Durgapur Steel Factory, Durgapur Coke Oven Plant and a host of other industries. In its lower stage from Durgapur downstream the river has to carry domestic, industrial and agrarian pollutants from a thickly populated agricultural-cum-industrial belt.

(b) *Effluences from fields:* The crop fields too add to their share of pollutants by drawing nitrogen and ammonia rich waste water which incidentally increases the BOD level of the water, beyond permissible limits.

(c) *Urban Sludge:* The solid waste disposals from human dwellings as also from cattle sheds are also creating constant pollution load which degrades the water quality considerably.

All these loads together have given a sum total of pollutants which has reached an alarming figure for environmental degeneration.

The different types of pollutants from different industrial sectors are presented in the Table 25.1.

The Water Quality of the Damodar

A systematic chemical analysis of the river water has been organised by the Central Pollution Board and also by the Damodar Valley Corporation at different intake points of effluences. The water quality of the river together with the tolerance limit, as fixed by the Central Pollution Board according to the Indian Standard are presented here with (*vide* Table 25.2).

It appears that the water quality particularly deteriorates during winter and summer. The total coliform organisms, Bio-Chemical Oxygen (BOD), Ammonia and Lead content of the river are well above the tolerance limit for drinking as also for outdoor bathing purpose. BOD itself is not a pollutant but an indicator to signify the propagation of the existence of dangerous bacteria. These micro-organisms release enzymes which in contact with oxygen decompose chemical substances of sulphite, sulphides from paper mills and ferrous iron. High BOD level thus indicates a high pollutant state of water.

Coliform pollution is a more serious and difficult problem to handle in maintaining water quality. A majority of the coliform species (bacterial) are pathogenic. Their presence in water is an indication of focal contamination. These come in contact with the water through untreated water of the drains, sewers etc. and pollute the water of the river.

Pollution Status in the River

Prof. A. K. De, Department of Chemistry, Visva Bharati has conducted a detailed chemical analysis of both water and sediment samples of the Damodar at about 50 sites of effluence intake points in the Asansol Durgapur industrial conurbation. His findings indicate that IISCO, CCIL, Carew and Co., Bengal Paper Mill and the DSP plants discharge pollutants which accumulate in the Durgapur barrage. The concentrations of these toxic pollutants exceed their respective tolerance limits. The barrage water is the principal source of domestic water supply for Durgapur city, but the existing conventional water

Table 25.1: Source-wise disposal of pollutants in the Damodar

Industries	Water intake/ waste water(mgd)	Water products
Durgapur project (D.P.L)	21/3.7	Phenol, ammonia, cyanide, benzene, solvent naptha. sulphide
Durgapur Chemical Ltd. (DCL)	1.0 (waste water)	Mercury, phenol, hydrochloric acid
Durgapur Steel Plant (DSP)	12.75/2.5	Ammonia, cyanide, phenols, tar, sulphide, oil and grease
Hindusthan Fertiliser Corpn.	2.0 (waste water)	Urea, ammonia, arsenic, chromate oil and grease
Bengal Paper Mill	13/11	Tannin, lignin, sulphide, phenolic compounds
Carew and Co. Ltd.	0.6/0.12	Nitrate, molasses
Indian Iron and Steel Co.	—	Ammonia, phenols, cyanide, sulphide, tar
Cycle Corporation of India Ltd.	0.2 (waste water)	Cyanide, chromate
Indian Explosive Ltd.		Lead
Bokaro, Chandrapur and Patratu Thermal Power Station		Fly ash
Gidde, Jamadova, Chashala, Kargali coal waste etc.		Oil, grease, coal fine, shale, dirt.

Source: D. V. C. Report

Table 25.2: Water quality of the Damodar against tolerance limit

Characterisation	Tolerance Limit					Water quality of the Damodar		
	A Drinking water source without conventional treatment	B Out door bathing	C Drinking water source with conventional treatment	D Propagation of fisheries	E Irrigation Industrial cooling	Summar	Rainy season	Winter
P.H.	6.5-8.5	6.5-8.5	6.0-9.0	6.5-8.5	6.0-8.5	8.1	7.3	8.4
Dissolved Oxygen (mg/1)	6.0	5.0	4.0	4.0	—	6.2	7.2	7.4
BOD (mg/1)	2.0	3.0	3.0	—	—	4.6	3.3	5.6
Total coliform organism (MPN/100 ml)	50.0	. 500.0	5000.0	—	—	340-1666	1060-1560	350-1800
Total dissolved solids (mg/1)	500	—	500	—	—	211.4	2136.2	131.6
Chlorides (mg/1)	250	—	600	—	600	17.4	42.3	16.8
Nitrates (mg/1)	20	—	50	—	—	7.6	9.2	10.1
Ammonia (mg/1)	—	—	—	1.2	—	0.11	0.05	0.23
Iron (mg/1)	0.3	—	50	—	—	0.62	0.67	0.83
Lead (mg/1)	0.1	—	0.1	—	—	0.05	0.03	0.04

Source : IS-1982 and D.V.C. report

treatment methods fail to remove these undesirable pollutants and about 5 lakh people of Durgapur and surrounding localities are constantly exposed to these health hazarding toxic pollutants.

The Tamla Nalah delivers a heavy pollution load to the river which are deposited in the river bed. The sediments containing these pollutants are 10-100 fold in excess and act as store house of these pollutants which continuously pollute the entire river course.

The water effluences from Hindusthan Fertilizer Corporation show chromium and arsenic which are both highly toxic elements. The concentration of arsenic both in sediment and effluent samples is much above the threshold limit of toxicity. The concentration of arsenic sample in sediments is about 50-100 times higher. His findings emphatically put forward the fact that the Damodar river bed has already stored fairly high level of arsenic which slowly leaches these toxic elements into the river.

Water use Zone of the Damodar

The Damodar is the principal source of water in its neighbourhood. Most of the towns and rural settlements find no other way but to depend on the river for drinking, bathing, other domestic purposes, agricultural and industrial uses.

A field survey conducted by the worker (Fig. 25.1) revealed that from its confluence with vera at Rajrappa upto Bokaro the Damodar water is not polluted in a strict sense and can be used as drinking water after conventional treatment. In this stretch the Damodar is a sacred river known as Deonad and is widely used for holy dips. From Bokaro down stream the water gets contaminated with various chemical effluences, oil slicks, domestic sludge materials and water quality deteriorates rapidly. From Asånsol down stream the water of the river is extremely polluted. But a huge population of this region upto Burdwan depends on the river for their drinking water. The water quality of the river down stream from Burdwan is not suitable even for fish culture and is mainly used for irrigation. This too is not desirable as the pollutants leave their traces on the field outturns.

The Central Pollution Control Board, together with the Bihar and West Bengal state boards is monitoring the water quality of the river and has classified it into a few zones according to its water quality

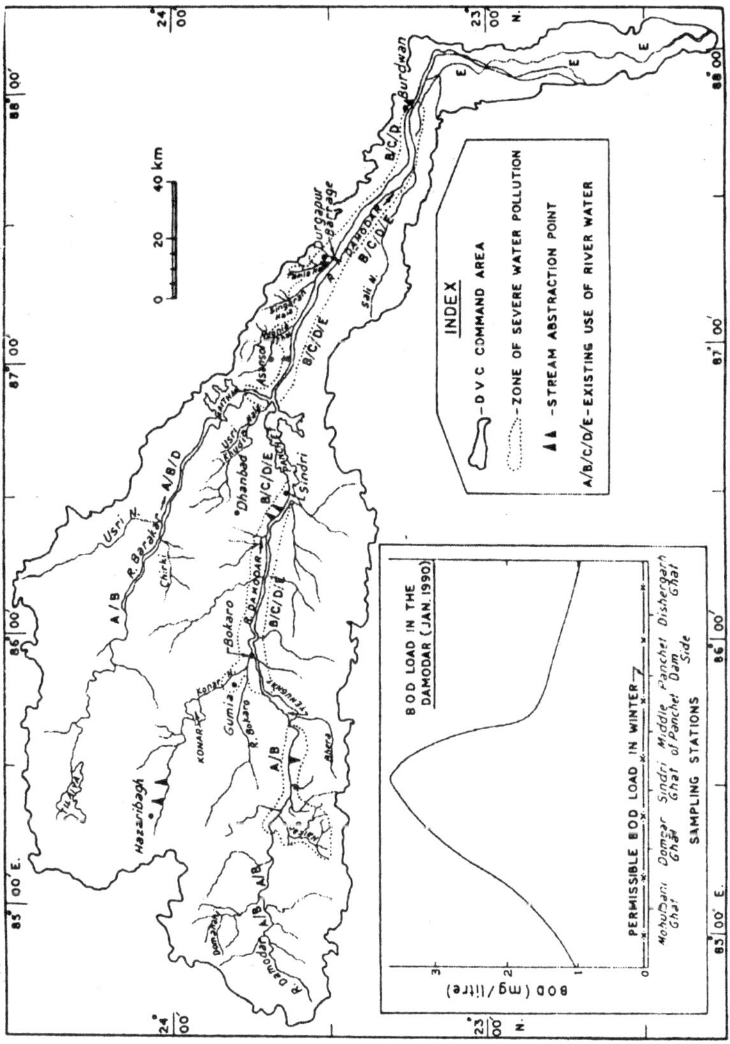

Fig. 25.1 : Damodar Valley Region : Zone of Severe Water Pollution in the River due to Industrialisation

(Fig. 25.1). According to the findings, the Damodar is predominantly in a state of class-D quality that is suitable for propagation of wild life only. There is only a small stretch between Giddi and Bokaro thermal power station which may be put to class C, that is water, after conventional treatment may be utilised for drinking purposes. Its major tributaries like Bokaro, Konar, Barakar, Nalkari etc. are also polluted and down graded to class-D. Other small tributaries situated in Asansol, Durgapur belt, Nuniakhal, Singaran, Tamla etc. are extremely polluted and can not be put to any class. The present state of water from utilising point of view may be tabulated in Table 25.3.

Consequences of Pollution

The above facts and findings hold a brief in favour of the fact that the Domodar water pollution, for most part of the river has crossed the threshold limits.

The concentration of undesirable elements result in propagation of various diseases to which the people of the valley are continously exposed. The intake of sulphides causes headache, nausea, lassitude, bad attack on nervous system, kidney, gastro intestinal track. The concentration of coliform in water results in the outbreak of serious diseases like typhoid, cholera, polio-myleitis, meningitis, hepatitis etc. Arsenic on the otherhand bring disasters by creating blindness, skin diseases etc. Although no systematic hospital records have yet been kept in this respect, increasing number of patients in the region, suffering from these ailments are regularly being reported of. It is hightime to make a keen observation as also to put a check on the situation to avoid future disasters.

Remedial Measures

The water quality of the Damodar, which is the life line of the entire catchment area through which it flows, has undoubtedly been polluted to a level of concern. It is imperative to restore back the state of water quality to stabilize the condition for future.

Following measures may help in improving the present state of condition.

(a) All washeries and collieries that throw effluences in the river must have to install clariflocculator plants to separate solids

Table 25.3: Zoning of the Damodar

Stretch	River length (in kms)	Critical primary water quality	River classification Existing	Desired
Mainstream				
Upstream of Giddi to downstream of Giddi	4	Total coliform BOD	C	C
Downstream of Nalkari to upstream of Bhairavi	24	BOD	D	C
Downstream of Bhairavi to upstream of Bokaro Thermal Plant	38	Total coliform BOD	C	C
Downstream of Bokaro-Konar to upstream of Barakar	125	Total coliform BOD	D	C
Downstream of Barakar to Burnpur	18	Total coliform BOD	D	C
Burnpur to Burdwan	125	Total coliform BOD	D	C
Tributaries				
Nalkari		Total coliform BOD	Extremely polluted	—
Bhairavi		BOD	C	C
Nuniukhal		Total coliform	Extremely polluted	—
Singaran		BOD	Extremely polluted	—
Tamla Nallha		BOD	Extremely polluted	—

Source: DVC Report

followed by chemical treatment to restore the original standard of the water;

(b) The cities and towns situated on the river should treat their municipal discharge effluents to remove solid content and bacteria before discharging the water back into the river;

(c) All the industries thriving on the Damodar water should be enforced to install treatment plants before draining their effluences into the river;

(d) Since water scarcity is a problem all the washeries and large scale industries should recycle the drawn water which is possible with modern technical knowhow;

(e) The original species of fishes, tortise etc. which are disappearing fast in contact with polluted water are to be grown to check biological pollution;

(f) A reasonable good flow of water in the lean period is to be ensured by (i) regular dredging in the river course, (ii) arrest of silt load before draining in the reservoirs and (iii) large scale afforestation in the upper valley; and

(g) Economic and judicious utilisation of the river water.

All these effective measures taken together, along with a drive for mass awarness circulated through mass media programmes is likely to bring back a healthy state of condition in the river water as well as in the socioeconomic status of the region.

Acknowledgement

The author expresses her deep sense of gratitude to Professor B. Banerjee for his valuable guidance during preparation of the paper.

References

Chakrabarti, S. (1990), Damodar River System, Steps for its water purification, Workshop on water quality management for the Damodar river system, D.V.C. Calcutta.

De, A. (1985), The saga of the Damodar River, *Journal of Indian Chemical Society*, Vol. LXII, Dec. pp. 1038-42.

De, A. *et al.* (1985), Pollution profile of Damodar river sediment in Raniganj-Durgapur Industrial Belt, West Bengal, India, *Environment International*. Vol. 11, pp. 453-8.

Mukhopadhya, D.P. (1990), Identification of River Stretches of the Damodar requiring quality improvement, Workshop on water quality management for the Damodar river system, D.V.C., Calcutta.

26

Population and Harvested Area in Ethiopia:

A Temporal Perspective

KEFIALEW ABATE and H. PRASAD

The average rate of population growth from 1990-1995 was estimated at 3.2 per cent per annum (UN, 1997). This high rate of population growth has been taken as one of the most important factors for inducing desertification through expansion of cultivated land. Desertification as defined by UNCED (1992) is land degradation in arid, semi-arid and dry sub-humid areas resulting from various factors, including climatic variations and human activities. Among several human activities, crop cultivation is one of the potential factors which may induce desertification.

As population increases, cultivable land will be cropped year after year (without fallowing) and/or less productive (marginal) land will be brought under cultivation (expansion of cultivated land). In the absence of modern agricultural inputs, reduced fallowing practices and expansion of cultivated land will both induce desertification. When everyone cannot be accommodated any longer in the rural areas because of population growth, the extra people emigrate to urban areas in search of alternative livelihood, which in turn leads to rapid growth in urban population.

The purpose of this study is, therefore, to examine if the rapid population growth in Ethiopia from 1950-1985 was associated with expanding cultivated land, rapid urban growth, increased crop yield and/or increased per capita crop productivity.

Background Information

Ethiopia, covering an area of approximately 1,104,300 km^2 is the second largest country (next to Sudan) in eastern and the Horn of Africa. The total population was estimated at 5,66,77,000 in 1995

(UN, 1995). The crude population density in 1995 indicated an average of fifty one persons per square km. The crude density, of course, does not tell much about availability or non-availability of enough natural resource base for the number of people theoretically expected to be found in each square kilometer, neither does it tell anything about the nature of the population distribution in the country.

Some sources (Ethiopia/UNCED National Report, 1992) indicated that about 90 per cent of the population lived in the highlands (above 1500m) which account for 45 per cent of the total area of the country. The remaining 10 per cent of the population lived in the lowlands (below 1500m) which covered 55 per cent of the country. The highlands are humid and sub-humid, whereas, the lowlands are arid and semi-arid except along the Sudan-Ethiopia border. The climatic conditions seem to have a major role in deciding the distribution of the population and the type of economic activities carried out in the country.

Small scale subsistence crop cultivation is the main economic activity in the highlands. The highlanders also raise livestock for draught power, for supplementing income from sale of animals such as sheep and goats, for purchasing agricultural implements like chisel, plow tips, axes and hoes, for buying clothing, soap, paraffin and salt. The highlanders are, therefore, mixed farmers practicing both crop and livestock production. About 74 per cent of the total livestock in the country is raised in the highlands while the pastoralists in the lowlands raise in the remaining 26 per cent of the total livestock population (Ethiopia/ UNCED National Report, 1992). The lowlanders use their livestock as a source of their staple food and social prestige. With the exception of a few large scale irrigated farms in the Awash River Basin, the extensive arid and semi-arid areas are used for grazing 26 per cent of the country's livestock population owned by about 10 per cent of the population.

The amount of cultivable land in the highlands in which 90 per cent of the Ethiopian population is found is by far less than 45 per cent if mountainous areas with frequent forest occurrences and very steep slopes are excluded. The land resource is not, therefore as big as the total area of the country suggests. The widely held view is that as the population grows at a high rate, the increasing pressure on land may force people to cultivate marginal and less productive areas. That is to say more of the mountainous and the semi-arid areas currently considered marginal areas, will be brought under cultivation and hence expansion of desertification if conservation measures are not taken. Similar conditions were experienced in Thailand between 1950 and 1988 when 108 million *rai* of forest land was cleared to alleviate rural poverty and to meet the demands of the rapidly growing population for land, by expanding agricultural land into the national forest reserves (David, 1993). In view of the virtually stagnating crop yields in Thailand and the finite forest land, the strategy of bringing forest land under cultivation was not sustainable; Thailand, instead, has to face the consequences of deforestation in the form of national disasters (e.g. floods) which are very costly. The high cost and severe damages of national importance will continue to occur unless the trend of deforestation is reversed. Do we experience similar expansion of cultivated land in Ethiopia?

Statement of the Problem

Rapid population growth may result in one or some of the following methods of adjustment:

(a) Additional land in the marginal mountainous areas and in the semi-arid areas will be brought under cultivation;
(b) Productivity per farmer and per unit of land will increase to support the additional number of people; and
(c) There will be an influx of rural people to urban areas in search of jobs and hence rapid increase in urban population.

The marginal areas which might have been brought under cultivation due to population pressure will be impoverished; the quality and diversity of life forms including the fertility of the soil will be adversely affected by the encroachment of the cultivators. The pastoralists will also be pushed to poorer and poorer pastures; or their grazing grounds will be gradually shrinking, hence over grazing. The wild life will also be negatively affected when their habitat is disturbed by the encroaching cultivators and pastoralists. Most of contemporary urban problems witnessed in many developing countries, including Ethiopia, are attributed, among other factors, to rapid urban population growth due to immigration of rural people to urban areas looking for better opportunities.

Research work (Berhe Wolde Aregay 1996, IUCN 1993, Adrian Wood and Michael Stahl 1990, Humi 1988) gave gloomy prospects in food production and inadequate pasture by the year 2000 and beyond if the population continued to grow at the same rate and if concerted conservation measures were not taken. On the other hand. unfavorable government policies, poor implementation strategies, poor popular participation and minimal efforts in raising environmental awareness witnessed in Ethiopia have always contradicted the deep concerns stated by the researchers.

Objectives

The overall objective of this paper is to examine if population growth in Ethiopia has been associated with an increase in the size of cultivated land which indicates encroachment of people to marginal lands and hence likelihood of the formation of desertification. More specifically the paper tries to:

— see if land under cultivation has significantly increased with the growth of population in Ethiopia from 1950 to 1985.
— examine if productivity per unit of land has increased over the years since 1950.
— Investigate if population pressure in the rural areas is reflected by rapid urban population growth.

In view of the high rate of population growth reflected in various literature (including *UN Demographic Year Books*), the following hypotheses have been formulated:

(a) Harvested area has increased significantly since 1950.
(b) Crop productivity per unit of cultivated land has increased significantly since 1950.
(c) Urban population has significantly increased in the country since 1950.

Materials and Methods

Secondary data was obtained from *UN Demographic Year books* and *World Crop and Livestock Statistics*. Population data has been taken from the estimates made by Asmerom Kidane (1987). The data have been summarised in the form of tables and graphs. Some statistical methods have also been employed to see relationship in the temporal changes between population, cultivated land and yield. The paper focuses only on harvested area, cereal production, production of pulses, root and tuber crops. Cereals include wheat, rice, barley, maize, rye, oats, millet and sorghum, mixed grain. Pulses include dry beans. Dry broad beans, dry peas, chick peas, lentils, cow peas, pigeon peas, vetch and soybeans. Potatoes. sweet potatoes and cassava fall under roots and tubers.

This paper excludes areas under sugarcane, ground nuts, oil seeds such as sesam, caster oil. sunflower and flax, stimulants and perennial plants such as coffee, tea, tobacco and chat. These crops were excluded because there was no concurrent data with the other crops over the period under consideration (1950-1985). The hectarage under these crops could be considerable, but the author believes that there will be very little change in the hectarage in the period under consideration (1948/50—1985). With the exception of oil seeds which are grown in the central and northern parts of Ethiopia, most of the other crops that are excluded are found in the south and southwest where environmental degradation is not yet a serious concern. It should also be noted that the Hararghe Plateau which has been subject to environmental degradation is an important coffee and tea growing area. The paper should, therefore, be read bearing in mind these limitations.

Results and Discussion

Harvested Area

Table 26.1 shows that an average of 6,123,889 ha of land had been brought under cereals, pulses, root crops and tubers every year from 1948/50—1985. The use of a one tailed Student's distribution indicates that the average harvested area will vary between 5,966,970 ha and 6,280,808 ha at 95 per cent confidence level. Cereals accounted for about 81 per cent of the total area under harvest, while pulses covered around 14 per cent of the harvested area. Root crops and tubers accounted only for 5 per cent of the total harvested area. Harvested areas did not increase consistently over the period in question. There were inter-annual fluctuations in the size of harvested areas.

As compared to pulses, root crops and tubers, the hectarage under cereal crops showed higher inter annual variations. Harvested area under cereal crops consistently increased from 1948/50 to 1964 when area under cereals reached its maximum (Fig. 26.1a, 1b). Since 1965, there had been a general decline in land under cereal crops until 1985. The period between 1972 and 1979 shows a conspicuous depression (fall in areas under cereals). The temporal changes of harvested areas under pulses resembles that of cereals. The 1970s are again characterized by a depression in the area under pulses (Fig. 26.1c). In the case of root crops and tubers, there has been a steady increase in harvested area, except in 1984-85 when production, yield and harvested areas all declined probably because of the widespread 1984 drought (Fig. 26.1d).

Population and Harvested Area

Population has shown considerable growth over the years since 1948 as shown in Fig. 26.2. Unlike rapid population increase, the harvested areas have not shown much change from 1948/50 to 1985 (Fig. 26.2). The total harvested area remained the same with wave like inter-annual changes in its amplitude. The same figure (Fig. 26.2) shows that the difference between 1948/50 and 1985 'in magnitude of harvested area is small. The correlation coefficient between population and total harvested area ($r = 0.134$) is very low..In other words, only around 1.8 per cent of the changes in the size of harvested area can be explained by population increase as indicated by the coefficient of determination ($r^2 = 0.018$). This relationship appears to contradict the widely held view that high rate of population growth leads to increasing cultivated areas and expansion of desertification due to encroachment of marginal lands.

The poor relationship between population change and harvested area signals the presence of several other factors (more important than rapid population growth) which has always been taken as a major

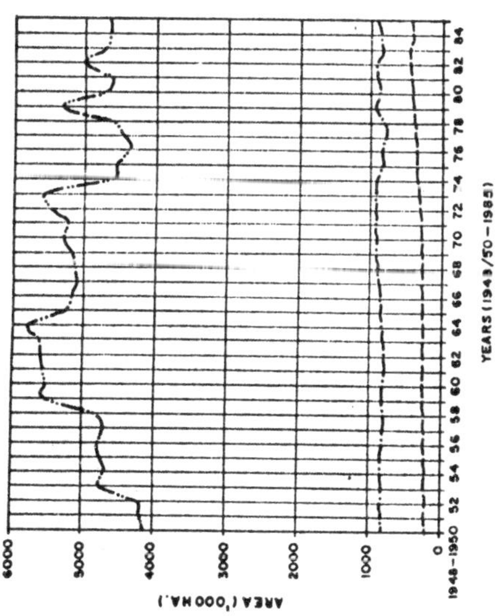

Fig.: 26.1(a) : Area harvested ('000 ha) in Ethiopia (1948/50—1985)

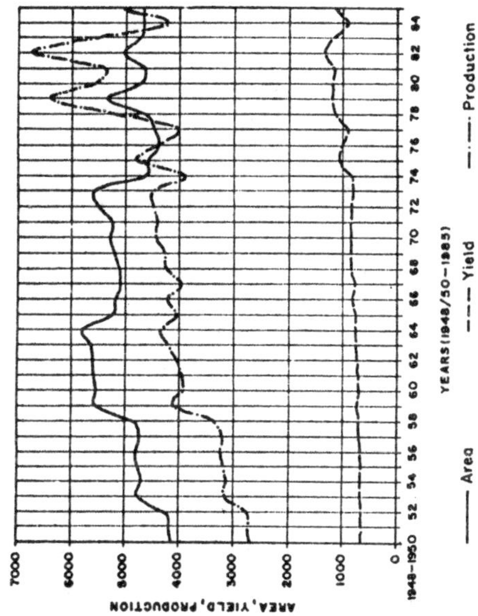

Fig. 26.1(b) : Area harvested ('000 ha). Cereal yield (kg/ha) and cereal production in Ethiopia (1948/50—1985)

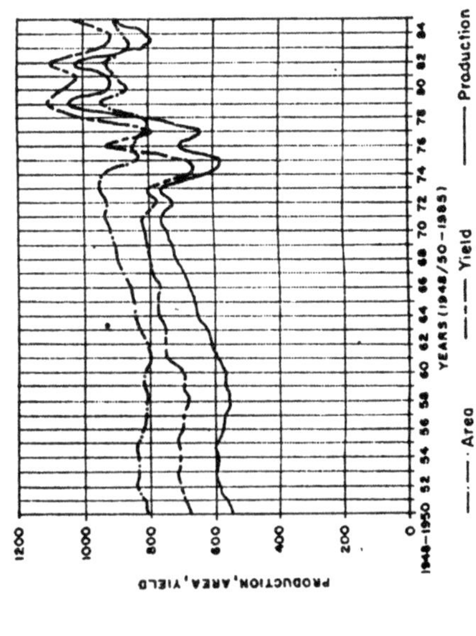

Fig.: 26.1(c) : Production of Pulses ('000 mt) area Harvested ('000 ha) and yield (kg/ha) in Ethiopia (1948/50—1985)

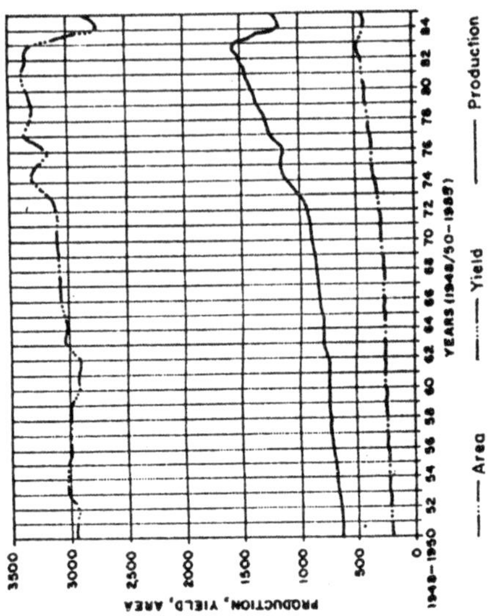

Fig.: 26.1(d) : Production of Root Crops and Tuber ('000 mt) yield (kg/ha) and area Harvested ('000 ha)

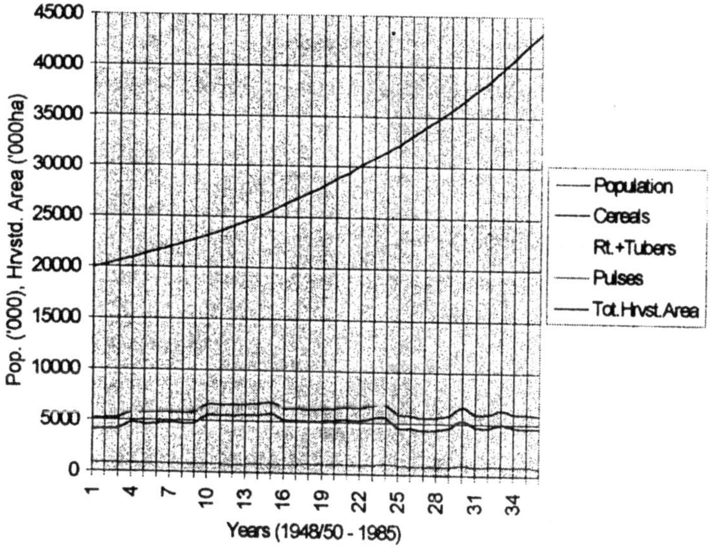

Fig. 26.2 :Temporal changes in Population ('000) and Harvested areas ('000), 1948/50—1985

factor for expansion of cultivation to semi-arid and mountainous areas and hence inducing land degradation (desertification). Cultural, social, economic, technological, political and physical factors such as poor soil, scarcity or lack of moisture, and hence poor land productivity may weigh more in the expansion of cultivated fields than mere increase in the number of people. These factors, unfortunately are not examined in this paper. Whether or not one single factor or a combination of factors affects size of harvested area, the fact remains that the increase of harvested area for a period of about 36 years (1948/50—1985) has been very small (Fig. 26.2). The difference in harvested area between 1948/50 and 1985 (6,022,000-5,154,000) is only 868,000 ha (see Table 26.1). That gives an average rate of increase in harvested area of about 24.111 ha per year.

In 1950 the population was estimated at around 19,900,000 (Asmerom, 1987) and the harvested area in the same year, according to FAO (1987), was about 5,154,000 ha which gives a harvested area of about 0.26 ha per person. In 1985 the ratio of harvested area to population (6,022,000/43,300,000) dropped to 0.14 ha/person. In a period of 36 years (1948/50 - 1985) harvested area per person

Table 26.1: Population ('00) and Harvested Area ('00)

Year	Population	Urban Pop.	Cereals	Rt. +Tubers	Pulses	Total Hrvst. Area
1948/50	19,900	—	4,130	216	808	5,154
1951	20,200	—	4,180	220	816	5,216
1952	20,600	—	4,222	222	838	5,282
1953	20,900	—	4,762	226	835	5,823
1954	21,300	—	4,685	230	837	5,752
1955	21,700	—	4,752	232	837	5,821
1956	22,000	—	4,790	237	818	5,845
1957	22,400	—	4,714	241	811	5,766
1958	22,800	—	4,823	245	810	5,878
1959	23,200	—	5,558	247	818	6,623
1960	23,600	—	5,533	253	816	6,602
1961	24,100	—	5,561	255	796	6,612
1962	24,600	—	5,593	256	807	6,656
1963	25,100	—	5,625	258	825	6,708
1964	25,700	—	5,787	261	842	6,890
1965	26,300	1,525.4	5,234	264	851	6,349
1966	26,900	1,702.3	5,172	268	858	6,298
1967	27,500	1,980	51,092	273	877	6,242
1968	28,100	2,108	5,112	278	897	6,287
1969	28,800	2,245	5,175	284	905	6,364
1970	29,400	2,391	5,292	289	918	6,499
1971	30,200	2,546	5,225	293	937	6,455
1972	30,900	2,699	5,481	299	927	6,707
1973	31,600	2,861	5,550	313	955	6,818
1974	32,300	3,075.7	4,611	334	936	5,881
1975	33,100	3,276	4,571	356	834	5,761
1976	34,800	3,488.9	4,368	358	854	5,580
1977	34,800	3,580.8	4,458	370	807	5,635
1978	35,700	3,794.1	4,631	389	818	5,838
1979	36,700	4,112.9	5,331	411	946	6,688
1980	37,700	4,224.9	4,712	422	867	6,001

(contd.)

Table 26.1: *(contd.)*

Year	Population	Urban Pop.	Cereals	Rt +Tubers	Pulses	Tot. Harvst. Area
1981	38,700	4,472.3	4,646	433	905	5,984
1982	39,800	4,719.6	5,029	454	932	6,415
1983	40,900	4,749.5	4,716	475	860	6,051
1984	42,200	4,779.4	4,651	428	878	5,957
1985	43,300	4,880.7	4,660	447	915	6,022
Sum			1,78,432	11,037	30991	2,20,460
Average			49,56,444	306.5833	860.86116	,123.889
Per cent			80.93,622	5.00635	14.05743	100
Corr. Pop. & total harvested area						0.133735
Stdev.			454.1824	78.8403	48.17774	463.7614
Sqrt.n			6	6	6	6
S.E.			75.69706	13.14005	8.029623	77.29356

decreased by (0.26 –0.14) = 0.12 ha. In other words each person lived on less harvested area in 1985 than in 1948/50. When harvested area is examined by crop type over time, we do not get a different picture. Neither cereals nor pulses have expanded over larger hectarages over the 36-year period. Root crops and tubers always covered relatively small areas and showed less inter-annual changes.

All the three crop categories seem to be affected by some common factors, because we do not see a shift from one crop type to another, which might occur when a certain crop category is affected by unfavourable weather conditions or by policies biased toward increased hectarage of specific crop groups. The question that may be raised at this juncture is that; if harvested area has not expanded and if there was no shift to the cultivation of some crop types when other crops could not be grown could it be possible that yield per unit area has increased over time?

Crop Production and Yield

(a) *Production*

There had been a general upward trend in the total crop production m the period 1948/50 and 1985. The highest amount of crop production in the period under consideration was obtained in 1982 (Table 26.2). The period 1978-1982 was characterized by the highest amount of crop production. Crop production in 1984 dropped to below the 1970s due to the effect of the 1984 drought. The 1984 crop production was, however, higher than any year before 1970 (Table 26.2).

Figure 26.3 shows the temporal changes, in crop production, of the three crop categories. Out of the three crop categories under treatment, pulses increased the least over the 36-year period. Production of root crops and tubers gradually grew except in 1984 when it went down slightly (Fig. 26.3).

The overall trend of crop production since 1948/50 has shown a consistent growth, hence a high correlation (r=0.884) beween population growth and crop production. This high correlation does not, however, indicate enough crop production for the increasing population. The rate at which crop production grew was much lower than the rate of population growth. The correlation is, therefore, only an indicator of change in crop production (irrespective of its magnitude) and population growth in the same direction. The production increase does not guarantee adequate food for the growing population for it can be far below what can support the population. We also do not expect stagnation or negative trend in crop production while population increases.

(b) *Yield and Productivity*

Yield refers to production per unit area of cultivated land. Crop yield in this paper is given in kilograms of crop per ha. As shown in Figs. 26.1b to 26.1d and 26.4a, crop yield remained almost stagnant except in 1984 when yield declined. Only in the case of pulses, we observe a period of relatively high yield from 1978 to 1983 (Fig. 26.1c). Even in 1984 when drought was blamed for crop failure and for low productivity, the yield of pulses remained much higher than pre-1977 period.

Productivity is taken to mean the ratio of crop production to the total number of people. It shows the per capita crop productivity. The per capita crop production had been declining since 1959 until 1978 when it started to go up again (Fig. 26.4b). The highest per capita

Table 26.2: Population ('000), Crop Production ('000 mt) and per capita productivity

Year	Pop.Crop	Prdctn.Crp.	Prd./pr	Cerl.prd./pr	Plse.Prd./p	Rt.+Tub./pr
1950	19,900	3,888	0.195	0.136	0.028	0.032
1951	20,200	3,940	0.195	0.135	0.028	0.032
1952	20,600	4,000	0.194	0.134	0.029	0.032
1953	20,900	4,457	0.213	0.152	0.029	0.033
1954	21,300	41,423	0.208	0.147	0.028	0.033
1955	21,700	4,510	0.208	0.148	0.028	0.032
1956	22,000	41,533	0.206	0.148	0.026	0.032
1957	22,400	4,506	0.201	0.144	0.025	0.032
1958	22,800	4,652	0.204	0.148	0.024	0.032
1959	23,200	5,430	0.234	0.178	0.024	0.032
1960	23,600	5,215	0.221	0.166	0.024	0.031
1961	24,100	5,300	0.22	0.164	0.025	0.031
1962	24,600	5,418	0.22	0.165	0.025	0.03
1963	25,100	51,024	0.224	0.168	0.025	0.031
1964	25,700	5,805	0.226	0.17	0.025	0.031
1965	26,300	5,475	0.208	0.152	0.025	0.031
1966	26,900	5,729	0.213	0.157	0.025	0.031
1967	27,500	5,493	0.12	0.144	0.025	0.031
1968	28,100	5,814	0.207	0.151	0.026	0.031
1969	28,800	5,879	0.204	0.148	0.025	0.03
1970	29,400	6,060	0.206	0.15	0.025	0.03
1971	30,200	6,075	0.201	0.146	0.025	0.03
1972	30,900	6,140	0.12	0.145	0.023	0.03
1973	31,600	6,217	0.12	0.141	0.024	0.031
1974	32,300	5,593	0.173	0.12	0.02	0.034
1975	33,100	6,539	0.198	0.145	0.02	0.035
1976	34,000	6,281	0.185	0.13	0.021	0.033
1977	34,800	5,883	0.17	0.115	0.02	0.036
1978	35,700	7,318	0.205	0.146	0.023	0.036
1979	36,700	8,805	0.24	0.174	0.029	0.037
1980	37,700	7,945	0.211	0.149	0.024	0.038
1981	38,700	7,768	0.201	0.139	0.024	0.038
1982	39,800	9,264	0.232	0.169	0.026	0.038
1983	40,900	7,912	0.193	0.135	0.02	0.038
1984	42,200	6,174	0.146	0.1	0.019	0.028
1985	43,300	7,539	0.174	0.123	0.022	0.029

Corr. Pop. & Crp. Prdctn. 0.884

Corr. Pop. & Crp.Prdctivity. – 0.357

Corr. Pop. & per capita Cerls. Prdctivity. – 0.359

Corr. Pop. & per capita productivity of pulses – 0.617

Corr. Pop. & per capita productivity of Rt. crps. and Tubers 0.429

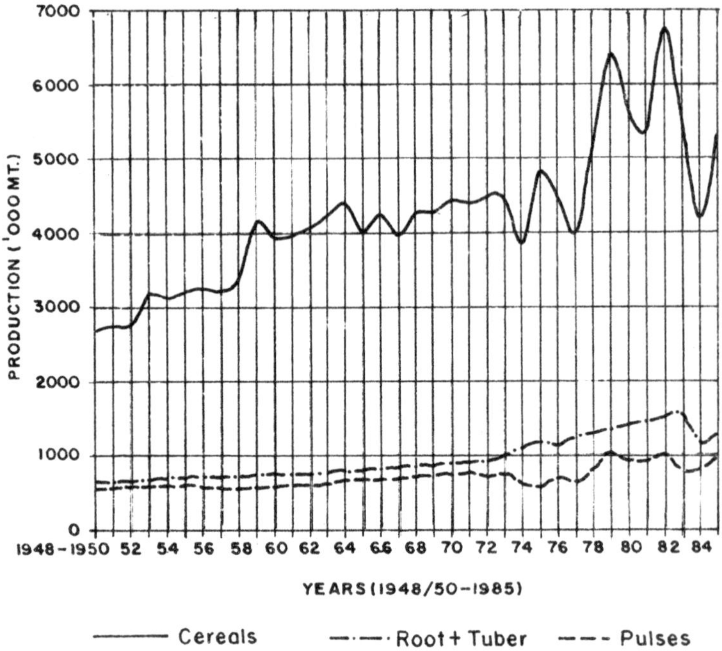

FIG. 26.3: Production of Different Crops ('000 mt)

crop productivity was experienced in 1979. The period 1978 to 1983 was relatively a period of high productivity; but a low productivity soon resumed in 1984.

The low and constant crop productivity per person was reflected by negative correlation coefficients when correlated with the population change over the given period (1948/50—1985). Negative correlation is an indication of increasing in opposite directions. The following table shows the negative relationships of per capita productivity and population change in Ethiopia.

	Crop production per person	Cereal production per person	Pulses production per person	Root + Tubers production/ person
Population	– 0.3575	– 0.3600	– 0.6177	0.42930

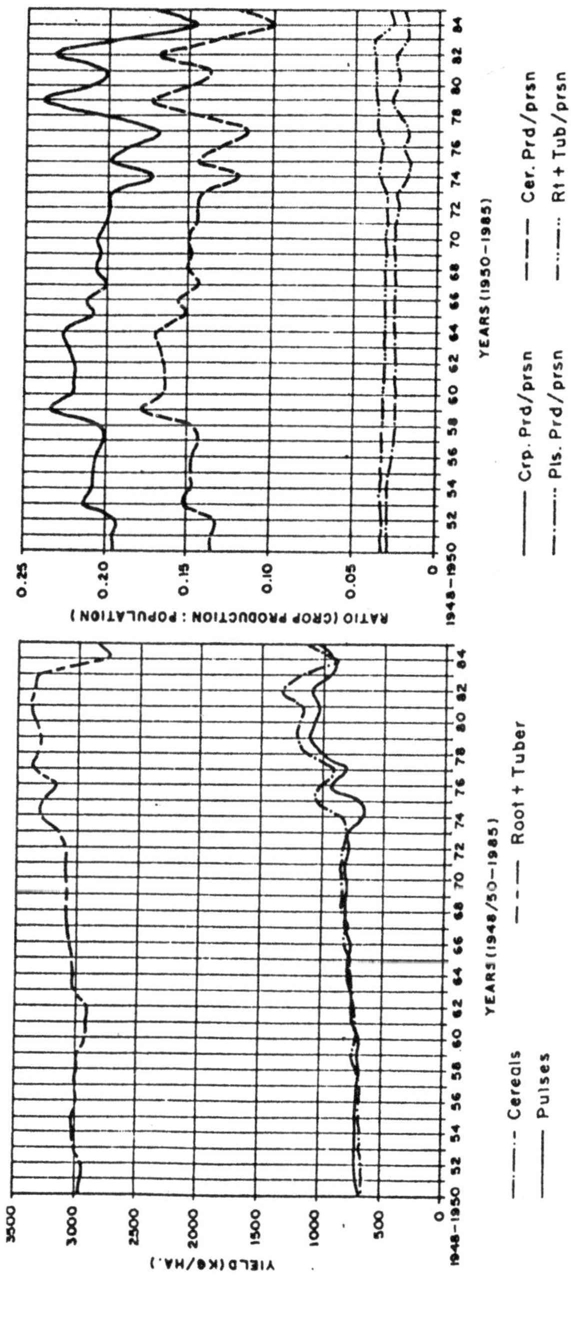

Fig. 26.4(a) and 26.4(b)

The productivity per person on cereals and pulses do not seem to respond to population growth; if they do, the response seems on the opposite direction. This holds true especially for the per capita productivity on pulses which show a negative correlation of –0.62. Although the correlation is low (0.43), only the productivity of root crops tends to increase in the same direction with population growth.

Urban Population

Urban population growth over the period (1965-1985) was examined to see if the additional rural people who could not get land emigrated to urban areas in search of job opportunities, and hence to contribute to rapid urban population growth. Urban areas in Ethiopia are defined as localities in which there are 2,000 people or more. As given by *UN Demographic Year Book* (1973-1988), the urban population increased steadily from 1965 to 1982. From 1982 to 1985, the urban population increase became even more flattened (Fig. 26.5). The period 1982, 1983 and 1984 was a period of wide spread droughts in eastern and southern Africa including Ethiopia and in north central India (Martin and Timothy, 1988). That period was a period of pronounced El Nino or Southern Oscillation phenomena, a periodic fluctuation in the intensity of the inter-tropical atmospheric and, oceanic circulation (southern oscillation) that is usually coincident with an anomalous warming of the, eastern tropical Pacific Ocean (El Nino).

The sea suface temperature changes and the resulting anomalies in atmospheric and oceanic circulation are all believed to cause droughts and substantial decrease in productivity, large aggregate falls in production and in extreme (though not frequent) instances of starvation and death (Martin and Timothy, 1988). Extensive hardship and migration from farms to towns would normally be expected when crop production failed or decreased because of drought conditions. In spite of the wide spread sufferings of most rural Ethiopia in that period, it was not reflected by rapidly growing urban population. Instead the data shows the contrary (i.e. a nearly leveling off of urban population during that period). The authors have found the relatively slow growth of the urban population (especially during hard times) very difficult to explain.

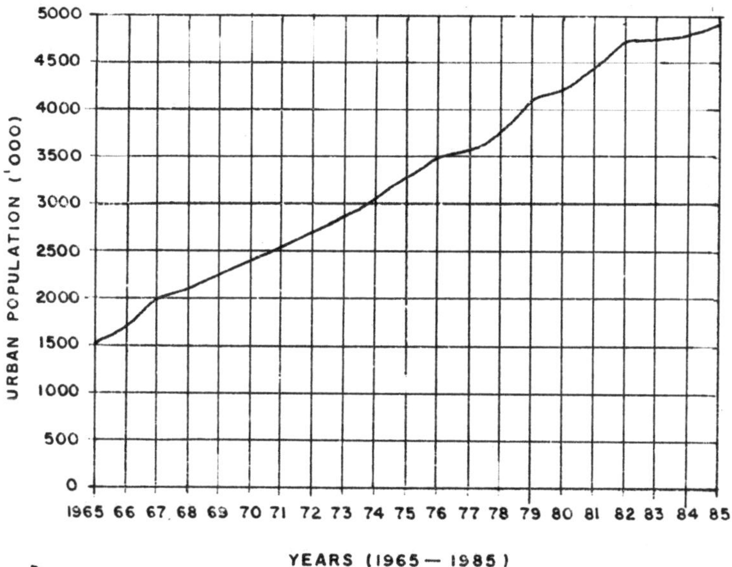

Fig. 26.5 : Temporal changes of Urban Population ('000) 1965—1985)

Conclusion and Recommendations

The authors are aware of some case studies conducted in northern Shewa, northern Gondar, Bale and in Hararghe where cultivated areas were found to expand upwards (to mountainous areas) and downwards (to semi-arid and arid lowlands) as a result of population growth. These case studies were not, however, reflected when the changes in population and in harvested areas over 1948/50 to 1985 were examined in this study at a national level.

The analysis based on the secondary data obtained from the *World Crop and Livestock Statistics* (1987) and from *UN Demographic Year Books* indicated that harvested area did not expand significantly from 1948/50 to 1985 although population increased at a high rate over the same period. Yield and per capita crop productivity did not also show any significant growth. The steady growth of urban population obtained in this study defies the idea of attributing rapidly growing urban population to population pressure in rural areas. These findings, therefore, lead to the following conclusions:

(i) Population growth did not lead to expansion of harvested areas. neither did it lead to expansion of desertification (due to expanding cultivated land) over the time period considered in this paper;

(ii) More and more people depended on the same harvested area and on the same crop yield. In other words each person got less and less of crop products from the decreasing per capita hectarage as population increased. That indicates the deepening of poverty from time to time;

(iii) The practice of fallowing probably decreased as population increased without a corresponding increase in harvested area or in yield, and hence a general decline in the productivity of the soil or the occurrence of desertification *in situ;*

(iv) The deepening poverty either led to hunger mass starvation, famine and even death, or it led to emigration of people out of the country since the steady urban population growth did not reflect an influx of rural people to urban centers. There were either prohibitive policies to prevent rural-urban migration or rural people did not expect anything better in urban areas. Movement of people probably took place from deficit to excessive areas within rural Ethiopia instead of to urban areas; and

(v) As the population grew, the additional people who could not cultivate more land probably engaged themselves in petty trade and other informal non-agricultural activities.

We have always taken it for granted that population increase leads to expansion of cultivated land and hence encroachment of marginal areas like mountains, arid and semi-arid lands. There has also been a popular view that population growth results in population pressure on the available land and leads to the exodus of people to urban centers which, in most cases, are not prepared to accommodate the influx of rural people in terms of jobs, housing, health, education and transport facilities. These popular views are not, however, supported by this paper.

The findings in this paper gave a new insight for further investigation on the identification of major factors. other than population growth, that affect expansion of cultivated land in Ethiopia. Further studies, on how the increasing number of people has adjusted to the somewhat fixed cultivated land, should be conducted and encouraged. Ways and means must be sought to increase yield and per capita crop produtivity.

Attributing the development of slums and many social problems in many urban centers in Ethiopia to rural population influx must be re-examined since these problems can prevail due to political, policy, transparency, accountability, and other related problems other than rural emigration. More rigorous studies are, therefore, recommended to understand and tackle urban problems.

References

Asmerom, K. (1987), Re-estimating the Ethiopian population by age and geographical Location, *NEAS,* (9)3, pp. 59-73.

Berhe, W. A. (1996), *Twenty Years of Soil Conservation in Ethiopia: A Personal Overview,* Regional Soil Conservation Unit/SIDA.

David, R. (ed.) (1993), *Structural Adjustment and the Environment,* Earthscan Publication Ltd.. London.

FAO (1987), *World Crop and Livestock Statistics (1948-1985)*: Processed Statistics Series 1. Rome.

Hurni, H. (1988), 'Ecological Issues in the Creation of Famines in Ethiopia'. Paper presented at the National Conference on a Disaster Prevention and Preparedness Strategy for Ethiopia.

Martin. A. J.W., and Robert, C. B. Jr. (1996), *Interactions of Desertification and Climate.* Arnold, London.

Martin, P. and Timothy, C. (1988). Assessment of Effects of Climatic Variations on Agriculture: A Summary of Results from Studies in Semi-Arid Regions; *The Impact of Climatic Variations on Agriculture,* Vol.2, Assessments in Semi-Arid Regions, Kluwer Academic Publishers, London, pp. 9-60.

UN (1997), *Demographic Year Book 1995,* 47th issue, New York.

List of Acronyms

corr.	Correlation	Sqrt.	Square root
Pop.	Population	Plse. or Pls.	Pulses
Tot.	total	Cerls.or Cerl.	Cereals
Rt.	Root crops	Stdev.	Standard deviation
Hrvst.	Harvested area	Prn.or Prsn	Person
mt.	Metric tonnes	Prdctn.	Production

27

Conclusion:
A Critique of Professor Kayastha's Work

Dr P. Nag

Environment cannot be viewed in isolation. Population and development vitally affect the state of environment. Therefore, it would be necessary to consider population, environment and development as interacting aspects of the global phenomena. Development, to be meaningful, has to be sustainable developments. This is the opening paragraph of the first chapter of the book of Professor Kayastha entitled *Geography of Population*. The model proposed by him involves the issues related to environment, population and development (EPD). Such approaches have a greater acceptability in geography from very early stage. In fact this disciple has been considered as the study of human-environment relationship. The hidden factor is the resources. The concept of development has undergone drastic changes. Its reincarnation as 'sustainable development' has close links with population, environment and resources. Professor Kayastha has highlighted the changing equation very effectively while dealing with issues at any level. While there is general agreement on the need for the protection and optimization of 'sustainable', according to Kuhnen (1992) means survival, that is, keeping the community alive. 'Sustainability' also means 'an ecologically acceptable production', where everything removed is then replaced so as not to harm the ecological system. Sustainability can also be understood in a social sense, meaning a thriving economic and social order where productive structures are relationships which ensure a fair distribution of income, power and opportunities, thus providing basis of social peace.

Sustainability finally environment, there is repeated emphasis on sustainable development. The term stands to mean long-term carrying capacity of a region or regions, where there is no negative impact on environment and sustainable development (Kayastha, 1998). Though, 'sustainable' has been considered as a comprehensive expression of acceptable balance between EPD issues, there are several terms now in currency, such as 'sustainable development', 'sustainable economy', 'sustainable society' and 'sustainable use'. The World Commission on Environment and Development (WCED) defined sustainable development as development that means the needs of the present without compromising the ability of future generations to meet their own needs. Further, the term has been criticized as ambiguous and open to a wide range of interpretations, many of which are contradictory. The confusion has been caused because of using these terms interchangeably, as if their meanings were considered almost the same. They are not. 'Sustainable growth' is a contradiction in terms : nothing physical can grow indefinitely. 'Sustainable use' is applicable only to renewable resources : it means using them at rates within their capacity for renewal.

In addition, 'sustainable development' is used in this strategy to mean : improving the quality of human life while living within the carrying capacity of supporting ecosystems. A 'sustainable economy' is the product of sustainable development. It maintains its natural resource base. It can continue to develop by adapting, and through improvements in knowledge, organization, technical efficiency, and wisdom. Furthermore, a 'sustainable society' lives by the nine principles outlined below (IUCN/UNEP/WWF, 1991).

— Respect and care for the community of life
— Improve the quality of human life
— Conserve the earth's vitality and diversity
— Minimize the depletion of non-renewable resources
— Keep within the earth's carrying capacity
— Change personal attitudes and practices
— Enable communities to care for their own environments
— Provide a national framework for integrating development and conservation
— Create a global alliance

After scanning the contributions on EPD, it appears that he was able to understand the significance of studying in integrated manner from the very beginning of his career and that too at regional level. Geographers have always considered oregano as a handy tool for spatial analysis. In his case it was the Himalayan Beas Basin. His early contributions including his Ph.D. and the first important publication entitled *The Himalayan Beas Basin: A Study in Habitat, Economy and Society* (1964) highlighted the significance of a river basin concept which now has been sò widely accepted approach for development. Nevertheless, he did not confined to regional level for dealing with EPD issues. He was one of the foremost geographers to deal with policies, legislation and philosophy. This forced him to operate at the global and national levels. On the other hand, he mostly engaged his students and research scholars to deal with EPD issues at local level. Perhaps the mechanism such developed allowed him to have first hand information in this process which in turn could be generalized and tested at regional level; and that again in turn contributed to policy oriented research. Such feedback mechanism worked very efficiently for the benefit of not only the professor but his students as well.

Professor Kayastha's Model

A close study of the work will indicate that Professor Kayastha had implicitly put the environment, population and development at the vertex of an equilateral triangle with resources at the centre. His in-depth study of Indian philosophy and frequent reference to Mahatama Gandhi's views were meant to support this model. His studies and contributions at different levels were suppose to revalidate this model, partly or wholly. The best part of his model is that it is based on qualitative statements. He was able to shrug off the so-called quantitative revolution in geography when it was at its peak, which incidentally coincided with the peak of the contributions of this professor. As a result, his contributions are easily understood with out traversing through the associated juggler. In fact he was able to swim against the tide.

Another important aspect of his contribution was to give due importance on social issues instead of economic aspects. This was due to his inclination on population related issues. With the award to Nobel Prize to Dr. Amartya Sen, social issues gained importance recently, but Professor Kayastha has been discussing these aspects from the very beginning of his academic life. In fact, again he was able to push aside the much-accepted economic models in geography in preference to social and behavioral models that too at the peak of the former.

Professor Kayastha at National Level

Geographers have hardly made contributions to work at global and national levels. Further there are rarer contributions in policy oriented researches particularly on EPD issues. His earliest contributions on water, forest and food security are testimony of strong views that were to come later in his career. In 1981, he made an appraisal of water resources of the country and pleaded for national water policy. In the following year he studies forests of the country and the impact of economic development (1982). In 1986 he analyzed the environmental policies of India. Recently, he made a valuable contribution on the water resources, their conservation and management (1998d). His attention was also on environmental problems such as floods, their occurrence, forecasting and control (1983). He was not only sensitive to environmental problems but also their mitigation strategies and relationship with human being (1998c). Population aspects have been closely studied in an integrated manner. These issues were sustainability of metropolitan centres, fertility and migration (1988a, 1991a; with Mukherjee, 1979). In fact, in his recent book entitled, *Geography of Population*, he has discussed several spatial aspects of population and that too in relation to environmental and developmental issues (1998a). Though, Professor Kayastha has written several times latter issues (1989, 1990), but at conceptual level his views on environment, development and quality of life has broken new grounds (1993). While writing the foreword for our earlier book he raised a very pertinent question: What kind of world shall the coming generation inherit? The answer largely depends on the actions taken now by the governments and the individuals all over the world. It is of prime importance that sincere efforts are made towards conservation of environment and development of a conscience for sanctity in life (1993). Regarding the latter aspect, he has widely quoted from *Gita* and contributions of Mahatama Gandhi. Perhaps, he wanted to evolve an oriental view of conservation and way of life that revolves around the concept of limited wants and universal welfare deeply imbedded into Indian philosophy and culture.

Professor Kayastha at Regional Level

As mentioned earlier, Professor Kayaṣtha's earlier contributions have been on Himalayan Beas basin. He has extensively written on the issues related to population, tourism, horticulture, forest and ecology (1956, 1983a, 1990b, 1992). In fact, the study of Himalaya was very close to his heart, but tourism in this region was closest (1975, 1983a, 1998; with S.N. Singh, 1983). According to him, tourism for various purposes is largely based on the use of environment. The state of environment depends on the manner of use of environment. Tourism on large scale makes heavy demand on environmental resource and hence the need for greater concern for necessary measures for its protection and conservation (Kayastha, 1997). He has also studied forestry at regional level, particularly concerning this mountainous region. His contribution has also been valuable in the volume *India: A Regional Geography*, particularly the chapter on Himachal Region (1971). Regional studies were also made concerning environmental problems such as floods (with Yadav, 1985, 1990). Further in the researches conducted under his guidance during the early period were also on regional issues related to EPD. This trend continued till later period of his life.

Professor Kayastha at Local Level

During the later part of the researches conducted under his influence, the local issues caught his attention. This was due to the nature of problems faced by the people at large. Most of the issues were urban in nature like slums, air, noise, water and solid waste problems. Kumra conducted an extensive survey on urban environmental problems of Kanpur city (1978, 1979, 1980, 1983). Attempts were also made to study urban ecosystems and quality of human environment of selected cities. Other studies on Varanasi was also carried out based on perception studies (with S.N. Singh, 1977). He was able to make one of the earliest study on Bhopal disaster with Nag (1989). Nevertheless, the study at the local level was not limited to urban areas only. Aspects related to energy resources and technology transfer in rural scene were given due importance (with R.B. Singh, 1983). He feels that environmental degradation and bad planning have lead to poor, lop-sided development. All these need proper planning, environmental protection and sustainable development. Location-specific planning and co-operation of voluntary agencies and the people will help to remove the perception gabs between the people and the policy-makers (Kayastha, 1988a).

Conclusion

Professor Kayastha has very coordinated way of studying EPD issues at all the three levels. The studies were complimentary to each other. Migration, tourism, hazards and similar issues were discussed at a greater length. Though maps and tables were used to support the qualitative statements, mathematical equations and analysis at decimal levels were omitted. This made reading extremely lucid. Perception studies have widely used for studying EDP issues, which made analysis more humane. The contributions at the later part of his life were more delightful in reading as they reviewed different aspects of environment, population and development in more integrated fashion. They also include national reports submitted to the International Geographical Congresses or presidential addresses of important scientific events (1998b,1998c; with Raza, 1984). As mentioned earlier he falls back on meta-philosophical considerations to deal with the EDP issues. According to him, the Indian traditions of peace and harmony are very comprehensive, and extend to all living and non-living matter. We must evolve the concept of not only living in harmony amongst ourselves, but also in harmony with our environment. Through much harmony alone are peace and progress possible (Kayastha, 1997).

References

Clarke, John I. (1997), *The Future of Population,* Predictions, Phcenix.

IUCN, UNEP & WWF (1991), *Caring for the Earth, The World Conservation Union,* United Nations Environmental Programme & World Wide Fund for Nature, Gland, Switzerland.

Kayastha, S.L. (1956), Tourist industry of Kangra, Kullu and Mandi in the

Himalayan Beas basin, *The National Geographical Journal of India,* vol. II, pt. 3, pp. 128-48.

Kayastha, S.L. (1971), Himachal region, in R.L. Singh *et al* (eds), *India: A Regional Geography,* National Geographical Society of India, Varanasi.

Kayastha, S.L. (1975), Tourism, in Walter Imber and Hans Boesch (eds), *India,* Kiminerly & Frey.

Kayastha, S.L. (1981), An appraisal of water resources of India and need for national water policy, *Geo-Journal,* vol. 5, pp. 563-74.

Kayastha, S.L. (1982), Forest in India and the impact of economic developments, in the *Proceedings of the IGU Commission on Environmental Problems,* VI Symposium, Sao Paulo, Brazil, pp. 1-12.

Kayastha, S.L. (1983a), Impact of tourism in the Himalayan region, in the *Proceedings of the IGU Symposium of Commission on Environmental Problems,* No. VII, Palma de Mallorca, Spain, December 20-28, pp. 1-110.

Kayastha, S.L. (1983b), Floods in India : A study of their occurrence, forecasting and control, *The National Geographical Journal of India,* vol. 29, pp. 1211-4 1.

Kayastha, S.L. (1986), Environmental policies in India, in C.C. Park (ed), *Environmental Policies: An International Review,* Croom Helm, London, pp. 223-58.

Kayastha, S.L. (1988a), Sustainability of metropolitan centres in Third World countries, in *Population in Transition in India,* vol. II, B.R. Publishing Corporation, Delhi, pp. 327-34.

Kayastha, S.L. (1988b), Environment and rural development, *Rural System,* vol. VI, Pt. 1.

Kayastha, S.L. (1989), Environment and development, in Hempriya Devi (ed), *Development and Environment,* Varanasi, pp. 1-7.

Kayastha, S.L. (1990a), Environment and development, in R.S. Doroa *et al* (eds), *Man, Development and Environment,* Ashish Publications, New Delhi, pp. 67-76.

Kayastha, S.L. (1990b), *Population pressure in the Himachal Himalaya: Characteristics and Problems,* Vikas Publishing House, New Delhi, pp. 27-55.

Kayastha, S.L. (1991a), Some aspects of fertility in India, in J. Bahr & P. Gans (eds), *Geographical Approach to Fertility,* Geographical Institute, Kiel University, Keil, pp. 369-77.

Kayastha, S.L. (1991b), Review article on Kullu valley: Impact of tourism in mountain areas, *Indian Journal of Social Sciences,* vol. 4, No. 3, pp. 466-71.

Kayastha, S.L. (1991c), Foreword, in Onkar Singh, P. Nag, V.K. Kumra & J. Singh (eds), *Frontiers in Environmental Geography,* Concept Publishing Company, New Delhi, pp. 5-7.

Kayastha, S.L. (1992a), Forest and ecology in the Himalaya, in R.B. Singh (ed), *Dynamics of Mountain Geosystems,* New Delhi, pp. 2-30.

Kayastha, S.L. (1992b), Forests and ecology in the Himalaya, in *Dynamics of Mountain Geosystems,* Ashish Publishing House, New Delhi, pp. 26-39.

Kayastha, S.L. (1993), Environment, development and quality of life, *Annals of NAGI,* vol. 12(2), pp. 54-75.

Kayastha, S.L. (1998a), *Geography of Population: Selected Essays,* Rawat Publications, Jaipur & New Delhi,

Kayastha, S.L. (1998b), Frontiers on Environment : Reappraisal of Environmental Problems and Policies, Presidential Address, Geology & Geography Section, Indian Science Congress (Pune), Indian Science Congress Association, Calcutta, pp. 1-36.

Kayastha, S.L. (1998c), Population, disaster and disaster-reduction strategies, *Geographical Review of India,* vol. 53, pp. 206-18.

Kayastha, S.L. (1998d), Water resources of India : Their appraisal, conservation and management, *Acquatic Biology,* Nainital, pp. 20-5.

Kayastha, S.L. (1998e), Tourism and environment in the Himalayan region, in P. Nag, V.K. Kumra & J. Singh (eds), *Geography and Environment: Regional Issues,* Concept Publishing Company, New Delhi, pp. 31-45.

Kayastha, S.L. & V.K. Kumra (1978), Air pollution problem in Kanpur City, *Indian Journal of Air Pollution Control,* vol. 1, No. 1, pp. 38-42.

Kayastha, S.L. & V.K. Kumra (1979), Scientific study and control of ecological deterioration in urban ecosystem: A case study of Kanpur city, in the *Proceedings of the IGU Commission on Environmental Problems, Study and Control of Anthropogenic Transformation of Natural Ecosystems,* Moscow, pp. 234-63.

Kayastha, S.L. & V.K. Kumra (1980), Noise as a factor in environment, *Philippine Geographical Journal,* vol. XXIV, No. 3, pp. 127-42.

Kayastha, S.L. & V.K. Kumra (1985), Quality of human environment : A case study of Kanpur city, in *Spectrum of Modem Geography,* Concept Publishing Company, New Delhi, pp. 447-59.

Kayastha, S.L. & S. Mukherjee (1979), Spatial disorganization and internal migration in India: Some strategies for restructuring the space economy and development, *Canadian Studies in Geography,* vol. 6, pp. 45-61.

Kayastha, S.L. & P. Nag (1989), The Bhopal disaster, in John I. Clarke, P. Curson, S.L. Kayastha & P. Nag (eds), *Population and Disaster,* Basil Blackwell, Oxford and Massachusetts, pp. 206-18.

Kayastha, S.L. & Mehdi Raza (1984), *Environmental Studies in India, in Country Report: Trends in Geographical Research in India,* 1980-83, Presented to the 25th IGC Paris, Indian National Science Academy, New Delhi, pp. 47-55.

Kayastha, S.L. & R.B. Singh (1983), Energy resources and technology transfer in the management of rural environment, in *Environmental Management,* Allahabad Geographical Society, Allahabad, pp. 379-89.

Kayastha, S.L. & S.N. Singh (1977), A study of preferences and behaviour pattern of tourists in Varanasi: A geographical analysis, *The National Geographical Journal of India,* Vol. XXIII, pts. 3 & 4, pp. 143-50.

Kayastha, S.L. & S.N. Singh (1983), Tourism in the U.P. Himalaya, in O.P. Singh (ed), *The Himalaya: Nature, Man and Culture,* Rajesh Publications, New Delhi, pp. 205-11.

Kayastha, S.L. & R.P. Yadav (1985), Flood induced population migration in India: A case study of Ghaghara Zone, in L.A. Kosinski & K.M. Elahi (eds), *Population Redistribution and Devèlopment in South Asia,* D. Riedel Publication Company, Dordrecht, pp. 79-88.

Kayastha, S.L. & R.P. Yadav (1990), Flood Hazard in Lower Ghaghara Plain: A study in perception and impact on socio-economic development, *National Geographical Journal of India,* vol. 26, pts. 2&3, pp. 19-28.

Kuhnen, Frithlof (1992), Sustainability, regional development and marginal locations, *Applied Geography and Development,* vol. 39, Tubingen, pp. 101-5.

Singh, H.H., P. Nag, V.K. Kumra, J. Singh, H. Prasad & S.K. Singh (1986), Professor S.L. Kayastha : A profile, in H.H. Singh *et al.* (eds), *Geography and Environment : Issues and Challenges,* Professor S.L. Kayastha Felicitation volume. Concept Publishing Company, New Delhi, pp. v-vii.

INDEX